GENE MANIPULATION
IN PLANT IMPROVEMENT
II

STADLER GENETICS SYMPOSIA SERIES

CHROMOSOME STRUCTURE AND FUNCTION
Impact of New Concepts
Edited by J. Perry Gustafson and R. Appels

GENE MANIPULATION IN PLANT IMPROVEMENT
Edited by J. Perry Gustafson

GENE MANIPULATION IN PLANT IMPROVEMENT II
Edited by J. Perry Gustafson

GENETICS, DEVELOPMENT, AND EVOLUTION
Edited by J. Perry Gustafson, G. Ledyard Stebbins, and Francisco J. Ayala

GENE MANIPULATION IN PLANT IMPROVEMENT II

II

19th Stadler Genetics Symposium

Edited by

J. Perry Gustafson

USDA–ARS
University of Missouri
Columbia, Missouri

PLENUM PRESS • NEW YORK AND LONDON

The editor would like to dedicate his effort in this
publication to Ed, Kathleen, Anna, Doug, Hajrial, Hassan,
Mohammed, and Zong-Min. They kept the lab running. In
addition, he would like to express a special appreciation to
Chris, Katy, and Nick, who kept him running.

Library of Congress Cataloging-in-Publication Data

Stadler Genetics Symposium (19th : 1990 : University of Missouri-
    -Columbia)
    Gene manipulation in plant improvement II : 19th Stadler Genetics
Symposium / edited by J. Perry Gustafson.
        p.   cm. -- (Stadler genetics symposia series)
    "Proceedings of the 19th Stadler Genetics Symposium ... held March
13-15, 1989, at the University of Missouri, Columbia, Missouri"-
    -Verso t.p.
    Includes bibliographical references and index.
    ISBN-13: 978-1-4684-7049-9   e-ISBN-13: 978-1-4684-7047-5
    DOI: 10.1007/978-1-4684-7047-5
    1. Plant genetic engineering--Congresses. 2. Plant breeding-
    -Congresses.   I. Gustafson, J. P. II. Title. III. Series.
SB123.57.S73   1990
631.5'23--dc20   90-36593
                                                                  CIP

Proceedings of the 19th Stadler Genetics Symposium,
Gene Manipulation in Plant Improvement II, held March 13-15, 1989,
at the University of Missouri, Columbia, Missouri

© 1990 Plenum Press, New York
Softcover reprint of the hardcover 1st edition 1990

A Division of Plenum Publishing Corporation
233 Spring Street, New York, N.Y. 10013

ACKNOWLEDGEMENT

The editor gratefully acknowledges the generous support of the following contributors: College of Agriculture, Department of Agronomy, Division of Biological Sciences, Graduate School and School of Medicine, University of Missouri-Columbia; United States Department of Agriculture-Agricultural Research Service; CIBA-GEIGY Corporation; DeKalb Pfizer Genetics; Del Monte Foods, USA; Garst Seed Company; Illinois Foundation Seeds, Inc.; Monsanto Company; Northrup King Company; and Pioneer Hi-Bred International, Inc. who made the 19th Stadler Genetics Symposium a success.

The speakers, who spent a tremendous amount of time preparing their manuscripts and lectures are gratefully acknowledged. Without their expertise and dedication the Symposium could not have taken place.

I wish to thank the local chairpersons for their efforts to see that everyone in the respective sessions were well taken care of during the Symposium.

The behind-the-scene and on-site preparation was excellently handled by Joy Williams from Conferences and Specialized Courses, University of Missouri, who tirelessly handled all of my peculiar requirements and made sure everything was extremely well organized.

Many thanks are due to Joyce Reinbott, University of Missouri, for her excellent secretarial help in handling all the correspondence and typing. A special thanks goes to Kathleen Ross and Ed Butler for keeping the lab running.

J. P. Gustafson

February 6, 1990
Columbia, Missouri

CONTENTS

Plant Breeding and the Value Contributed to
    Cereal Grain and Oilseed Production in
    Western Canada   1
    D. F. Kraft

Self-pollinated Crop Breeding: Concepts and
    Success 21
    R.H. Busch and D.D. Stuthman

The Romance of Plant Breeding and Other Myths 39
    D.N. Duvick

Targeting Genes for Genetic Manipulation in
    Crop Species 55
    J.W. Snape, C.N. Law, A.J. Worland,
    and B.B. Parker

Incompatibility Barriers Operating in Crosses
    of *Oryza sativa* With Related Species and
    Genera 77
    L.A. Sitch

Wheat x Maize and Other Wide Sexual Hybrids:
    Their Potential For Genetic Manipulation
    and Crop Improvement 95
    D.A. Laurie, L.S. O'Donoughue,
    and M.D. Bennett

Induced Mutations – An Integrating Tool in
    Genetics and Plant Breeding 127
    M. Maluszynski

In Vitro Culture of Rice: Transformation and
    Regeneration of Protoplasts 163
    T.K. Hodges, J. Peng, L. Lee,
    and D.S. Koetje

In Vitro Manipulation of Barley and Other
    Cereals   185
    H. Lorz, R. Brettschneider, S. Hartke,
    R. Gill, E. Kranz, P. Langridge,
    A. Stolarz, and P. Lazzeri

Transformation and Regeneration of Non-
    Solanaceous Crop Plants   203
    M.A.W. Hinchee, C.A. Newell, D.V. Connor-Ward,
    T.A. Armstrong, W.R. Deaton, S.S. Sato, and
    R.J. Rozman

Haploids in Cereal Improvement: Anther and
    Microspore Culture   213
    K.J. Kasha, A. Ziauddin, and U.-H. Cho

Transgenic Plants   237
    R. Dekeyser, D. Inze, and M. Van Montagu

Transformation and Regeneration of Important
    Crop Plants: Rice as the Model System For
    Monocots   251
    R. Wu, E. Kemmerer, and D. McElroy

Genetic Transformation of Maize Cells by
    Particle Bombardment and the Influence of
    Methylation on Foreign-Gene Expression. . . .   265
    T.M. Klein, L. Kornstein, and M.E. Fromm

Non-conventional Resistance to Viruses in
    Plants - Concepts and Risks   289
    R. Hull

Plant Transformation to Confer Resistance Against
    Virus Infection 305
    R.N. Beachy

Using Plant Virus and Related RNA Sequences
    to Control Gene Expression   313
    M. Young and W. Gerlach

Mapping in Maize Using RFLPs   331
    D.A. Hoisington and E.H. Coe, Jr.

RFLP Mapping in Wheat - Progress and Problems 353
    M.D. Gale, S. Chao, and P.J. Sharp

New Approaches for Agricultural Molecular
    Biology: From Single Cells to Field
    Analysis   365
    R.A. Jefferson

Regulation of Plant Gene Expression by Auxins. . . .   401
    T.J. Guilfoyle, B.A. McClure, G. Hagen,
    C. Brown, M. Gee, and A. Franco

The Molecular Basis of Variation Affecting Gene
    Expression: Evidence From Studies on the
    Ribosomal RNA Gene Loci of Wheat   419
    R.B. Flavell, R. Sardana, S. Jackson,
    and M. O'Dell

Index   431

PLANT BREEDING AND THE VALUE CONTRIBUTED TO

CEREAL GRAIN AND OILSEED PRODUCTION IN WESTERN CANADA

Daryl F. Kraft

Professor, Agricultural Economics and Farm Management
University of Manitoba, Winnipeg, Manitoba R3T 2N2,
CANADA

INTRODUCTION

Cereal grains and oilseeds account for two thirds of the
value of agricultural production in the Prairie Provinces of
Manitoba, Saskatchewan, and Alberta. Between 1978 and 1988 the
value of these crops ranged from $4 billion to $8 billion. The
extreme variation in the annual crop value is primarily related
to; droughts in 1980, 1984 and 1988; record production levels in
1981, 1982 and 1986; record high prices in 1980 and 1981 and
twenty year low prices in 1986 and 1987. Underlying the
economic and weather shocks has been a steady upward trend in
crop yields. Nevertheless bio-technology has had a significant
influence upon the financial affairs of 100,000 farmers in
western Canada.

The issue addressed in this paper is to account for the
income attributed to one technology namely plant breeding. Over
the past 25 years total cereal grain and oilseed production has
increased at an annual average of 2 percent (Kraft, 1980).
After removing the year to year influence of weather the level
of cereal grain and oilseed production predicted in 1990 was 46
million tonnes. The average for the 1980's was forecast to 42.5
million tonnes. So far the average between 1980 and 1988 has
been 42 million tonnes. The factors underpinning the long term
growth in production do not appear to have changed substantially
in the past decade relative to the 1960's and 1970's.

Land Related Increases in Production

In the span of 25 years (1961-1986) the area under
production increased from 15 million hectares to 23.6 million
hectares.

Gene Manipulation in Plant Improvement II
Edited by J. P. Gustafson
Plenum Press, New York, 1990

1

About half of the increase in seeded area took place in the last ten years. While the area in crop increased 1.3 percent annually the yearly contribution of land to total production was about 0.4 percent. Nearly all the additional area occurred because less land was summerfallowed. Stubble yields are lower than fallow yields and without additional expenditures on fertilizers and pesticides the total increase in output only ranged between 0.3 percent to 0.4 percent for a 1.3 percent increase in seeded area. Therefore land by itself contributed about 4 to 5 million tonnes of the 17 million tonnes of additional grain produced in the last 25 years.

Fertility Related Increases in Production

Fertilizer registered a six fold increase in use between 1968 and 1986. An additional 10 million tonnes of nitrogen, phosphorous and potassium were applied by Prairie farmers. The contribution of fertilizer to total cereal grain and oilseed production has been difficult to isolate. Flaten and Hedlin (1988) reviewed a number of studies on soil fertility and crop yields. However, they were unable to reach a conclusion on the combined effects of increased nutrients, pesticides and varietal improvements when all inputs were changing together. Traditional agronomic research normally does not design experiments to analyze the interactive relationships when a number of inputs are combined in different proportions.

Arthur and Kraft (1988), Arthur, Fields and Kraft (1986) and Kraft (1982) estimated the influence of varieties and fertilizer from farm level data collected between 1968 and 1980. The genetic yield potential of the crops depended upon the varieties seeded and the yield characteristics of the variety. An estimate of these yield attributes were taken from Cooperative Tests of Varieties sponsored by Agriculture Canada. Annual surveys undertaken by the Pool Elevator Companies identified the varieties seeded by farmers. Fertilizer applied and crop yields were obtained from surveys of farmers conducted by the provincial crop insurance corporations. Throughout the period fertilizer use was correlated with increased application of herbicides. Therefore, the relationships estimated between fertility and crop yields probably jointly reflect the use of pesticides. The elasticities estimated for wheat (Triticum ssp.), barley (Hordeum vulgare), oats (Avena sativa) and flax (Linum usitassimum) with respect to fertilizer ranged between 0.1 and 0.15. In other words a 10 percent increase in fertilizer would increase crop yields between 1 percent and 1.5 percent. Since fertilizer use increased 600 percent over the period 1960 to 1985 the crop production increase attributed to fertilizer would vary from 7.7 million tonnes to 11.5 million tonnes. Crop production was estimated to increase from 24 million tonnes in 1961 to 41 million in 1986. The 17 million

tonne increase which could be attributed to fertilizer/
pesticides ranged between 7.7 million and 11.6 million tonnes.
The additional area seeded was estimated to account for between
4.3 million tonnes and 5.7 million tonnes. The lower end of the
combined contribution of fertilizer/pesticides and land range
was 12 million tonnes. Given the total increase was estimated
to be 17 million tonnes the larger land-fertilizer/pesticide
contribution of 17.3 million tonnes was too high since varietal
improvement and mechanization added something to production.
The 12 million tonne estimate is a lower bound for the
fertility/pesticide and land contribution, and the task remains
to identify the additional output attributed to genetic
improvements.

The analysis of Arthur, Fields and Kraft (1986) and Kraft
(1982) estimated the elasticity for spring wheat yields with
respect to the weighted yield index of the varieties seeded to
be one. In other words a one percent increase in the yield
index of varieties grown by farmers resulted in a one percent
increase in farm spring wheat yields. Whereas the absolute
yield improvement cited in the Cooperative Tests did not occur
on farm, the relative change did occur. In the case of barley,
the variety elasticity was greater than one but not
statistically different than one. The yield relationship
between varieties and farm yields for oats and flax were not
statistically significant. This lack of significance can
probably be attributed to the paucity of new oat and flax
varieties grown by farmers between 1968 and 1980. canola
(rapeseed: _Brassica_ _napus_) displayed the largest number of new
varieties with an improvement but the statistical measurements
of the genetic contributions to farm production were
indistinguishable from the trends in fertilizers/pesticides.
For wheat and barley a range of fertilizer application rates
occurred with a different combination of varieties. This did
not occur for canola as the varietal changes were concurrent
with using more fertilizer. The relationship that farm yields
increased at the same rate as the genetic potential of the
varieties seeded was confirmed for wheat and barley and could
not be rejected for the other crops.

Given the change of varieties seeded and the potential
yield improvement, the ultimate effect on crop production and
value should be a straight forward arithmetic exercise.
Equation 1 determines the annual production improvement index;

$$I_{jt} = \frac{\sum_{i=1}^{n} (Y_{ij} \cdot P_{ijt})}{(Y_{ij} \cdot P_{ij}, \ 1961)}$$

where: I_{jt} -   production improvement index for the jth crop
                       in the tth year (t = 1962, 1963 .... 1986).

          Y_{ij} -   yield index for the ith variety of the jth
                       crop.

          $P_{i,j,t,}$ -   the proportion of the ith variety of the
                            j^{th} crop seeded in year t.

       n - the number of varieties available in year t.

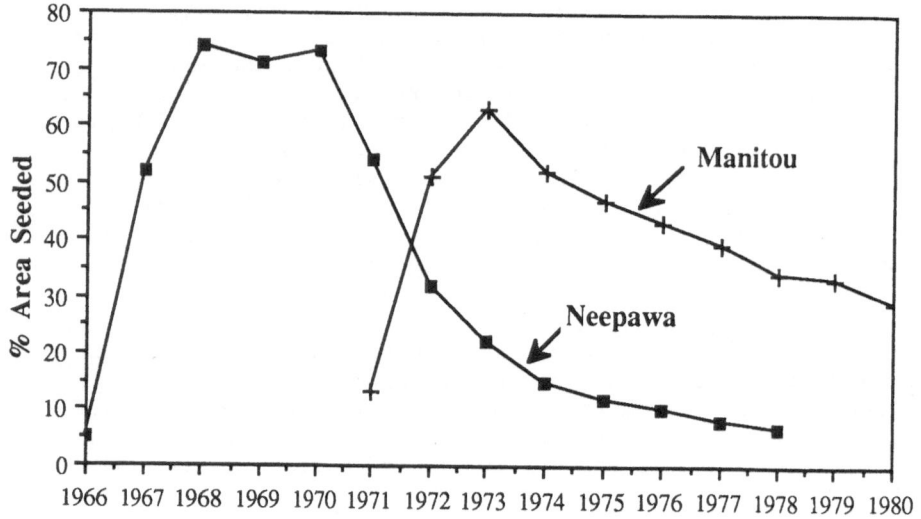

Years from Licensing

Fig. 1 Share of land seeded by a wheat variety

    The proportion of a variety seeded in a year will depend
upon its availability and productivity relative to other
varieties. Fig. 1 illustrates the relationship of two hard

spring wheat varieties in terms of their relative share of the
area seeded. The wheat 'Manitou' was released in 1964 and by
1968 exceeded 70 percent of the area seeded. However, the area
in 'Manitou' declined steadily after 1970 as a new variety
'Neepawa' appeared on the scene. The share of 'Neepawa' seeded
peaked at 62 percent in 1973 and has declined ever since. In
1973 the production index for hard spring wheat was comprised
almost entirely of 'Neepawa' (62 percent) and 'Manitou' (22
percent).

Yield indices for the crops grown in Manitoba appear in
Table 1. The crop indices determined from equation 1 were each
set equal 100 in 1961. Between 1961 and 1986 canola production
increased 15 percent while barley ranked second at a 10 percent
increase. Spring wheat oats and flax only registered marginal
productivity related improvements of between 2 percent and 4
percent during the 25 years. Spring wheat only registered a
weighted average increase of 2 percent however some areas in the
Provinces grew varieties which caused the productivity to
increase more.

Table 1. Production Indices for Crops Seeded in Manitoba

Year	Spring Wheat	Barley	Oats	Canola[1]	Flax
1961	100	100	100	100	100
1971	100	105	100	105	100
1981	101	107	101	107	102
1986	102	110	102	115	104

[1] Index change reflects new varieties and switch from
Polish (B. campestris) to Argentine (B. napus)
varieties.

Previous studies that analyzed the effects of genetic
developments upon crop output have merely shifted or pivoted the
supply function for the crop by the reciprocal of the
productivity index (Nagy and Furtan 1978, Akino and Hayami 1979,
Edwards and Freebairn 1984). Hertford and Schmitz (1977) point
out that the different methodologies in measuring the benefits
from genetic research do not differ substantially from just
determining the change in the value in crop production. This
was the methodology followed by Brennan (1988) in studying the
economic impact of wheat breeding programs in Australia.

The approach followed in this paper was to estimate the change in the value in crop production from plant breeding for all the major crops produced in the Canadian Prairie Provinces. Rather than analyze just one crop, all the major cereals and oilseeds were included because each one represents an alternative to the other. Given the variable costs are similar for each crop, a yield improvement in one, represents higher expected profits and will likely result in a decrease in the area seeded to the areas. Therefore given that prices, as well as crop yields, are changing at different rates from year to year over the 25 year interval studied it was necessary to analyze all crops together in terms of their relative profitability. Previous studies have estimated the value contributed by genetics by determining a supply equation for a particular crop on the basis of the prices for the crop, in question, alternate crops and other economic variables. Once the supply function has been estimated, then it is shifted or pivoted to the left by the reciprocal of the production index to represent the reduction in supply as if the genetic development had not happened. For example, given the 1986 Manitoba production index for canola was 1.15 (see Table 1) the procedure was to reduce the quantity supplied by 0.87 (1/115). This approach underestimates the impact of higher yields since the supply function not only shifts because yields are lower but also because less land would be seeded to the crop. The debate on how to represent the shift of the supply is secondary to omitting the magnitude of the change if other crops represent reasonable substitutes for the crop in question. Whereas previous studies underestimated the reduction in crop supply without the technical improvement they also underestimated the valued added because of the genetic development.

Previous studies cited, either assumed the change in the value of production was a fixed percentage cost reduction, or a fixed yield increase in the form of a percentage increase in output. However, the supply function shifted by more than the yield increase because more land was allocated to the crop with the higher yielding variety. The difference in the total crop receipts between the new variety and the total revenue from the displaced crops must be determined in addition to the added revenue from producing more of the crop for the same land base.

ECONOMIC METHODOLOGY

Every spring, prior to planting Prairie farmers evaluate their cropping alternatives. Past experience in terms of the success in growing and selling the crop along with the perception of upcoming conditions all have a bearing on the seeding decision. In western Canada the five major crops are hard spring wheat, oats, barley, canola and flaxseed. Since the operating costs involved with seeding, tillage, fertilizer,

pesticide use and harvesting are similar for each crop the
comparison of which crop to grow involves examining their
expected yields and prices. A decision to plant canola means
not planting wheat, barley, oats or flax. Therefore the cost of
growing canola is the opportunity lost in producing another
crop.

     The primary factors influencing the planting decision are
illustrated in Fig 2. Beginning with the improved crop land
base at the top of Fig. 2 places limits upon how much land can
be seeded to crops or summerfallowed. The total area available
for crop production changed slowly over time and has not been
directly linked to specific technological events. The area
planted every spring however, is affected by economic events
such as grain prices and stocks. They influence the decision to
summerfallow or reseed the land cropped the previous year. The
decision with respect to the mix of crops to seed is influenced
by three primary factors, namely crop prices, the grain
currently in storage on the farm and crop yields. As the
relative expected total value of the crop rises the tendency
will be to seed more of that particular crop. For the
individual farmer rotational consideration and past experience
influence specializing in a particular crop and therefore
modelling the economic and agronomic setting must account for
the changing conditions and how farmers have responded to these
events in the past.

     Varieties of each crop have a bearing upon how much crop
will be seeded in terms of the expected yields, harvestability
and disease resistance. As varieties are released they expand
the options available to farmers and have a direct bearing upon
which crop is seeded.

<u>Model for the Determination of Seeded Area</u>

     The total cropland base is defined as the area seeded to
cereal grains and oilseeds plus the area summerfallowed. The
share of the area seeded each year is estimated by equation (2).

$$Y_t = f(Y_{t-1}, T, S, PW \text{ and } L_{1970}) \qquad\qquad (2)$$

where:

      Y_t = share of the cropland area seeded to cereal grains
           and oilseeds (%)
      T = trend where t_1 is 1961 and t_{25} - 1986
      S = stocks of grain in storage in farm (tonnes)
      P_w = real price of wheat when the nominal price is
           deflated by the farm input price index
      L_{70}= a dummy variable to account for the Lower Inventory
           for Tomorrow program in 1970

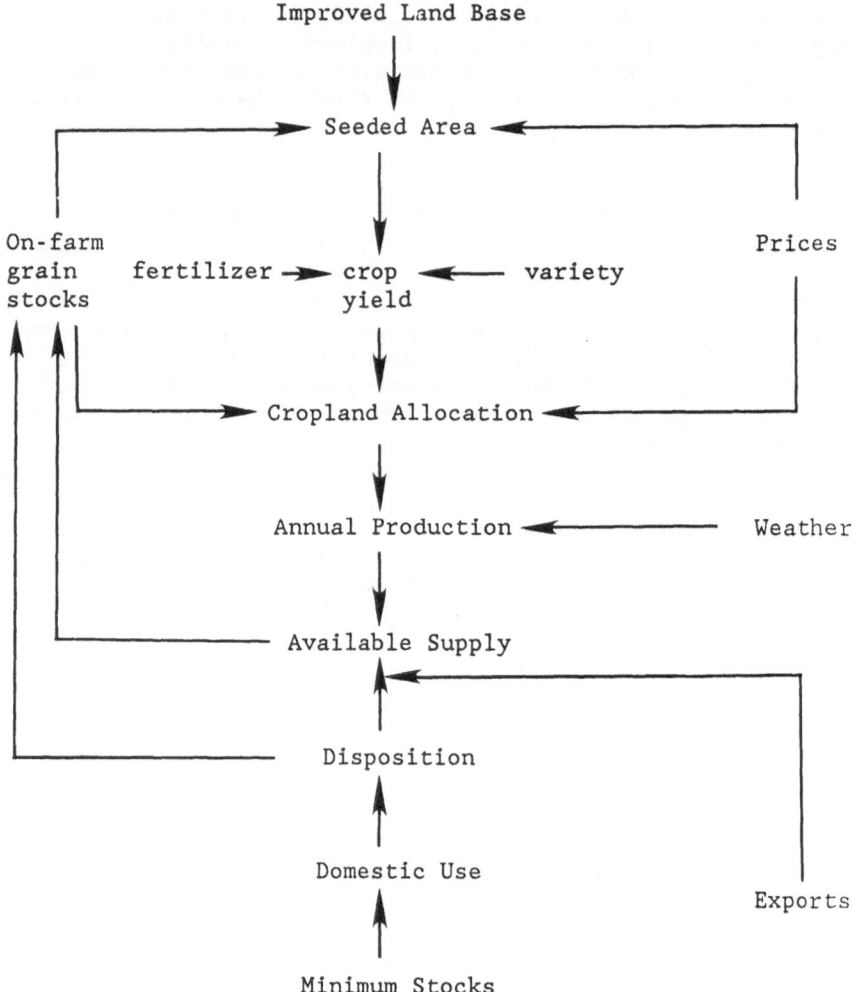

Fig. 2.   Schematic of economic model

Last years share (Y_{t-1}) of the area seeded accounts for the rigidity of cropping decisions because of rotations, weed control and soil moisture limitations. The trend reflects the technological advancement in weed control and soil moisture conservation machinery which result in less reliance upon summerfallow being part of the rotation. As the stocks of grain accumulate the tendency is for farmers to reduce the area seeded because of limited storage capacity, and pessimistic expectations with respect to the volume of crop which may be sold. Since wheat represents 60 percent of the area seeded its the real price is a good indicator of the expected profitability of seeding more land. Equation 2 forecasts the area seeded between 1961 and 1986 and the only difference between the simulation with and without new varieties was the grain stocks. Surprisingly without new varieties, the amount of grain in terms of total tonnes stored on farms was, on occasion, greater than with newer varieties because more wheat and less canola was produced. However over the course of the 25 years, the net effect upon the seeded area was negligible.

Economic Model for the Determination of Area Seeded to Each Crop

The areas seeded to wheat, barley, oats, canola and sunflower (Helanthus annus) were estimated by a system of simultaneous equations. The estimated equations are listed below:

$$\frac{A_W}{A_F} = B_w^o \left(\frac{A_W}{A_F}\right)_{-1} e^{\left(B_w^1 + B_w^2\frac{RW}{RF} + B_w^3\frac{RW}{RB} + B_w^4\frac{RW}{RO} + B_w^5\frac{RW}{RC} + B_w^6\frac{RW}{RS} + B_w^7 IW + B_w^8 IF\right)} \tag{3}$$

$$\frac{A_B}{A_F} = B_b^o \left(\frac{A_B}{A_F}\right)_{-1} e^{\left(B_b^1 + B_b^2\frac{RW}{RF} + B_b^3\frac{RB}{RW} + B_b^4\frac{RB}{RO} + B_b^5\frac{RB}{RC} + B_b^6\frac{RB}{RS} + B_b^7 IB + B_b^8 IF\right)} \tag{4}$$

$$\frac{A_O}{A_F} = B_o^o \left(\frac{A_O}{A_F}\right)_{-1} e^{\left(B_o^1 + B_o^2\frac{RO}{RF} + B_o^3\frac{RO}{RW} + B_o^4\frac{RO}{RB} + B_o^4\frac{RO}{RC} + B_o^5\frac{RO}{RS} + B_o^7 IO + B_o^8 IF\right)} \tag{5}$$

$$\frac{A_C}{A_F} = B_c^o \left(\frac{A_C}{A_F}\right)_{-1} e^{\left(B_c^1 + B_c^2\frac{RC}{RF} + B_c^3\frac{RC}{RW} + B_c^4\frac{RC}{RB} + B_c^5\frac{RC}{RO} + B_c^6\frac{RC}{RS} + B_c^7 IC + B_c^8 IF\right)} \tag{6}$$

$$\frac{A_S}{A_F} = B_s^o \left(\frac{A_S}{A_F}\right)_{-1} e^{\left(B_s^1 + B_s^2\frac{RS}{RF} + B_s^3\frac{RS}{RW} + B_s^4\frac{RS}{RB} + B_s^5\frac{RS}{RO} + B_s^6\frac{RS}{RC} + B_s^7 IS + B_s^8 IF\right)} \tag{7}$$

where:

A_W - area of wheat seeded

A_F - area of flax seeded

A_B - area of barley seeded

A_C - area of canola (rapeseed) seeded

A_O - area of oats seeded

A_S - area of sunflowers seeded

RW - expected total receipts per hectare for wheat

RF - expected total receipts per hectare for flax

RB - expected total receipts per hectare for barley

RC - expected total receipts per hectare for canola
    (rapeseed)

RO - expected total receipts per hectare for oats

RS - expected total receipts per hectare for sunflowers

IW - inventory of wheat stored in farms

IF - inventory of flax stored in farms

IB - inventory of barley stored in farms

IO - inventory of oats stored in farms

IC - inventory of canola (rapeseed) stored in farms

IS - inventory of sunflowers stored in farms

The lagged endogenous variable represents the tendency of farmers to continue with the previous years combination of crops. Besides, the relative profitability of crops, cropping rotations and confidence to realize the expected returns will dampen the year to year switching from one crop to another. The availability of a new variety will see it replacing older varieties of the same crop first and once the revenue potential becomes more widely known it will replace other crops. Therefore it may take up to five or six years after a variety is released before the potential area seeded is attained.

The expected returns per acre rather than simply crop prices or yields are the most relevant indicators in terms of profit a farmer will anticipate when growing any crop. When expected returns from one crop are divided by the expected returns for each of the other cropping options their relative profitabilities have a direct bearing on the share of the land

seeded to each crop. Therefore each equation includes a set of
variables to reflect the relative profitability. The expected
crop prices for barley, oats, flax and canola (rapeseed) were
the average cash prices in the months of January, February and
March 1989. The current Canadian Wheat Board initial prices and
the most recent final price became the expected price for wheat.
Expected crop yields are the three year moving average.

In addition to prices and yields the grain stored on farms
prior to seeding was important to forecasting the area planted
to each crop. Higher crop inventory prior to seeding would tend
to reduce the area planted to a specific crop. Approximately 75
percent of the cereal grain and oilseed crop production is
shipped out of the Prairie region and the limited grain handling
capacity often results in lower shipments relative to the
available supplies. Therefore relatively high inventories for
some crops increases the probability that not all the grain will
be sold during the year following harvest.

Elasticities Estimated from the Economic Model

The elasticities determined from equations 3 through 7
appear in Table 2 and provide a preliminary indication of how
farmers changed the area seeded in relationship to an increase
in revenue from a particular crop. A higher yielding spring
wheat variety adding 1 percent more revenue and assuming all
other factors remain unchanged will be expected to increase the
area seeded to wheat in the following year by 0.19 percent in
Manitoba, 0.15 percent in Saskatchewan and 0.14 percent in
Alberta. The weighted combined increase in Prairie total wheat
production from the new variety (1 percent) and the added area
(.16 percent) is 1.16 percent. The short-run elasticities only
measure increases in the area planted in the year following the
increase in crop yields. However crop yields represent a
permanent change in the potential to realize higher total
receipts from a particular crop and the annual increases in area
planted will grow until an equilibrium in seeded area is
established. In the case of wheat the long-run elasticity is
.54 percent. This indicates a one percent increase in wheat
yields will increase the area seeded by .54 percent before the
new equilibrium in seeded area is attained. In the longrun the
combined influence of yields and seeded area for a 1 percent
increase in wheat yield is an increase in wheat production of
1.54 percent.

The influence of a new canola (rapeseed) variety with a 1
percent higher yield potential is more dramatic in the short run
than the other crops given farmers inclination to seed more
canola (rapeseed) with a higher revenue potential. Given a 1
percent increase in canola (rapeseed) revenues farmers in
Manitoba have increased the area seeded to canola (rapeseed) by

1.16 percent, Saskatchewan 0.94 percent and Alberta 1.44 percent
(Table 2). A combined Prairie wide increase of 1.24 percent
occurs in just one year and the long term increase in the area
seeded to canola (rapeseed) is 2.25 percent. Therefore, the new
variety increasing canola yields by 1 percent would result in an
additional area seeded of 1.24 percent in the following year for
a combined production increase of 2.24 percent. In the long run
the combined increase in production of canola (rapeseed) is 3.25
percent. Studies focusing solely upon the increase in
production from the higher yields have assumed the revenue
elasticity is zero and only measured the contribution to
production attributed to the higher yields and not counted the
production from the increased area seeded.

Table 2.   Short and long run supply elasticities for a
           1% increase in crop revenue.

| Province | Crop | | | | |
	Wheat	Barley	Oats	Canola	Flax
Manitoba	.19	.70	.38	1.16	.64
Saskatchewan	.15	.67	.61	.95	.94
Alberta	.14	.08	.68	1.44	1.44
Prairies	.15	.37	.60	1.24	.80
	$(.54)^1$	$(3.9)^1$	$(1.96)^1$	$(2.25)^1$	$(1.80)^1$

[1] Long-run elasticities for a permanent increase in
prairie crop revenue.

    In the case of canola (rapeseed) Table 3 illustrate the
source of the reduction to other crops. The percentage
reduction in oats is the largest for all Provinces. The area
seeded to oats has been declining steadily between 1961 and
1986. The increasing share of canola (rapeseed) appears to have
replaced some of the area formerly growing oats. Wheat ranked
second in being replaced by canola (rapeseed) barley and flax
showing much lower reductions when the canola (rapeseed) area
are expanded. Therefore the net revenue contributed from the
additional area seeded involves substracting the total receipts
of the crop not grown from canola (rapeseed) revenue.

Table 3. Short run cross elasticity for a 1% increase
in canola (rapeseed) crop revenue.

Province	Crop			
	Wheat	Barley	Oats	Flax
Manitoba	-.15	-.13	-.28	-.20
Saskatchewan	-.05	-.04	-.40	-.07
Alberta	-.19	-.01	-.40	-.03

Validation of the Economic Model

Prior to recreating history the validity of the model
should be established. The capability of the model to track the
events between 1961 and 1986 in terms of the total revenue from
crop production and the areas seeded to each crop is important
because the simulated history becomes the bench mark for
comparing different crop variety scenarios. Validation is
evaluated in terms of three measurements, namely root mean
percentage error (RMS% error), mean percent error and R2. The
RMS% error is determined by equation 8.

$$\text{RMS\% error} = \frac{(A_t - P)^2}{A_t^2} \tag{8}$$

where: A_t - the actual value of a variable in periood t
        P_t - the predicted value of a variable in period t

A perfect forecast would result in a RMS% Error of 0 (Table
4). The overall forecasting performance of the model was judged
to be acceptable with the lowest error being recorded for wheat.
The model tended to overestimate the total area in crop
production as all forecasted revenues except flax tended to
exceed the historical revenue. Given that prices were assumed
not to be influenced by the level of crop production the only
variable which caused the forecasted revenue to exceed the
actual was the quantity of production. The mean percent error
for barley was the lowest, however, the forecast errors about
the mean were larger than wheat and oats. The oilseed crops
appeared to be the most difficult to forecast. The largest mean
percent error was -5.11 percent for flax. The model simulation
appeared to be sufficiently sound to represent and compare two
different historical scenarios.

Table 4.   Forecasting performance of the model.

Variable	R^2	RMS% Error	Mean %Error
		Performance Statistic	
Wheat Revenue	.99	3.6	.55
Barley Revenue	.98	7.4	.05
Oats Revenue	.96	6.5	1.40
Flax Revenue	.90	17.1	-5.11
Canola Revenue	.97	17.3	2.84

Scenarios Analyzed by the Economic Model

The benchmark scenario became the simulated history with farmers adopting the new varieties as they had in the past. One of the alternate scenarios assumed the plant breeding programs released no new varieties after 1961. The other scenario assumed the canola (rapeseed) breeding program were never started but the varieties of wheat, oats and barley and flax were released and adopted at the same rate as they were historically.

RESULTS

Value of Production

During a twenty-five year time interval (1961-86) Prairie farmers were simulated to produce $101.5 billion of cereal grains and oilseeds. The actual value was $101.9 billion. If no new plant varieties were released after 1961 then the value of twenty five year accumulated cereal grain and oilseed production was simulated to be $95.2 billion. Over the period of 25 years the value of cereal grains and oilseeds produced was simulated to be $5.6 billion lower without new varieties of 5.3 percent of the total value of production. The income contributed by the new varieties is nearly equal to the added value of production since the differential cost of seeding newer varieties diminishes as their availability expands over time. In 1980 an estimate of the cost of the new varieties purchased that year equaled 10 percent of the value contributed. Assuming the relationship was comparable throughout the 25 years then the net farm income contributed by new varieties would equal 90 percent of $5.6 billion or $5.0 billion dollars.

Table 5 shows the annual revenue from all crops for each of the three scenarios. The difference between the revenue produced from all crops and either of the scenarios with fewer crop varieties is an estimate of the income of new varieties

contributed. During the 1960's the newer varieties contributed
$0.2 billion.

Table 5. Simulated value of Priarie cereal grain and
oilseed production and attributed to yield
related varietal development.

	Total Crop Revenue			Income Lost	
Years	Past Variety Development All Crops	No Variety New Development All Crops	No Variety Development Canola	No Variety Development All Crops	No Canola Variety Development
	($-Billion)				
1962-69	14.6	14.4	14.6	.2	.0
1970-79	37.7	36.0	37.3	1.7	.4
1980-86	49.2	45.5	48.0	3.7	1.2
1962-86	101.5	95.9	99.9	5.6	1.6

Canola (rapeseed) varieties added very little additional
income during the 1960's. By the 1970's all new varieties were
simulated to add $1.7 billion. Canola (rapeseed) varieties
alone were determined to contribute $0.4 billion to farm income
during the 1970's.

During the 1980's $1.2 billion of $3.7 billion crop variety
related income was associated with the canola varieties released
since 1961. By 1986 10.5 percent of the total value of all
crops produced was linked to varieties released in the past 25
years. Canola (rapeseed) accounted for over half of the added
value from varieties in 1986.

The relative importance of canola's (rapeseed) contribution
through new varieties is measured in terms of the total added
income between 1961 and 1986. New canola (rapeseed) varieties
contributed $1.6 billion of the $5.6 billion from all varieties.
The importance of canola (rapeseed) developments will be
examined in more detail later.

Crop Rotations

The significance of variety developments is registered
through the diversification of the crops produced. Table 6
shows that if no new varieties were released after 1961 the
relative importance of wheat, the most dominant crop in the
Prairies, would have been even greater. Wheat production was
simulated to contribute $59.7 billion between 1961 and 1986 with
historical variety developments but would have equaled $61.4
billion with no new varieties. In other words low yielding
wheat varieties would have produced more wheat because more land
would have been seeded to wheat. During the 1960's there was
little difference between the historical simulations and the no

new variety simulations with respect to the area producing
wheat, however simulations of the 1970's resulted in at least 5
percent more land growing wheat and up to 15% more in 1978.   The
1980's averaged 10 percent more land in wheat if 1961 varieties
were the only choices available to farmers.

Table 6.   Simulated value of wheat production in
                the Canadian Prairies.

Time	Historical Variety Development All Crops	No Variety Development All Crops	No Variety Development Canola
	($-Billion)		
1962-69	9.4	9.4	9.4
1970-79	20.8	22.0	22.0
1980-86	29.5	31.1	31.2
1962-86	59.7	62.4	62.7

If canola (rapeseed) variety developments had not occurred
but the historical varieties of wheat, barley and oats and flax
were available wheat production over the 25 years was simulated
to be 5 percent higher than it was historically.   During the
1960's the influence of canola (rapeseed) varieties was
negligible in terms of wheat production.   By the 1970's the
value of wheat production was lowered by $1.27 billion or 6.1
percent because land was seeded to canola (rapeseed) instead of
wheat.   Between 1980 and 1986 wheat production was $2.00 billion
lower (6.7 percent) then it would have been without the option
of seeding the varieties of canola (rapeseed) released in the
past 25 years.

The growth of canola (rapeseed) production attributed to
the variety development after 1961 was simulated to occur
predominantly in the 1970's and 1980's.   The first new varieties
'Tarka' (B. napus) and 'Echo' (B. campestris) were released in
1964.   By 1969 they were grown upon 46 percent of the land (Nagy
and Furtan, 1978).   By 1970 only 5 percent of the area was
seeded to varieties available prior to 1962 and in two years
(1972) the older varieties were no longer seeded.   The simulated
canola (rapeseed) revenue without the varieties was $1.7
billion, but with the varieties the revenue was $4.2 billion or
243 percent more.   By themselves, the newer varieties yielded 7
percent more each year or 70 percent of the added output.   The
remaining 173 percent was attributed to additional land seeded
to canola (rapeseed).   The land seeded to canola (rapeseed) was

200 percent higher with variety development but since some of the additional land was on stubble the increase in production was less than the increase in area. For the seven years simulated in the 1980's the canola (rapeseed) varieties averaged 10 percent per year or a 50 percent cumulative increase in production due to the varieties grown. The area seeded to canola (rapeseed) was simulated to be 2.3 times greater than with no variety improvements but the added area contributed just two times the production because of the growing share of canola (rapeseed) seeded on stubble land. In the absence of the yield improving varieties of canola (rapeseed) the crop would be comparable to flax production in the Prairies.

Limiting the value of plant breeding too just higher yields ignores the contribution of quality changes. This is particularly true for canola (rapeseed) with the improvements in the erucic acid, glucosinolates and fiber content of the newer varieties. The economic model, however, provides some insight with respect to the magnitude of the benefits. Qualitative changes are reflected in the price of the crop. Varieties of canola (rapeseed) with higher concentrations of erucic acid or glucosinolates would be discounted in terms of lower prices. This in turn represents lower expected total receipts from growing these varieties. Lower revenues can be incorporated into the economic model in the same manner as lower yielding varieties. Assuming that lower quality canola (rapeseed) would sell for 10 percent less, then the model indicates that the 25 year (1961-86) canola (rapeseed) production with the older varieties would be approximately $1.8 billion. This represents a further reduction of canola (rapeseed) output from the 25 year cumulative level of $10.7 (Table 7). Canola (rapeseed) production was simulated to be only $4.2 billion when just yield was influenced by the newer varieties. Added the loss of quality to yield suggests the value of canola (rapeseed) would have declined further to $1.8 billion. The latter estimate may in fact be too high because if the quality improvements prevented only a 10% price discount then without the yield and quality improvements canola (rapeseed) would have returned an income comparable to the other crops. Without the yield and quality occurring together, canola (rapeseed) would likely only be a minor specialty crop in the Prairie Provinces. Assuming this to be the case then the income attributed to genetics would be the difference between canola (rapeseed) receipts and the mix of crops it replaced. In Table 5 the income attributed to canola (rapeseed) variety development was $1.7 billion between 1962 and 1986. If canola (rapeseed) remained at a level of production it had in 1961 because the varietal yield and quality changes were nonexistent then the loss in farm income would be $2.2 billion or nearly a hundred million per year.

CONCLUSIONS

Plant breeding through increasing the yields of cereal
grains and oilseeds contributed $5.6 billion between 1962 and
1986. The $225 million average annual increase in income should
be compared to the annual expenditures on research and
development for cereal grains and oilseeds. Specific studies on
canola (rapeseed) and malting barley have suggested the annual
returns exceed expenditures by a two to one margin. The
findings of this study would suggest the returns estimated may
exceed other studies. Previous analyses have limited the
benefits from plant breeding to just the higher yields on the
existing area growing the crop. The benefits, however extend

Table 7.   Simulated value of Canola (rapeseed)
                production in the Canadian Prairies.

Time	Historical Variety Development All Crops	No Variety Development All Crops	No Variety Development All Crops
	($-Billion)		
1962-69	.4	.3	.3
1970-79	4.3	1.8	1.7
1980-86	6.0	2.4	2.2
1962-86	10.7	4.6	4.2

further as less of the crop would be seeded if the genetic
improvements had not occurred.

The long run elasticities estimated for barley, oats,
canola (rapeseed) and flax indicate the increase in production
from the added area producing a crop exceed the increase due to
the higher yields. In the case of the Canadian Prairie
Provinces the major genetic improvements have been in barley and
canola (rapeseed). The effect of these developments has been to
diversify the mix of crops grown by farmers as without the
higher yields fewer hectares would be seeded to barley and
canola (rapeseed) and more to wheat. In fact the simulations
suggested that more wheat would be produced in western Canada if
farmers only had the 1961 varieties available to them during the
following twenty-five years.

The analysis indicates that in 1986 the value contributed
by genetics was equal to 10 percent of total crop receipts
although the additional volume produced was only 5 percent. In
other words genetics influenced the switch from lower valued
crops to higher valued crops.

REFERENCES

Akino, M., and Hayami, Y., 1974, Efficiency and Equity in Public
    Research: Rice Breeding in Japan's Economic Development,
    American Journal of Agricultural Economics, 57:1-10.
Arthur, L. M., and Kraft, D.F., 1988, The Effects of
    Technological Change on the Economic Impact of Agricultural
    Drought in Manitoba, Canadian Journal of Agricultural
    Economics, 36: .
Arthur, L. M., Fields, V. J., and Kraft, D. F., 1986, "Toward a
    socio-economic assessment of the implications of climate
    change for agriculture in the Prairie Provinces." Final
    report to Environment Canada. Winnipeg University of
    Manitoba, Department of Agricultural Economics.
Brennan, J. P., 1988, An Economic Investigation of Wheat
    Breeding Programs, Agricultural Economics Bulletin No. 35,
    Department of Agricultural Economics and Business
    Management, University of New England, Armidale, N.S.W.
Edwards, G. W., and Freebairn, J. W., 1984, The gains from
    research into tradeable commodities, American Journal of
    Agricultural Economics, 66:41-49.
Flaten, D., and Hedlin, R., 1988, Influence of Fertility,
    Pesticides and Genetics on Cereal Grain Yields, contributed
    paper annual meeting, Agricultural Institute of Canada,
    Department of Soil Science, University of Manitoba.
Hertford, R., and Schmitz, A., 1977, Measuring economic returns
    to agricultural research, in: "Resource Allocation and
    Productivity in National and International Agricultural
    Research," Arndt, T. M., Dalryple, D. G. and Ruttan, V. W.,
    eds., University of Minnesota Press, pp. 148-167.
Kraft, D. F., 1988, Production Potential for Cereal Grains and
    Oilseeds in Western Canada, Crop Production Symposium, The
    Canadian Wheat Board, Winnipeg, Manitoba.
Kraft, D. F., 1982, Crop Yield Model, Study Element No. 4,
    Saskatchewan Drought Studies, Prairie Farm Rehabilitation
    Administration, Regina, Saskatchewan.
Nagy, J. G. and Furtan, W. H., 1978, Economic Costs and Returns
    from Crop Development Research: The Case of Rapeseed
    Breeding in Canada, 26:1-14.

SELF-POLLINATED CROP BREEDING: CONCEPTS AND SUCCESSES

R.H. Busch and D.D. Stuthman

ARS-USDA and Department of Agronomy and Plant Genetics
University of Minnesota
St. Paul, Minnesota 55108

Self-pollinated crop breeding has changed greatly from the
time of plant domestication to present. Classical breeding procedures have been developed, but are rarely employed in their original form. Modifications have been adopted as well as various
combinations of these classical procedures are used as appropriate
in given situation for a particular crop. The topic of breeding
self-pollinated crops is quite broad, and is discussed in the
following sections as: 1) successes of breeding self-pollinated
crops, 2) equipment modernization, 3) impact of recurrent selection, 4) implementation of rapid generation advance, 5) a recent
contribution from cytogenetics, and 6) a contribution from cereal
technology.

SUCCESSES

As an art, plant breeding has been practiced for thousands of
years. However, during the last 80-100 years science has made an
increasingly greater contribution to plant breeding with the
greatest impact during the last 35 years. This impact has occurred
through application of scientific principles in disciplines related
to plant breeding. Statistics, genetics (qualitative, quantitative,
cytogenetic, and now molecular), pest management (plant pathology
and entomology), plant physiology, and soil science have contributed
to greater efficiency as well as impressive improvements in many
traits including large gains in yield potential.

Yield improvement in self-pollinated crops is relatively easy
to document, but obtaining unbiased estimates of the contribution
due solely to genetic improvement, without confounding by agronomic

Gene Manipulation in Plant Improvement II
Edited by J. P. Gustafson
Plenum Press, New York, 1990

improvements, is difficult and usually requires certain assumptions.
Often these assumptions are made in reference to the performance
of check cultivars under contemporary production systems. These
systems use modern growing conditions and an old check may perform
relatively poorly because of lodging, disease reaction, or response
to fertility. Special studies under carefully controlled growing
conditions have helped reduce some of the assumptions required for
accurate interpretation. Breeding successes in wheat (Triticum),
oats (Avena sativa), barley (Hordeum vulgare) and soybean (Glycine
max) will be discussed here although good success has also been
documented in other self-pollinated crops.

Genetic gains in yield have varied depending on crop and
location. Usually larger yield gains have been obtained in environments more favorable for production. Austin et al. (1980) compared newly released semidwarf winter wheats with cultivars released
in 1908, 1935, 1953, and 1964. All wheat genotypes were grown in
both high and low soil fertility with fungicide and netting to
minimize the lodging and disease variables. Yields of the new
semidwarf cultivars were about 40% higher than the older check
varieties in both low and high soil fertility. Yield gain over
the entire 70 years averaged 0.6% per year (Fig. 1), however the
greatest genetic gain in yield occurred from cultivars released
after 1964. If lodging had been allowed to occur the yield advantage of the newer cultivars would have been considerably greater.

Schmidt (1984) studied the genetic contribution to yield gain
in wheat in the U.S. from 1958 through 1980 using data from the
USDA-ARS Wheat Uniform Regional Nurseries. Over all geographical
areas and types of wheat grown in the U.S., he estimated an
increase in yield of 17% due to genetic improvement, with an
average increase in yield of 0.9% per year (Fig. 2). From 1958-
1980, wheat grown in the U.S. showed an average increase in yield
from 1688 to 2226 kg/ha, or about 32%. Based on the 17% genetic
yield increase found in the Uniform Regional Nurseries, genetic
improvement was estimated to have contributed 53% of the gain
total increase in yield of the wheat crop in the U.S.

Riggs et al. (1981) studied the grain yield of spring barley
cultivars released in Great Britain from 1880 to 1980. All cultivars were grown under high and low fertility with fungicides and
netting to prevent disease and lodging. Over all environments, a
quadratic regression better fit yield versus year of release than
did a linear regression (Fig. 3), suggesting an ever increasing
rate of progress with time. From pre-1900 to 1952, genetic gain
averaged 0.27% per year but from 1953 to 1980 it averaged 0.84%
per year or about three times more.

Wych and Rasmusson (1983) studied genetic gain in yield of
malting barley cultivars released in the north central U.S. from

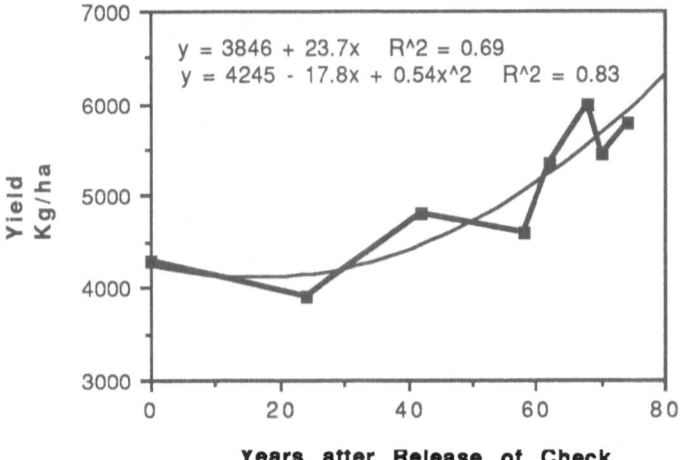

Fig. 1.   Gain in yield of winter wheat in Great Britain, 1908-80.
          Regression of winter wheat varieties yields (kg/ha) on
          years after release of the oldest variety in the trial,
          by Austin, et al. (1980).

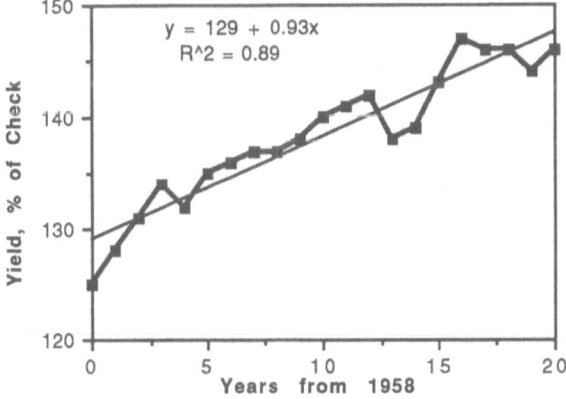

Fig. 2.   Gain in yield of wheat in the USA, 1958-80.   Regression
          of wheat yield from new lines expressed as a percent of
          check over all wheat classes grown in Uniform Regional
          Nurseries, 1958-1980, by Schmidt (1984).

1920 to 1978. High fertility environments with fungicide as
necessary were used to evaluate performance. Annual increase in
grain yield averaged 0.9% (Fig. 4), but during the last 40 years,
the time period in which serious breeding effort has taken place,
the genetic increase in yield has been nearly linear and averaged
2% per year.

Genetic gain for yield increases in spring oats have also
been substantial in both England and the U.S. Lawes (1977) found
the rate of 0.56% gain per year in Great Britain when cultivars
released from 1908 through 1962 (Fig. 5) were compared in the same
trials. Wych and Stuthman (1983) compared oat cultivars adapted
to the upper-midwestern U.S. and released from 1923 to 1980. An
annual genetic gain in yield (Fig. 6) of about 0.8% was found when
no diseases were present.

Luedders (1977) reported that genetic gain in yield for various
soybean maturity groups (Fig. 7) from 1920 to 1970 averaged about
0.84% in midwestern U.S. Based on year of release, four cycles of
cultivars were considered (pre-1933, 1938-43, 1947-53, 1961-68,
and 1971). Yield increases above previous cycle cultivars were
21, 9, 14, and 6%, respectively.

The gain in yield of these self-pollinated crops was a result
of a combination of improvements including better ability to: 1)
respond to favorable growing conditions, 2) tolerate unfavorable
growing conditions, 3) resist or tolerate diseases and/or insects,
4) resist lodging, and 5) partition a larger share of carbohydrates to the grain.

Possibly the most significant agronomic improvement which has
accompanied yield increases has been lodging resistance. In wheat
and barley, yield increases have been accompanied by reduced
height with a corresponding increase in harvest index. Height has
not been reduced in spring oats, but even so increases in harvest
index have been similar to those for wheat and barley. Modern
soybean cultivars, like modern oats, are slightly taller than
older cultivars. Enhanced disease resistance and more timely
harvesting of all crops due to better mechanization and ease of
harvesting may also have resulted in greater yield stability and
higher quality grain with less weather damage. Modern cultivars
often possess higher grain quality, such as increased groat yield,
kernel weight and test weight in oat (Wych and Stuthman, 1983) and
increases in percent plump kernels, malt extract, alpha amylase,
and diastatic power in barley (Wych and Rasmusson, 1983).

EQUIPMENT

Mechanization of plant breeding related work has had and will

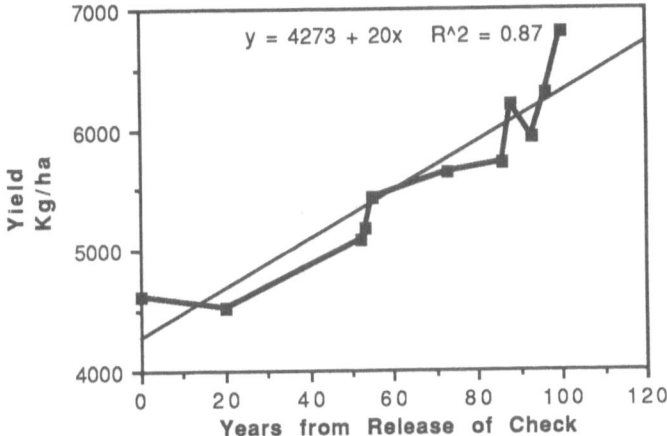

Fig. 3. Gain in yield of barley varieties in Great Britain, 1880-
         1980. Regression of barley varieties yields (kg/ha) on
         years after release of the oldest variety in the trial,
         by Riggs et al. (1981).

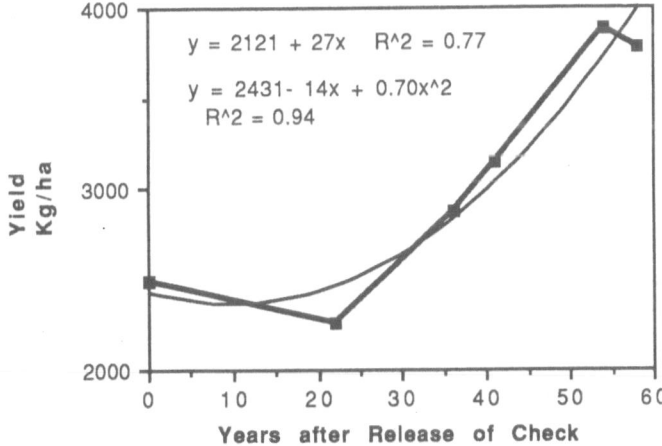

Fig. 4. Gain in yield of malting barley varieties in Minnesota,
         1920-78. Regression of barley varieties yields (kg/ha)
         on years after release of oldest variety in the trial, by
         Wych and Rasmusson (1983).

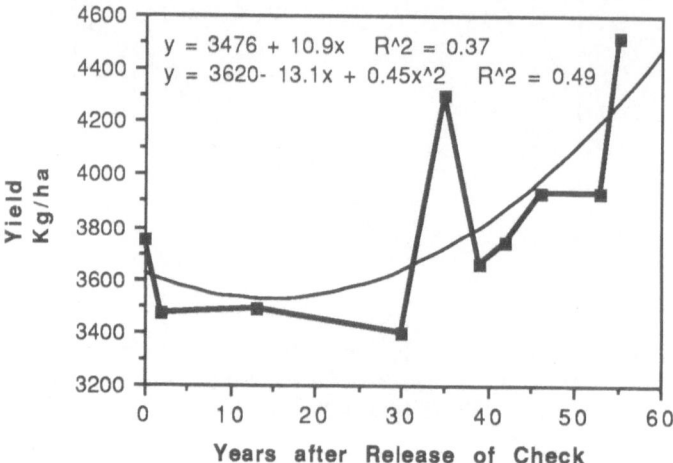

Fig. 5.   Gain in yield of oat varieties in Great Britain, 1908-
          1962. Regression of oat varieties yields (kg/ha) on
          years after release of the oldest variety in the trial,
          by Lawes (1977).

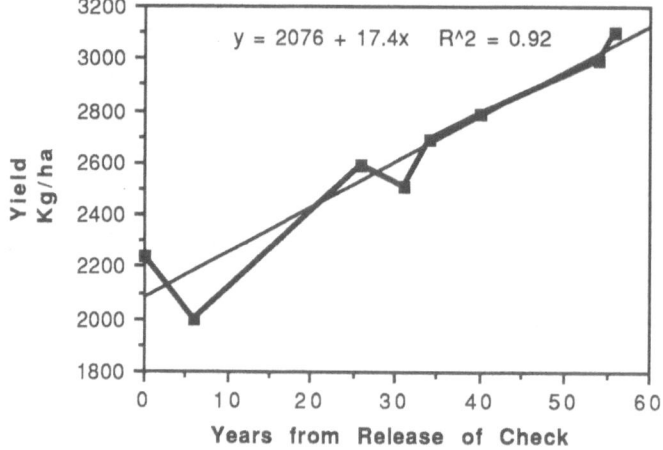

Fig. 6.   Gain in yield of oat varieties in Minnesota, 1920-80.
          Regression of oat varieties yields (kg/ha) on years after
          release of the oldest variety in the trial, by Wych and
          Stuthman (1983).

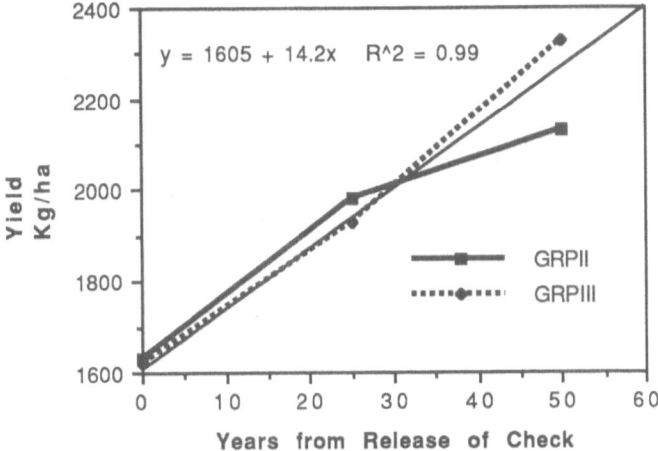

Fig. 7. Gain in yield of soybeans in the USA, 1925-75. Regres-
         sion of soybean varieties yield (kg/ha) on years after
         release of oldest varieties from maturity group II and
         III, Luedders (1977).

continue to have a major impact on the success of plant breeding
programs. Planters using tray systems can seed over 20,000 rows
of early generation materials per day. The trays used for the
planting can be used to store the seed before planting as well.
Yield trial plots can also be planted rapidly and accurately
without major physical effort using multiple row dividers.

Disease reactions, height, heading date and other field data
may be entered directly into hand held or portable computers and
later downloaded into either maintrame or personal computers for
data storage and analysis. Harvesting, once a time consuming
labor intensive job, can now be accomplished quickly by combines
which can be fitted with computers and scales to allow automatic
data entry in the field. Field sheets, randomizations, data
analyses, and seed inventories can be done in the laboratory on
personal computers with great time savings and improved accuracy.
Because of the ability to rapidly analyze data from field trials,
performance of earlier generation lines can be easily assessed
before they are used as parents in crosses which provide more
rapid recycling of germplasm in the breeding program.

Careful use of herbicides has greatly reduced the need for
hand labor to control weeds in nurseries, although weed control

continues to be a challenge in most breeding programs. With
herbicides, the much larger programs, made possible by mechanized
planting and harvesting, can be sustained.

Capabilities to access quality of grain have been enhanced as
well. Two such developments will be mentioned here. Protein
determination is now relatively easy, less hazardous, and more
rapid since near infra-red analyzers have been developed. Nuclear
magnetic resonance similarly provides an easy and rapid method to
measure oil content in grains. These two examples along with many
other developments have resulted in much greater efficiency for
evaluating genotypes in a rapid, systematic fashion. A plant
breeder can now manage a program at least 4-5 times larger with
better definition of the evaluated genotypes than was possible 30
years ago.

Breeding programs have grown in size by increasing the number
of crosses, increasing the number of lines per cross, or combinations of both. The added efficiency has also allowed more traits
including resistance to more pests to be evaluated.

RECURRENT SELECTION

New plant breeding procedures have been developed to take
advantage of the new equipment and instrumentation technology.
Similarly, programs have benefitted from the application of quantitative principles to plant breeding and the use of better statistical analyses in evaluation. Approaches developed and successfully used in cross-pollinated crops for many years have recently
been adopted by self-pollinated crop breeders as well. These
include cyclic selection which will be discussed in some detail
here.

Recurrent selection consists of three basic steps: 1) selection of parents, 2) intercrossing selected parents, and 3) progeny
testing. For self-pollinated crops, most testing is conducted on
a phenotypic basis and selfed progeny are used directly as parents
for the next cycle, as opposed to testing progeny to determine
which parents of the progeny will be utilized in the next cycle.
Many self-pollinated breeding programs have been conducted as
nonstructured recurrent selection programs, with parents not
restricted to the recurrent population (open). Only recently have
structured recurrent selection programs with the parentage being
restricted to the population under selection (closed) been conducted
and reported for self-pollinated crops.

Rickey (1927) is generally credited with first introducing
the idea of cyclic improvement in crop plants. Jenkin's (1940)
suggested alternatives to mass selection and early methods of

recurrent selection. His ideas on the advantage of selection among rather than just within lines, and testing for yield potential were relevant to both self- and cross-pollinated crops. Hull (1945) outlined a similar procedure and was the first to introduce the term recurrent selection. Comstock et al. (1949) introduced a method called recurrent reciprocal selection, which was intended primarily to improve two heterozygous populations concurrently. The goal of this procedure was mutual improvement of the two populations so that extracted lines from each, when crossed, will produce improved hybrids.

Recurrent selection has proven to be highly effective in improving corn (Zea mays) populations for yield if sufficient genetic variability existed in the initial populations and adequate testing of progeny allowed for effective recombination. Hallauer (1985) has summarized recurrent selection methods and their application. Rodriguez and Hallauer (1988) reported that yield was increased for all types of cyclic selection, except one population undergoing reciprocal recurrent selection. The rate of gain was probably slower than if single trait selection had been employed, because these germplasm sources were being developed for use in applied breeding programs so attention was devoted to other traits as well as yield.

Most of the recurrent selection methods have been developed and long term studies have been conducted in corn. Because of favorable results in corn, recurrent selection has also been adopted by breeders to self-pollinated crops, but the methods used in corn have required modification to fit the pollination and production mode of the crop. The problems involved using recurrent selection in a self-pollinated crop usually involve either random intermating between cycles of selection, and/or methods of progeny evaluation. Male sterility has been proposed for use in wheat (Sorrells and Fritz, 1982), soybeans (Brim and Stuber, 1973) and other crops but most reports of recurrent selection have been conducted using extensive hand crossing. The methods used for crossing in self-pollinated crops are important but less expensive than progeny testing.

Strategies for utilizing recurrent selection in self-pollinated crops have not been extensively evaluated, but Piper and Fehr (1987) conducted empirical studies in a soybean population to determine the most effective amount of testing and the optimum number of random intermating generations required. One year of hill plot testing was compared to one year of hill plots and a reduced number of lines selected from hills tested in rows a second year. From one to three cycles of random intermating were included between each cycle of selection. Their results indicated that multiple generations of intermating did not enhance genetic gain and that two year testing was superior to one year of testing to improve yield per cycle and per year.

Recurrent selection has been utilized in wheat improvement primarily in breeding studies. Loeffler et al. (1983) and McNeal et al. (1978) evaluated recurrent selection for increased grain protein content of spring wheat with success reported in both reports. Both evaluations followed only two cycles of selection and both used F3 rows to screen progeny for protein to identify parents for the next cycle. McNeal et al. (1978) did not report gain per cycle but indicated that several lines met the requirements for agronomic traits in order to qualify as parents for the next cycle. Loeffler et al. reported a gain of 0.5% protein for random lines and 0.35% for bulk populations per cycle. Height and yield values gave indications of becoming less desirable in both studies. Loeffler et al. (1983) stressed the need for multiple trait selection if the resulting germplasm is to be useful in applied breeding.

Busch and Kofoid (1982) reported that recurrent selection effectively increased kernel weight in spring wheat by 3% per cycle, measured on two cycles of lines, and 7% per cycle, measured over four cycles of bulks. F2 plant evaluations were the basis for selecting parents to intercross for the next intercrossing cycle. Avery et al. (1982) conducted three cycles of recurrent selection for early heading in winter wheat with considerable success, since gain averaged 1 day earlier heading per cycle.

There are more reports on utilizing recurrent selection in soybeans than for any other self-pollinated species. The method has increased grain protein (Brim and Burton, 1979), yield (Sumarno and Fehr, 1982; Piper and Fehr, 1987), resistance to iron chlorosis (Prohaska and Fehr, 1981), oil quality (Carver et al., 1986) and tolerance to Phytophora root rot (Walker and Schmitthenner, 1984). Recurrent selection appears to be an effective method of providing populations of improved germplasm for further soybean improvement.

Grain yield in oats has been improved using recurrent selection techniques as well. At Minnesota, five cycles of recurrent selection for grain yield in oats has been completed during the last 20 years. Payne et al. (1986) reported on gain after the first three cycles. Their procedures for cycle 1 used 12 parents intercrossed to provide 64 crosses. Single seed descent was used to advance to the F4 followed by a year for seed increase. The highest yielding line from each of 21 best yielding crosses was recombined to form cycle 2. Cycle 3 was formed similar to cycle 2. Gain in yield of 3.8% per cycle for three cycles was obtained. Almost 1% gain per year was found over the cycle 0 parents. This was a rapid increase in grain yield since the cycle 0 parents were

high yielding genotypes. Bregitzer et al. (1987) reported no
increase in harvest index for the same cycle 3 parents when they
were compared to cycle 0 parents for yield and correlated responses
in morphological traits. Most morphological structures measured
increased in size when yield increased. Pomeranke and Stuthman
(1988) reported 50% yield increase after five cycles of selection
but there were undesirable correlated responses for later maturity
and plant height.

In barley, results from studies of recurrent selection are
very limited. Patel et al. (1985) reported on one cycle of selection for increased yield using doubled haploid progeny for evaluation. Using a very intense selection differential, a small increase
in mean yield of cycle 1 lines was observed when compared to cycle
0. McProud (1979) evaluated three major barley breeding programs,
which can be considered as open parent, non-structured recurrent
selection from different areas of the world. He assigned base
populations to each, noted germplasm additions and constructed a
chronological order of selection cycles. The evolution of all
programs was considered as phenotypic recurrent selection with
cycle times varying from 6.5 to 10.5 years. All three programs
have been highly successful, but McProud suggested that an expan-
sion of genetic variation and a reduction in cycle time would
increase the efficiency of each program.

To summarize, recurrent selection can be considered a proven
system. It has been shown to be effective in both cross and self-
pollinated crops. Recurrent selection's effectiveness is influenced
by methods of screening of progeny, amount of genotypic variance
present, and length of time per cycle (Piper and Fehr, 1987). The
knowledge gained from quantitative genetics has allowed improvements and modifications which have increased its effectiveness and
usefulness.

RAPID GENERATION ADVANCE

Additional selfing generations before each cycle of selection
allows easier selection because the germplasm is more homozygous
and more rapid incorporation of desirable traits. Several methods
have been used to obtain additional generations per year, using
greenhouses and off-season nurseries. Goulden (1939) proposed
using a method, now called single seed descent, to provide a rapid
and broad genetic base approach to homozygosity. Single seed
descent can be applied by randomly advancing a single seed from
each plant and repeating the process until the desired level of
homozygosity is reached. With the use of single seed descent and
greenhouses, up to three generations per year of spring wheat,
oats, and barley can be obtained. The technique has been used as
suggested by some breeding programs or has been modified by other

programs by using selection for certain highly heritable traits
before applying single seed descent.

An example using rapid generation advance can be drawn from
the spring wheat breeding program at the University of Minnesota.
Fl (crosses) seed is produced in the fall greenhouse, with the F2
seed produced on F1 plants in the spring greenhouse. F2 plants
are planted in the field in the summer permitting selection for
disease reaction and plant type. One seed from each selected F2
plant is planted in the fall greenhouse to produce the F3, and a
seed from each F3 plant is grown in the following spring greenhouse. All seed from the F4 plants are harvested and F5 plant
rows are grown in the field with selection for disease and agronomic traits. Heads are harvested from each selected F5 row, and
one to two heads from each row are sent to the winter increase
nursery in Texas as F6 rows. Sufficient seed is produced to grow
a preliminary yield nursery of F7 lines at two locations the
following summer. Thus, generation advance to the first yield
evaluation in F7 is accomplished in three years after crossing.
Assuming the selection intensity and heritability remain constant,
the shorter the cycle time, the higher the rate of expected gain.
This time reduction likely contributes to the greater genetic
gains for yield documented in the last 30 years as compared to
previous periods.

CYTOGENETICS

In the past, cytogenetics has contributed both genetic knowledge and useful genes for resistance to disease and insects
(Sears, 1981). An alien transfer into present-day wheats that
appears to involve more than disease resistance is the substitution of rye chromosome 1R for wheat chromosome 1B, or a translocation involving 1RS and 1BL. These transfers have been reported to
be advantageous for yield as well as disease resistance (Zeller,
1973; Mettin et al., 1973; Rajarm et al., 1983; Gustafson, 1988).
The 1BL/1RS translocation has been of such significance that some
European breeding programs are screening progeny to ensure the
presence or absence of this translocation. Problems of sticky
dough have been reported for this 1BL/1RS translocation, but its
use is widespread and new cultivars containing it recently have
been released in both the hard winter and spring wheat, and soft
wheat areas in the U.S. and in Australia and Europe. Systematic
studies of 1R effects on wheat after its incorporation into wheat
are still lacking, however.

CEREAL TECHNOLOGY

Cereal technology is also beginning to impact on wheat breeding

in the area of wheat quality. High molecular weight subunits of
glutenin have been correlated with rheological and bread-making
properties in European wheats (Payne et al. 1979; 1980; 1981;
1987). The locations of the high molecular weight subunits of
glutenin were related to chromosomal locations and the presence or
absence of these subunits was related to good or poor bread-making
quality differences (Ng and Bushuk, 1988). In some of the worlds
wheat breeding programs the parents are assessed for their glutenin
subunit composition and crosses are made which provide the maximum
opportunity for progeny to exhibit good bread-making qualities.

Lawrence et al. (1988) has indicated that differences in
specific high molecular weight glutenin subunits may also be
important in determining quality differences among bread wheat
cultivars in Australia. In 44 U.S. hard red spring genotypes,
Khalil Khan (pers. commun.) found subunit 2 (1A) in all wheats
except 7 of 44, subunit 7 (1B) in all the wheat lines with subunits 8 or 9 (1B) in about equal frequency with each other, and
subunits 5 and 10 (1D) in all spring wheats except one. The high
molecular weight composition of the wheats grown in the upper
midwest indicated that the vast majority of these genotypes contained the subunit combinations that are characteristic of good
bread-making quality. Well over 90% of the spring wheat genotypes
contained the 5 and 10 subunit combination characteristic of
strong dough strength used for blending with weaker-type European
wheats while the presence of the 5 and 10 combination was present
in only 19% of the English wheats (Payne, 1987). Khan (pers.
commun.) also stated that subunit 8 tended to be associated with
lower protein content and had negative relationships with other
quality parameters. Subunit 9 was associated with higher protein
content and better baking qualities.

Although the vast majority of spring wheat genotypes contained the high molecular weight glutenin subunit composition
which characterize good bread-making quality, substantial differences exist in bread-making quality traits among these genotypes.
Lawrence et al. (1988) suggested that (quantitative differences)
in high molecular weight subunit composition may explain these
differences when similar subunits are present but bread-making
qualities differ. These differences may influence the interactive
properties of the gluten proteins. Although the use of SDS-PAGE
electrophoresis may not be useful for wheat classes with a long
history of high bread-making quality, many wheat breeders can use
the technique to predict parents to cross for improved quality in
the progeny if the parents are from diverse origin. These types
of techniques hold good promise for future breeding efforts for
use as markers for complex traits which are not readily observable.

SUMMARY

     As plant sciences have evolved, crop improvement progress has
become more sophisticated. In the future, there will be a need
for additional basic geneticists to improve germplasm sources, so
that genetic products can be used directly in breeding programs.
However, most applied plant breeders are unable or do not have the
time to make the detailed effort required to improve these basic
genetic materials for direct use. Therefore, the role of an
"intermediate" level geneticist will become increasingly important
for the basic geneticist to improve the raw germplasm for use by
the plant breeder. Frequently, genetic stocks, produced by basic
genetic manipulations, lack desirable agronomic traits. Even
worse, some genetic stocks have linkages of the favorable gene
with very undesirable trait expression. Ways to break these
linkages and studies of effective methodology to rapidly and
effectively improve germplasm will greatly aid future germplasm
development. The emerging areas of genetic manipulation will
increase the demand for intermediate stages of genetic improvement. The level of coordination and cooperation among basic,
intermediate and applied genetics will dictate future gain in crop
cultivar development.

REFERENCES

Austin, R.B., Bingham, J., Blackwell, R.D., Evans, L.T., Morgan,
     C.L., and Taylor, M., 1980, Genetic improvements in winter
     wheat yields since 1900 and associated physiological changes,
     J. Agric. Sci., Camb., 94:675-689.
Avery, D.P., Ohm, H.W., Patterson, F.L., and Nyquist, W.E., 1982,
     Three cycles of simple recurrent selection for early heading
     in winter wheat, Crop Sci., 22:908-912.
Bregitzer, P.P., Stuthman, D.D., McGraw, R.L., and Payne, T.S.,
     1987, Morphological changes associated with three cycles of
     recurrent selection for grain yield improvement in oats,
     Crop Sci., 27:165-168.
Brim, C.A., and Burton, J.W., 1979, Recurrent selection in soy-
     beans. II. Selection for increased percent protein in seeds,
     Crop Sci., 19:494-498.
Brim, C.A., and Stuber, C.W., 1973, Application of genetic male
     sterility to recurrent selection schemes in soybeans. Crop
     Sci. 13:528-530.
Busch, R.H., and Kofoid, K., 1982, Recurrent selection for kernel
     weight in spring wheat, Crop Sci., 22:568-572.
Carver, B.F., Burton, J.W., Wilson, R.F., and Carter, T.E., 1986,
     Cumulative response to various recurrent selection schemes in
     soybean: Oil quality and correlated agronomic traits, Crop
     Sci., 26:853-858.
Comstock, R.E., Robinson, H.F., and Harvey, P.H., 1949, A breeding

procedure designed to make maximum use of both general and specific combining ability, Agron. J., 41:360-367.

Goulden, C.H., 1939, Problems in plant selection, in: Proc. 7th Int. Genet. Congr., R.C. Burnett, Ed., (Edinburgh) Cambridge Univ. Press, p. 132-133.

Gustafson, J.P., Evaluation of a 1R/1D substitution in wheat, in: Proc. of Seventh Int. Wheat Genetics Symp., R. Koebnen and T. Millen, Eds., Cambridge, Great Britain, in press.

Hallauer, A.R., 1985, Compendium of recurrent selection methods and their application, CRC Crit. Rev. Plant Sci., 3:1-33.

Hull, F.H., 1945, Recurrent selection and specific combining ability in corn, J. Amer. Soc. Agron., 37:134-145.

Jenkins, M.T., 1940, The segregation of genes affecting yield of grain in maize, J. Amer. Soc. Agron., 32:55-62.

Lawes, D.A., 1977, Yield improvement in spring oats, J. Agric. Sci., Camb., 89:751-755.

Lawrence, G.J., MacRitchie, F., and Wrigley, C.W., 1988, Dough and baking quality of wheat lines deficient in glutenin subunits controlled by the Glu-A1, Glu-B1 and Glu-D1 loci, J. Cereal Sci., 7:108-112.

Loffler, C.M., Busch R.H., and Wiersma, J.V., 1983, Recurrent selection for grain protein percentage in hard red spring wheat, Crop Sci., 23:1097-1101.

Luedders, V.D., 1977, Genetic improvement in yield of soybeans, Crop Sci., 17:971-972.

McNeal, F.H., Mcguire, C.F., and Berg, M.A., 1978, Recurrent selection for grain protein in spring wheat, Crop Sci., 18:779-782.

McProud, W.L., 1979, Repetitive cycling and simple recurrent selection in traditional barley breeding programs, Euphytica, 28:473-480.

Mettin, D., Blunthner, W.D., and Schlegel, R., 1973, Additional evidence of spontaneous 1B/1R wheat-rye substitutions and translocations, in: Proc. of Fourth Int. Wheat Genetics Symp., E.R. Sears and L.M.S. Sears, Eds., Univ. of Missouri, Columbia, pp. 179-184.

Ng, P.K.W., and Bushuk, W., 1988, Statistical relationships between high molecular weight subunits of glutenin and bread-making quality of Canadian-grown wheats, Cereal Chem., 65:408-413.

Patel, J.D., Reinbergs, E., and Feger, S.O., 1985, Recurrent selection in doubled-haploid populations of barley (Hordeum vulgare L.), Can. J. Genet. and Cytol., 27:172-177.

Payne, P.I., Corfield, K.G., and Blackman, J.A., 1979, Identification of a high-molecular weight subunit of glutenin whose presence correlates with bread-making quality in wheats of related pedigree, Theor. Appl. Genet., 65:153-159.

Payne, P.I., Law, C.N., and Mudd, E.E., 1980, Control by homoeologous group 1 chromosomes of the high-molecular-weight subunits of glutenin, a major protein of wheat endosperm,

Theor. Appl. Genet., 58:113-120.

Payne, P.I., Cornfield, K.G., Holt, L.M., and Blackman, J.A., 1981, Correlations between the inheritance of certain high-molecular-weight subunits of glutenin and bread-making in progenies of six crosses of bread wheat, J. Sci. Food Agric., 32:51-60.

Payne, P.I., 1987, The genetical basis of bread-making quality in wheat, Aspects of Appl. Biol., 15:79.

Payne, T.S., Stuthman, D.D., McGraw, R.L., and Bregitzer, P.P., 1986, Physiological response to three cycles of recurrent selection for grain yield improvement in oats, Crop Sci., 26:734-736.

Piper, T.E., and Fehr, W.R., 1987, Yield improvement in a soybean population by utilizing alternative strategies of recurrent selections, Crop Sci., 27:172-178.

Pomeranke, G., and Stuthman, D.D., 1988, Evaluation and comparison of an open and closed system of recurrent selection for yield in oat, Agron. Abst., p.92.

Prohaska, K.R., and Fehr, W.R., 1981, Recurrent selection for resistance to iron deficiency chlorosis in soybeans, Crop Sci., 21:524-526.

Rajaram, S., Mann, C.E., Ortiz-Ferrara, G., and Mujeeb-Kazi, A., 1983, Adaptation, stability and high yield potential of certain 1B/1R CIMMYT wheats, pp. 613-621, in: Proc. Sixth Int. Wheat Genetic Symp., S. Sakamoto, ed., Kyoto, Japan, pp. 613-621.

Richey, F.D., 1927, Convergent improvement in selfed lines, Am. Nat., 61:430-449.

Riggs, T.J., Hanson, P.R., Start, N.D., Miles, D.M., Morgan, C.L., and Ford, M.A., 1981, Comparison of spring barley varieties grown in England and Wales between 1880 and 1980, J. Agric. Sci., Camb., 97:599-610.

Rodriguez, O.A., and Hallauer, A.R., 1988, Effects of recurrent selection in corn populations, Crop Sci., 28:796-800.

Schmidt, J.W., 1984, Genetic contributions to yield gains of wheat, in: Genetic contributions to yield gains of five major crop plants, W.R. Fehr, ed., Crop Science Society of America, special publication No. 7, Madison, WI.

Sears, E.R., 1981, Transfer of alien genetic material to wheat, in: Wheat Science--Today and Tomorrow, L.T. Evans and W.J. Peacock, eds., Cambridge University Press, Cambridge, pp. 75-89.

Sorrells, M.E., and Fritz, S.E., 1982, Application of a dominant male-sterile allele to the improvement of self-pollinated crops, Crop Sci., 22:1033-1035.

Sumarno, and Fehr, W.R., 1982, Response to recurrent selection for yield in soybeans, Crop Sci. 22:295-299.

Walker, A.K., and Schmitthenner, A.F., 1984, Recurrent selection for tolerance to Phytophthora rot in soybeans, Crop Sci., 24:495-497.

Wych, R.D., and Stuthman, D.D., 1983, Genetic improvement in Minnesota-adapted oat cultivars released since 1923, Crop Sci., 23:879-881.

Wych, R.D., and Rasmusson, D.C., 1983, Genetic improvement in malting barley cultivars since 1920, Crop Sci., 23:1037-1040.

Zeller, J., 1973, 1B/1R substitutions and translocations, in: Proc. Fourth Int. Wheat Genetics Symp., E.R. Sears and L.M.S. Sears, eds., Univ. of Missouri, Columbia, pp. 209-221.

THE ROMANCE OF PLANT BREEDING AND OTHER MYTHS

Donald N. Duvick

Pioneer Hi-Bred International, Inc.
700 Capital Square, 400 Locust Street
Des Moines, Iowa, 50309, U.S.A.

INTRODUCTION

Myths are essential to human existence. They give meaning to life, inspiration and strength for the daily struggle. They often express profound truths, embedded in elements of fantasy. Everyone deeply believes in some myths, while deprecating others as dangerous falsehoods. The origins of myths usually are unknown, but probably they are inspired by dramatic events, clearly important to those who first noted them, but with cause and effect imperfectly understood.

In my world, plant breeding, I have learned three myths. I believe in them and act on them, and at the same time I know they are fallacious. I shall describe them, and attempt to explain why I believe in them even as I understand — with ever deepening knowledge — that they also are untrue.

Myth Number One. The world's food supply depends absolutely on past achievements of plant breeders of the major food and feed crops. Without their contributions, food production of wheat and rice, or meat production based on maize and soybean meal, could provide only a fraction of our daily needs. Equally important, if plant breeders do not continually improve productivity of the important food and feed crops, our burgeoning world population soon will outstrip global food supplies.

This is myth number one. It is an interesting and perhaps a unique myth. Very few people believe it — or even know about it — except plant breeders. But this is not surprising, since almost no one knows about plant breeders.

Gene Manipulation in Plant Improvement II
Edited by J. P. Gustafson
Plenum Press, New York, 1990

Myth Number Two. Genetic engineering, applied to plants, uniquely will revolutionize our abilities to grow crops. New levels of pest and chemical resistance, new kinds of weather-proofing, and exotic new plant-produced chemicals will transform our food production systems, even to the point of causing excessive and undesirable over-production. But also possible is accidental release of uniquely dangerous, ecologically disastrous mistakes, spreading without possibility of recall.

This is myth number two. Everyone - or almost everyone - knows about it and believes at least parts of it. Only the plant breeders, at least some of them, express reservations about it.

Myth Number Three. Excessive genetic uniformity, dubbed "genetic vulnerability", and expressed as too few varieties, with too much uniformity, spread over too broad an expanse, is the greatest present threat to the world's food supply. New disease and insect pests, uniquely adapted to the widely planted, uniform varieties, inevitably will arise and devastate them. Our chief recourse is to diversify our major crop varieties, greatly increasing their numbers and variability. This can be done by drawing upon the wealth of already existing variability in the myriad peasant varieties (landraces) and wild relatives of these crops, still grown or found in nature throughout many parts of the world.

This is the third myth. It is believed by a small but select group of people, an interesting mixture of biologists, politicians and social activists. Plant breeders again have certain reservations, and the world at large, as with myth number one, has never heard of it.

DISCUSSION

Plant Breeding. Myth number one ("plant breeders are the indispensable basis of the world's food supply") is true to the extent that plant breeders, by developing improved crop varieties, indeed have contributed more than one-half of the 5- and 10-fold increases in productivity achieved in our major crops during the past 50 years. Without these gains in crop productivity the world could not be fed, today. Certainly, greatly increased inputs of chemical fertilizers, herbicides, and insecticides, and better planting and harvesting machinery, have made important contributions also, but without the plant breeder's contribution the other yield-raising inputs would have had much less impact. For example, only the newer cereal and maize varieties are capable of withstanding the increased levels of nitrogen fertilizer that support increased grain yields. The older varieties, when heavily fertilized, grow too tall, and fall down before harvest.

Interestingly, the newer varieties not only can better absorb the stresses and strains of intensive cultivation, they also are better at withstanding the problems of poor growing conditions, of under-fertilization, of heat and drought, and of excessively cool, wet conditions. The new high yield varieties, in general, are stronger and more broadly adapted. (Exceptions are to be found of course. For example, the new, short, high yielding rice varieties cannot survive in deep water conditions to which certain ancient varieties are superbly adapted, because they devote much of their energy to stem elongation. And some of the new maize hybrids, with upright leaf habit to accommodate thick planting, may be less able to shade out weeds in poorly weeded, sparsely planted maize fields.)

It also is true, however, that plant breeders now are on a treadmill - they must continue to turn out new varieties, frequently and with regularity, because (as stated in myth number three) new biotypes of disease and insect pests, uniquely adapted to the widely grown varieties, soon make varieties obsolete. New varieties, with new kinds of resistance, must replace the old ones. (Old varieties of course are replaced for other reasons as well. New varieties routinely displace older ones because they provide higher yields, greater reliablity of yield, or improved quality of product.)

Why are so few varieties planted so widely? The answer (or, at least, one of the answers) is simple. Today's commercial farmers, worldwide, are keenly aware of experiences of their compatriots over wide expanses of territory. They also can evaluate varietal performance with precision. They soon identify the best one or two of a series of new varieties, and plant only them. As a general rule, the more consistently a variety excels in any one locale, the more widely adapted it will be to many locales. Thus, many farmers over a broad expanse simultaneously choose the same one or two varieties as the best for them, at that particular time and place. Additionally, they quickly learn the special needs and strong points of their chosen varieties and typically carry them well beyond the capabilities envisaged for them, by the breeders who developed them.

For example, my company bred an early maturing maize hybrid for the very short but relatively warm season of southern Manitoba, in midwestern Canada. Farmers in Canada's Quebec province, in eastern Canada, soon found that the hybrid also was suited to their short, cool maritime seasons and began to plant it in Quebec. Farmers in a protected bay region of southeastern Sweden, through relatives in Quebec, then also heard of the hybrid, found that they could make it mature for ensilage purposes, and soon were planting it widely, buying the seed from Canada. In all cases, farmers, on their own, had heard about,

tested, and adopted the hybrid, carrying it to places and using it
in ways never envisaged by its breeders.

And this example points up the fallaciousness of myth number
one.   Plant breeders are not the sole mainstay of our food
production system.   They are only one part of a highly integrated
system.   Farmers, plant breeders, chemical and fertilizer
suppliers, and many others provide interactive ingenuity, power
and direction for the system.   Plant breeders may be indispensable
to the system's operation – but so are the other players.

And importantly, the effects of the other players constantly
pull and haul plant breeders to and fro.   For example, when
farmers in northern Germany decided to grow maize as a high energy
grain crop (even though present maize hybrids cannot produce fully
mature, dry grain in northern Germany), they simply ground and
ensiled the entire immature maize ear, using the most suitable
hybrids then on hand.   ("Corn-cob mix" is the common name for this
new kind of high energy feed.)   This new practice, in turn, has
pressured maize breeders to select new grain-type hybrids for
northern Germany, hybrids with a high ratio of grain to fodder.
This is a turnabout from earlier pressures from that region, to
select silage-type hybrids with large stature and a lower ratio of
grain to fodder.

Another example, again with maize.   In the 1960s U.S. farmers
increased planting rates of maize hybrids by 50% and more, to take
advantage of increased application rates of newly available low
priced commercial nitrogen fertilizers.   The crowded maize plants
suffered; poor root development and subsequent plant lodging, poor
ear development and subsequent yield loss were typical of the ills
brought on by over-crowding.   Maize breeders responded to the
challenge by developing new hybrids genetically able to withstand
the ill-treatment of high plant density and over-fertilization.
They discarded old breeding lines with alacrity and selected new
genotypes with new adaptations and new strengths – able to
withstand the new cultural conditions of the 1960s.

But U.S. farmers, switching en masse to the new hybrids,
continued to raise planting rates and fertilizer amounts, thus
subjecting the new generation of hybrid genotypes to stresses more
severe than had been expected by the maize breeders.   So breeders
once again were pushed to breed a still more resilient line of
hybrids, for the 1970s and 1980s.

Now, with fertilizer use steady or declining due to economic
and environmental reasons, maize breeders once again are faced
with new challenges.   They are creators of change but they also
are manipulated by their surroundings.   They are only one strand
in an intricate web of actions and reactions, with consequences

always changing, often unpredictable but certainly important.

Genetic Engineering. Myth number two ("genetic engineering for plants will overwhelm us with enormously productive new varieties, some of which may spawn runaway genes uniquely causing ecological disaster") has in it certain elements of truth. Molecular genetics, with genetic transformation, will allow identification of powerful genes of agronomic utility, their transfer into elite germplasm, and then fine-tuning the genes to maximum levels of performance, conducive to high yields of grain and forage. For example, genes with defensive utility – genes conferring disease and insect resistance, or heat and drought tolerance – will increase the reliability, the stability of performance, of the new genetically engineered varieties. Some of the engineered genes will come from other, distant, species, even other kingdoms, and will enable the production of new protective chemicals, the operation of useful new biochemical pathways never before present in our major crop species.

And some of those genes or gene systems potentially could spread into the natural biome, through hybridizations with weedy relatives or cross-fertile wild species, giving potential for new kinds of vigor, of pest tolerance not previously there, and with unpredictable ecological consequences.

But myth number two is fallacious in part, also. Although useful new genes and new processes will be introduced into our breeding pools, they will produce the same end results, in most cases, that plant breeders always have turned out, ever since the profession got underway some 75 or 100 years ago. Breeders always and consistently have turned out improved varieties with better disease and insect resistance, more heat and drought tolerance, and better ability to withstand too little or too much fertilizer. Genetic engineering's results, although they will be real and invaluable, will be introduced gradually and in general will not be recognized as uniquely different by most farmers, accustomed to seeing regular and frequent varietal improvements.

In one way, however, genetic engineering will give recognizably new traits. In time, genes for production of expensive specialty chemicals will be placed in some of our commodity-type field crops. Soybeans, for example, may be engineered to make highly specialized industrial grade oils, or sunflower to make rubber. (But most such speculations are far from even the drawing board stage today. Initial plans are more modest; rapeseed plants may be engineered to make higher percentages of saturated – or of unsaturated – fatty acids, for example.) When old crops make new products, or at least strangely different ones, the age of genetic engineering, spawned by molecular genetics, will be visible to all.

And yet there is fallacy even in this belief, at least as it
is interpreted by some people to say that genetic engineering is
the sole means of introducing radically new products into old
crops. Ingenious plant breeders, with mutagens or with exhaustive
searches through collections of diverse farmer varieties, of weedy
and wild relatives, already are finding unusual variants and
breeding with them, to put new products into old crops using
standard sexual breeding techniques. Rapeseed, for example, has
been reincarnated via standard plant breeding as "canola" – low in
erucic acid and glucosinolates, and safe for human and animal
consumption. (Erucic acid and glucosinolates, present in native
rapeseed, have anti–nutritional properties.) Thus, genetic
engineering's achievements in making new products with old crops
will not be unique, even though they will be startling, highly
useful, and the result of ingenious scientific endeavor.

What about the second portion of myth number two? Will
runaway genes or varieties, the products of genetic engineering,
induce ecological disaster, such as insect–proof, herbicide
resistant, wild sunflowers? (Sunflowers already are classified as
a noxious weed in the U.S., and one would not want to make them
even more aggressive.) Here, powers for precise prediction are
low, in my opinion, but chances for disaster seem small. Since
wild species (by definition) already are well buffered
genetically, I personally find it hard to imagine that one
additional gene – not selected for survival in the wild – will
move through a genetically well–balanced natural population with
large impact for either good or bad. And strict governmental
regulations about planned release of engineered organisms will be
administered on the side of safety.

But in the world of biology the unexpected always happens.
For example, a virus disease from Peru, called maize chlorotic
mottle, somehow teamed up with another virus called maize dwarf
mosaic (closely akin to the Asian–derived sugarcane mosaic virus)
and produced, in Kansas, U.S., a virulent new disease of maize
called, appropriately, corn lethal necrosis.

In future times we can be sure that additional unexpected and
unwished for biological phenomena will occur. Plant breeders,
including genetic engineers, could be connected with some of them,
simply because they do biology. But neither genetic engineering
specifically nor today's plant breeding in general will be
uniquely the cause of such problems.

Unforeseen consequences from human actions are not new in
biology, even to our imagination. Maize's original creators were
blamed for destruction of wild "proto–maize", for example, in one
scenario that pictures domestication of maize from a wild
progenitor species which in turn was swamped into extinction by

ROMANCE OF PLANT BREEDING AND OTHER MYTHS

non-adaptive genes from its domesticated progeny. (This
hypothesis is now generally out of favor, not because biologists
think the swamping was impossible, but rather because they think –
for the most part – that the proto-maize species never existed.)

So another fallacy of myth number two is the belief that
genetic engineering uniquely will bring about ecologically
undesirable consequences.

Genetic Vulnerability. Myth number three ("genetic
vulnerability, expressed as varietal uniformity, is too prevalent
and will bring on epidemic disaster; it could be prevented by
incorporating freely the great diversity available in third world
farmer varieties and in their wild relatives") is perhaps my
favorite. Certainly it is the most romantic of the three. It
throws into one pot, first world and third world farmers, plant
breeders, epidemiologists, systematists, ethnobotanists, first
world and third world politicians, social activists, genetic
engineers, geneticists of the Mendelian kind, and, occasionally,
interested bystanders.

Some of the believers of this myth note that it has political
utility. The belief that developing countries possess a valuable
store of genetic variability in their highly diverse farmer
varieties has led some developing countries, and social activists
speaking on their behalf, to express interest in a doctrine of no
release – except for a price – of the native germplasm in their
countries. (The problem really is more complicated than this
statement indicates, but it does hinge on a belief that plant
breeders, especially in the industrial countries, strongly and
uniquely need the rich variability that exists in some of the
developing countries.)

Biologists of ecological and systematics persuasions are
attracted to this myth perhaps because it fits so well with many
observations of nature in the raw – of great stores of diversity
within and between species, apparently mediating harmonious
homeostasis.

Epidemiologists are attracted to the myth for about the same
reason – they have seen or known of many epidemics that came about
when a uniquely adapted new biotype – a new strain – of an old
disease (or insect, or nematode) swept across fields of a variety
or population that was uniformly susceptible to the new biotype.
Often – although not always – the susceptible crop was of a single
variety, highly uniform for all traits, including the specific
pest susceptibility. The maxim suggested by these epidemics was:
uniformity means susceptibility; diversity means safety.

Plant breeders also have seen the same epidemics as the

epidemiologists and they, too, have noted the uniform
susceptibility of some varieties, or of families of varieties
connected by a thread of common ancestry.

And farmers, worldwide, have been first-line receivers of the
benefits or penalties associated with absence or presence of
genetic vulnerability. They have countered its threats with the
means at their disposal; countermeasurers used by subsistence
farmers in general are very different from those used by
commercial farmers of the industrial nations. Variability within
and among varieties, and planting many kinds of crops, are the
chief kinds of protection available to subsistence farmers.
Farmers of the developed nations, on the other hand, usually
depend on chemical and machinery aids to promote stability of
yields. Further, they are aided by the plant breeders, who
provide "diversity in time" by turning out a continuous stream of
new varieties sequentially countering the most recent insect or
disease variants.

Today's plant breeders can and do use some of the genetic
diversity available in farmer varieties and in wild relatives.
Rust resistance obtained from a wild grass relative (Agropyron),
has been useful in modern wheat varieties. Resistance to northern
corn leaf blight in maize is reported to have been transferred
from a related genus, Tripsacum. And soybean breeders depend
absolutely on repeated introductions, from primitive farmer
varieties, of genes for resistance to cyst nematode, or to
phytophthora root rot.

But, on the whole, plant breeders do not use large amounts of
such exotic germplasm in their breeding programs. They usually
depend on locally adapted, elite improved stocks as parents of new
varieties. Nor do they, in general, attempt to increase diversity
of the new varieties. They ordinarily are not trying to make
varieties less uniform, nor were they, until recently,
systematically trying to increase the total numbers of varieties,
nor to increase amounts of difference between varieties. For
example, in the 10 year span from 1970 to 1980, very little
reduction was seen, in the United States, in percent of total land
area planted to the leading half-dozen varieties of wheat and
soybeans, or of hybrids involving the six most widely used public
inbreds of maize. Concentration of farmer plantings on a few
varieties stayed about the same.

And, even greater restrictions in number and diversity of
varieties now are found in those developing countries that have
adopted the new high-yield rice and wheat varieties during the
past 20 years. Indonesia, for example, planted 66% of its rice
growing area to just two varieties, in 1984. And Bangladesh

planted just one variety of wheat on 67% of its wheat lands, in 1984.

We also have clear evidence of cases in which lack of diversity has been associated with onset of epidemic disease or insect attack. The most famous such event in recent times is the 1970 southern corn leaf blight epidemic in the United States. Nearly all maize in that year had a single cytoplasm ("T cytoplasm") tracing back to a single plant as ultimate seed parent. A new race of southern corn leaf blight, particularly virulent on that cytoplasm, swept across the land in 1970, lowering U.S. maize yields by an average of 15%.

So the main elements of myth number three are true: uniformity, low diversity, small effort to bring in diversity from outside (from exotic farmer varieties or from wild relatives), and epidemics associated with uniform and widespread susceptibility.

But myth number three has fallacious elements, also; it is not completely true.

Diversity does not necessarily guarantee safety. It if did, blight would not have destroyed the vast eastern American chestnut woodlands in the U.S. in the 1920s, nor the widespread, heterogeneous stands of American elm in the 1960s. Both of these native, wild species were highly heterogeneous, and variable for many traits – except for one: they were uniformly susceptible to the specific disease that killed them. So, simple diversity, "diversity in general", is not enough; diversity must include elements able to combat specific pest problems.

Another fallacious element in myth number three is the belief that farmer varieties or wild relatives are uniquely essential sources of diversity for improvement of modern crop varieties, particularly in the temperate zones. In fact, breeders say that improved locally adapted varieties are the most used and most valued sources of needed diversity, of needed improvements in disease and insect resistance, in stress tolerance, and in specific climatic adaptations. Breeders find that only rarely do they need to go to farmer varieties or wild relatives for needed new kinds of diversity. Interestingly, breeders in developing countries are as reluctant as breeders in developed countries, to use farmer varieties as the chief basis for varietal development once the first cycle of selection and varietal development is completed. They prefer to recycle their own elite new lines. And they consider elite lines from developed countries as their best source of exotic (non-local) germplasm. To them, the developed countries are "gene-rich". (In actuality, of course, all countries are "gene-rich" in one way or another; the world's store of improved varieties is one such treasure; the world's store of farmer varieties is another.)

Also fallacious is the belief that breeders, quickly and simply, can extract useful traits from exotic farmer varieties and wild relatives for use in development of new, highly diverse, modern varieties. Breeders do not ignore this possibility for genetic enhancement out of lassitude or because they don't see its theoretical advantages. Rather, they know from experience that such attempts at incorporating complexly inherited traits from exotic landraces or wild relatives are almost sure to be difficult, they will take many years to complete, and the results may well be unexpected and undesired.

To be sure, traits inherited as a single gene or as a tightly linked cluster of genes can be transferred with relative ease. But traits with complex inheritance bring in multitudinous linkages with undesirable qualities – qualities that likely were needed in the primitive variety or wild species, but that are detrimental to performance of modern varieties.

For example, I have used Zea diploperennis as a source of prolificacy – multiple-earing – to be added to the elite midwestern U.S. inbred, B73. (Zea diploperennis, diploid perennial teosinte, is a wild grass from western Mexico and a close relative of maize.) Several cycles of backcrossing and interpollination of multiple-eared plants have been used to make a population with the temperate day-length adaptation and desirable plant habit of B73, and the added ability to make two or three ears per plant, instead of one. I have selected plants able to make two or more ears but have not made conscious selection for any other diploperennis traits.

Recently, I have begun to make repetitive cycles of self-pollination in this population, intending to extract inbred lines uniform for the multiple-eared trait but like B73 in all other respects. After two generations of selfing, the progeny are reasonably prolific, but they also exhibit recessive traits not previously visible in the cross-pollinated population, nor in the inbred B73. Notable among these traits, in this past growing season, were: 1) extreme heat susceptibility, manifested by streaks of dead leaf tissue; 2) lack of root development, manifested by lodging of all plants in a progeny; 3) extremely small plant size; 4) white endosperm (B73 has yellow endosperm); 5) reduced numbers of kernel rows, resulting in undesirably small ears; and 6) premature plant death, resulting in severe stalk and root rot. Not all progenies exhibited all of these traits at once, but by the time prolific progenies were discarded, individually, for presence of one or more of these undesirable traits, very few select progeny remained for further selfing and selection in hybrid combination.

No doubt the diploperennis genes for prolificacy are linked

with the undesirable traits I have described; in time I should be able to select for individuals in which the linkages have been broken. But it is clear to me that at least one more cycle of intermating, selfing and selection will be needed to achieve this end. At a minimum of five years per cycle, plus another five years for extraction of inbred lines, and another five years to put them in successful hybrid combination and introduce them to the farm, I will not be finished - if I am successful - until another 15 years, at a minimum, have gone by. And this is just for a rather simply inherited trait, as quantitative traits go.

It also is highly likely that even if I do derive agronomically satisfactory lines with multiple-eared capability, they may have no real advantage over non-prolific lines derived from standard non-exotic materials. Prolificacy per se is no guarantee of increased total grain yield per unit area. Individual prolific maize plants simply might reduce the size of their ears, produced two or three at a time, to the point that the sum of their grain weight was no more than the grain weight of a single large ear. Or they might not. The only way to know what will happen is to try - on a 10 to 15 year time scale.

But, in the meantime, I could spend my time (were I a full time maize breeder) in developing new inbreds from superior, locally-adopted, non-exotic lines using tried and proven breeding methods, and with high probability of success, compared to my venturesome and somewhat erratic efforts to incorporate diploperennis traits into U.S. cornbelt maize.

If my success - and my job - depended on the number and frequency of useful inbred lines I produced in a professional lifetime, it might not take much time for me (were I a young maize breeder) to decide that I would devote most of my time and energy to standard breeding procedures.

The choice between genetic enhancement in the sense of incorporating traits from exotics and genetic advance through standard breeding with adapted lines nearly always comes down in favor of the latter course of action. Genetic enhancement, especially using landraces or wild species, is chancy, slow, has no broadly recognized standard operating procedures, no stand-alone professional literature - and almost no advocates. Plant breeding and varietal development through standard genetic advance is relatively fast, some degree of success always is guaranteed, its procedures are well outlined in standard textbooks, and it is a proud, well-recognized profession.

Genetic Enhancement. This brings me to a subject that I consider to be extremely important, and in need of attention: the subject of genetic enhancement of elite modern breeding stocks by use of exotic materials, particularly farmer varieties (landraces) and related weedy or wild species. This type of breeding – sometimes called "pre-breeding" – has many advocates but few practioners. A few outstanding biologists – Charles Heiser, Charles Rick, Douglas Dewey, Stanley Peloquin and Ernest Sears come to mind – have had success in such activity. But on the whole this activity is not recognized as a profession, there is no agreed-upon name for it, there certainly is no "Society for Germplasm Enhancement", and I know of no category for its funding in, for example, the U.S. Competitive Grants fund for agricultural science.

And yet, any thoughtful, experienced plant breeder will say that there is great and continuing need for introgression of exotic germplasm into elite breeding materials (whether from tropics to temperate zones, or from temperate to tropic zones). Such breeders will admit that in practice they usually do not go far afield for exotic breeding materials, but nevertheless they often will say, citing data, that most major advances in breeding of their crop occurred following successful introgression of quite different breeding materials into their standard stocks (i.e., following genetic enhancement with exotics).

But they also will say that because of the long-term, risky nature of genetic enhancement activities, we need to have a broadly based, reliable, long-term method of funding top quality geneticist/breeders, doing genetic enhancement. Such funding requirements rule out most (although not all) of the private breeding efforts, which typically are relatively short-term in outlook, narrowly focused, and have limited resources in ancillary scientific fields. And unfortunately, such funding requirements also run counter to present-day trends for public funding in countries such as the United States, where less and less money is placed in long-term, project-oriented plant breeding and genetics, and more and more emphasis is placed on competitive grants, peer-reviewed, and given for periods of two to three years (although with renewal possibilities).

How can any researcher hope to get reviewers' approval for a project of uncertain time scale and conclusion, with no likelihood of publishable results until at least 5 or 10 years have passed, and with little opportunity to lay out a definable, narrowly circumscribed hypothesis and the conclusive experiments needed to test its validity? This problem is especially daunting if there are no categories for this type of activity and few or no reviewers with experience and standing in the field. (I do not denigrate the great need for and utility of the competitive grants concept. But I do point out what the grants perhaps cannot accomplish, as presently administered, in a field where great need exists.)

How can we promote genetic enhancement as a needed activity, well-respected and well-funded? I have some suggestions:

1) Settle on a name for this discipline, and define its meaning.
2) Establish a society, or a division within a society, of its practioners.
3) Promote publication of papers precisely on this subject, treated as a discipline. By this means, build up a body of theory and experimental results.
4) Be sure to include, as respected parts of the discipline, work in molecular and cellular biology as well as in classical breeding, quantitative genetics and cytogenetics.
5) Publicize in appropriate media, and apprise appropriate public and private funding agencies of the existence of this discipline. Tell them why it is needed, its accomplishments to date, and its expected accomplishments in future.

As a preliminary to this agenda, I propose formation of a select committee, one that includes world-renowned experts in the field and also some enthusiastic young researchers. These people would have a vision of the practical and scientific possibilities of genetic enhancement, broadly defined. Their committee could be the spearhead for formation of the newly-recognized (not new!) discipline of genetic enhancement, and its promotion in the world of basic and applied science.

The discipline of genetic enhancement could be the most romantic of all the disciplines that support plant breeding. To create new pools of useful germplasm, new basic breeding stocks from diverse mixtures of ancient peasant varieties, modern high yield varieties, unlikely weeds, and even from unrelated, distant organisms, using the tools of cytogenetics, population genetics, molecular genetics, and anthropology, with goals set by plant pathologists, entomologists, physiologists, systematists and social scientists, is surely a heady challenge to any biological scientist, and a worthy goal for any public-spirited idealist. Our future existence really will depend on our plant breeding capabilities, everywhere in the world. The diversity needed for continued improvements in dependability and productivity of plant breeding products can come only from keen, assiduous and widespread application of the principles of genetic enhancement, to provide new basic breeding materials suitable for use by plant breeders, worldwide, as they develop a continuing stream of increasingly useful new varieties.

REFERENCES

Anderson, Edgar, and Brown, W.L., 1952, Origin and significance
    of corn belt maize, in: "Heterosis," Iowa State Univ.
    Press, Ames, Iowa, pp. 121-148.

Auston, R.B., Bingham, J., Blackwell, R.D., Evans, L.T.,
    Ford, M.A., Morgan, C.L., and Taylor, M., 1980, Genetic
    improvements in winter wheat yields since 1900 and
    associated physiological changes, Jour. Agric. Sci.,
    Cambridge, 94:675-680.

Cardwell, V.B., 1982, Fifty years of Minnesota corn production:
    Sources of yield increase, Agron. J., 74:984-990.

Carlone, M.R. and Russell, W.A., 1987, Response to plant
    densities and nitrogen levels of four maize cultivars from
    different eras of breeding, Crop Sci., 27:465-470.

Castleberry, R.M., Crum, C. W., and Krull, C.F., 1984, Genetic
    yield improvement of U.S. maize cultivars under varying
    fertility and climatic requirements, Crop Sci., 24:33-36.

Dalrymple, D.G., 1986, "Development and spread of high-yielding
    rice varieties in developing countries," Bureau for Science
    and Technology, Agency for International Development,
    Washington, D.C.

Dalrymple, D.G., 1986, "Development and spread of high-yielding
    wheat varieties in developing countries," Bureau for
    Science and Technology, Agency for International
    Development, Washington, D.C.

Dewey, D.R., 1977, The role of wide hybridization in plant
    improvement, in: "Genetics Lectures," Vol. 5, Oregon State
    University Press, Corvallis, OR, pp. 7-18.

Duvick, D.N., 1984, Genetic contributions to yield gains of U.S.
    hybrid maize, 1930 to 1980, in: "Genetic Contributions To
    Yield Gains Of Five Major Crop Plants," W.R. Fehr, ed.,
    CSSA Special Publication Number 7, Crop Sci. Soc. Amer.,
    Madison, WI, pp. 15-47.

Duvick, D.N., 1984, Genetic diversity in major farm crops on the
    farm and in reserve, Econ. Bot., 38:161-178.

Duvick, D.N., 1986, Plant Breeding: Past Achievements and
    Expectations for the Future, Econ. Bot., 40:289-297.

Duvick, D.N., 1988, Genetic enhancement and plant breeding, in: "Proceedings, First National Symposium for New Crops: Research, Development, Economics," Indianapolis, Indiana, October 23-26, in press.

Heiser, C.B., 1988, Aspects of unconscious selection and the evolution of domesticated plants, Euphytica, 37:77-81.

Hermundstad, S.A. and Peloquin, S.J., 1985, Germplasm enhancement with potato haploids, Journ. Hered., 76:463-467.

Hougen, F.W., Stefansson, B.R., 1982, Rapeseed: production, trade, composition, utilization, in: "Advances in Cereal Science and Technology, Vol. 5," University of Manitoba, Winnipeg, Canada, pp. 261-289.

Miller, F.R. and Kebede, Y., 1984, Genetic contributions to yield gains in sorghum, 1950 to 1980, in: "Genetic Contributions To Yield Gains Of Five Major Crop Plants," W.R. Fehr, ed., CSSA Special Publication Number 7, Crop Sci. Soc. Amer., Madison, WI, pp. 1-12.

Mooney, P.R., 1983, The law of the seed, another development and plant genetic resources, in: "Development Dialogue, No. 1-2," Uppsala: Dag Hammanskold Foundation.

National Research Council, 1972, Genetic Vulnerability of Major Crops, National Academy of Sciences, Washington, D.C.

Rick, C.M., 1976, Natural variability in wild species of Lycopersicon and its bearing on tomato breeding, Genet. Agr., 30:249-259.

Sears, E.R., 1977, Analysis of wheat-agropyron recombinant chromosomes, in: "Interspecific Hybridization in Plant Breeding," Proc. 8th Eucarpia Congress, Madrid, Spain, pp. 63-72.

Specht, J.E. and Williams, J.H., 1984, Contribution of genetic technology to soybean productivity - retrospect and prospect, in: "Genetic Contributions To Yield Gains Of Five Major Crop Plants," W.R. Fehr, ed., CSSA Special Publication Number 7, Crop Sci. Soc. Amer., Madison, WI, pp. 49-73.

U.S. Congress, Office of Technology Assessment, 1988, New
    developments in biotechnology: U.S. Investment in
    Biotechnology - Special Report, OTA-BA-360, U.S. Government
    Printing Office, Washington, D.C.

Wood, D., 1988, "Introduced crops in developing countries: a
    sustainable agriculture?", <u>Food Policy</u>, pp. 167-177.

TARGETING GENES FOR GENETIC MANIPULATION IN CROP SPECIES

J.W. Snape, C.N. Law, A.J. Worland, and B.B. Parker

Institute of Plant Science Research,
Cambridge Laboratory
Trumpington
Cambridge CB2 2JB
UK

INTRODUCTION

In broad terms, the genetic manipulation of crop species involves the creation of new allelic combinations, whether by the recombination of existing allelic variation within the species, or by the introduction of novel alleles or 'new' genes from outside of the species. In essence, the technologies to achieve this are already developed or are, presently, in an advanced state of development. First, the technologies for interplant, within species, sexual hybridization are used to generate new allelic combinations, and hence phenotypic variability, by exploiting the processes of random reassortment of chromosomes and genetic recombination within chromosomes. Such techniques are, in practical terms, operationally simple and have been the "standard fare" of crop improvement programmes for many years. Various forms of mutagenesis can also be used to create new allelic forms (Micke et al., 1987). However, now, in addition, techniques are available to augment this variation by the introduction of 'foreign' genes.

First the techniques of interspecific and intergeneric hybridization can be used successfully to introduce new homoeoallelic, 'alien' genes, into cultivated species (see for example in wheat, Triticum aestivum, Gale and Miller, 1988). This can be achieved either via sexual hybridization and embryo culture, or via non-sexual methods, such as protoplast fusion. Once interspecific hybrids are achieved, chromosome engineering techniques allow introgression of alien chromatin. Secondly, advances in molecular biology are enabling the isolation and introduction of individual, cloned genes into crop species from a wide range of

biological sources. It can be argued that there are no longer, or
soon will not be, any absolute barriers to the introduction of
genetic variation into crop species. Nevertheless, it is probable
that "standard" plant breeding techniques will still provide the
primary impetus and direction for crop improvement and the newer
techniques will be complementary rather than exclusive.

      In all of these types of manipulation it is becoming accepted
that progress rests not on limitations of the technology for genetic
manipulation but on the ability to identify, characterise, "isolate"
and then select for target genes controlling characters of interest.
In this context "isolate" means the ability to identify the
location(s) of genes in the genome which then provide a target
at the recombinational level of manipulation, but also ultimately a
target for genetic manipulation at the molecular level through
cloning and transformation techniques. Probably the biggest
challenge in crop improvement still remaining to be overcome is the
inability to translate phenotypic differences for characters of
interest into genes of known location and effect. This greatly
limits the efficient directed modification and selection for
important plant attributes.

      Although the introduction of foreign genes such as bacterial,
viral or even animal genes will provide a source of variation for
crop improvement it is likely that many, if not most, of the genes of
interest for genetic manipulation will be identified within crop
species themselves. Even for the transfer of genes across species
barriers it is likely that a major source of cloned alleles will
emerge within other cultivated species. For example, there is much
interest in putting genes for wheat glutenin endosperm proteins or
the very successful Norin 10 dwarfing genes into other cereals such
as sorghum, Sorghum bicolor and millet, Pennisetum americanum to change
end usage or to improve performance. To identify such genes
routinely in crop species requires a greatly enhanced understanding
of the genetic control of most agronomic characters than is presently
available. This can only be achieved by developing and applying
comprehensive methods of genetic analysis to translate phenotypic
variation into information concerning the locations of individual
genes of significant effect; the establishment of their linkage
relationships with other loci, particularly genetic markers; the
characterisation of their primary and phenotypic effects, including
interactions with other loci; and, in the longer term, the
characterisation of the underlying physiological, biochemical and
molecular basis of allelic variation. Only then will more efficient
selection of genes for improved agronomic performance be achieved,
thereby turning plant breeding into more of a science and less of an
art!

      This paper examines some of the issues involved in developing
and applying the methodologies of genetic analysis to important
agronomic characters in crop species. It attempts to focus on

approaches considered necessary to improve our understanding of the genetic control of such characters and thereby the identification of the targets for genetical manipulation. Examples from the small grained cereals, particularly wheat, will be used as illustrations.

DEFINING TARGETS

Genetical Considerations

Most characters of interest to plant breeders show phenotypic variability, and invariably genetic variation, within a particular crop species. The aim of genetic analysis is to understand the control of this variability and it is common as a first step to characterise it as discontinuous or continuous. This relates to whether discrete, separate, phenotypic classes are exhibited between different varieties, or, more importantly, between plants within segregating generations of crosses or families. Genetic analysis of discrete variation usually results in the conclusion of a simple genetic control arising from the segregation of single or a few major genes with a negligible effect of environment on expression. Such variation is easily amenable to defined genetic manipulation at most levels of complexity. On the other hand, much of the variation for important agronomic characters shows continuous variation where there are no discrete classes within a population but a continuous quantitative change between extreme phenotypes. It is often regarded that this variation is qualitatively different and is not controlled by single major genes, but is determined by, so called, polygenic systems. Classically, polygenic characters are described as being controlled by many genes, each of small individual effect, which behave in a mendelian fashion. Such loci have been termed quantitative trait loci (QTLs). Their summed segregational effects, in combination with an effect and/or interaction with the environment, produces the observed phenotypic variability. Consequently genetic manipulation of such characters is conventionally carried out by selection on the populations of genes as a whole and not on individual components. This, of course, defines the methodologies of quantitative genetics, a science which has been extensively developed since the classic paper of Fisher (1918) by various schools (Mather and Jinks, 1971; Falconer, 1960), and is still undergoing development and refinement (see Baker, 1984; Wier et al., 1988).

Conventional quantitative genetic methodologies assume that the effects of individual genes are not discernable in segregating populations but that phenotypic differences between individual plants and families can be expressed as statistical properties. Thus, the behaviour of genotypes and the genetical control of a character is described in terms of statistical parameters: means, variances and

covariances. This, nevertheless, provides a means for identifying
and selecting for improved individuals and populations and underpins
the strategies of most, if not all, crop improvement programmes. The
detection of the types of gene action: additive effects, dominance
effects, epistasis, linkage and genotype-environment interaction,
provides information that is descriptive of the "genetical
architecture" of a character and is predictive of the crosses,
families and individuals to be selected for genetic advance. The
contributions of this methodology to crop improvement has been and
will continue to be very large and should not be neglected. However,
it does not identify the effects of, or the location of, individual
gene components of variation in order to allow the defined genetic
manipulation of single component genes, a level of control which is
now desirable.

     For such an approach to be practicable it is, of course, a
requirement that a significant component of the genetical variation
of a continuously varying character is attributable to single genes
of identifiable effect. There is now a large amount of evidence to
indicate that genetic variation for such characters is mediated by
nuclear genes of variable, rather than equal effects. Further, a
small number of large effects can contribute a significant proportion
of the variation. For example in wheat, many characters such as ear
emergence time, plant height, bread making quality and even
components of yield have now been shown to have single loci with
alleles of large effect controlling a considerable proportion of the
variation between phenotypically diverse parents (Law et al., 1988,
Worland et al., 1988a). Obviously, there are also other genes with
small effects where the magnitude of allelic variation is less than
the environmental variation and where the level of detection of
single loci will always be poor because of background genetic and
environmental "noise". Nevertheless, if single gene components which
have a significant effect on the variation can be identified these
can be isolated, characterized and manipulated in a directed fashion.
Indeed, the distinction between a major gene and a QTL is perhaps
rather artificial since once a QTL is identified and located it
effectively becomes a major gene! In conclusion, the defined targets
of quantitative characters of agronomic importance should be single
gene components of discernable effect.

Character Considerations

     The choices of agronomic characters, particularly quantitative
characters which are the priorities for improvement and hence genetic
manipulation will obviously vary tremendously between crop species
and between crop improvement programmes within species. Generally
wide definitions of common objectives such as "yield" and "quality"
are too amorphous to be of real use in defining specific objectives.

For example, many phenotypic attributes under selection will
influence yield performance either through a direct effect on a
particular yield component or as responses to external stimuli which
limit yield such as disease, pests, stress or other environmental
factors. Undoubtedly "yield" and "quality" are target characters but
it is also probably more useful to define ways in which such targets
can be approached by combining character analysis together with
genetic analysis. It is often useful, therefore, to analyse the
phenotypic variation in terms of component characters, each of which
can be scored, and hence analysed in terms of its genetic variation,
and covariation, with other components.

    A classic example in crop plants of such character analysis is
the division of yield into components. In wheat, for example, yield
has been divided into three components: spikes per square metre,
grains per spike, and weight per grain. These characters themselves
can be subdivided further. Thus grains per spike can be divided into
spikelet number per spike and grain number per spikelet. Obviously
the opportunities for such reductionism are large (Rasmusson, 1984)
and the components identified for analysis can, and will, vary even
within a crop species. Ultimately a component should be chosen by a
consideration of the balance between ease of measurement, degree of
correlation with the primary character and identification of genetic
control by individual loci of large effect. These latter two
attributes of course, can only be determined by genetic analysis of
the available allelic variation. Another important consideration is
that the presence of covariation should be established in terms of
the pleiotropic effects on other components of the same or with
different primary characters, and also their interaction with the
environment. Such analysis is necessary to predict the consequences
of using identified alleles for individual loci and for
distinguishing between alternative loci controlling the same
character.

    For example, in considering the use of dwarfing genes in wheat
for lodging resistance, alternative sources are available for
directed genetic manipulation. The choice of which gene to use
depends on the environment to which varieties containing them are
targeted (Worland et al., 1988b). Thus in a winter wheat genetic
background in the UK, Norin 10 genes can, through pleiotropic effect
on fertility levels, increase yield. However, as high fertility
shows susceptibility to stresses such as high temperatures during
critical growth stages (meiosis to ear emergence), then in areas like
Southern Europe, where such natural stress conditions may be
encountered, the use of these dwarfing genes is less attractive
(Worland, 1986). Conversely in Southern Europe reduced height and
increased fertility is usually achieved by a combination of the
Akakomugi dwarfing gene Rht8, the photoperiod insensitive gene Ppd1
and a fertility factor linked to Yr16 (Worland et al., 1988c).
Whilst Rht8 reduces height by about 8 cms in the UK, in Southern
Europe with neutral pleiotropic effects on yield, Ppd1 reduces height

by a similar amount in both environments but with highly advantageous
pleiotropic effects on yield only in Southern Europe caused by
avoidance of dessicating summer conditions. In the UK Ppd1 reduces
yield by reducing the plants life cycle and time available for grain
fill under favourable conditions (Worland et al., 1988b).

     Similarly when considering a character such as ear emergence
time in wheat it is necessary to define the direct effects of
different loci in terms of their interaction with the environment to
manipulate the genes appropriately. Thus genetic variation for this
character can be divided into response to vernalisation on the one
hand and response to photoperiod on the other. Genes for these two
characters have now been shown to be under the control of separate,
unlinked, major gene loci which can, therefore, be manipulated
independently (Law et al., 1988). In this way the ear emergence time
of a genotype can be adjusted for different agroclimatic regions by
judicious use of alleles at the appropriate loci. The demonstration
of and the availability of a multiple allelic series at these loci
allows "fine-tuning" for both macro and micro environmental
differences. Only by defining the character in terms of these
components can this level of genetic manipulation be achieved. The
identification of the important components is critical to such an
analysis. Undoubtedly, defining target components is amongst the
hardest of tasks in applying genetic analysis to crop improvement.

     In defining actual target characters in a crop species it can be
useful to try to list those of interest to a plant breeder as a
preliminary to deciding priorities for genetical analysis. Clearly
different breeding programmes will have different objectives.
Nevertheless general lists of primary, secondary, tertiary (etc)
components can serve as a useful starting point to such an approach.
For example, Table 1 attempts to define the primary and secondary
component characters of interest for wheat improvement. Although
undoubtedly not exhaustive this list highlights the salient
objectives of most wheat breeding programmes. The identification of
genetic variation for these characters and its analysis and
interpretation can subsequently take place with specific objectives
in mind. Frequently within such an analysis it may also be possible
to identify qualitative differences controlled by major gene systems,
as well as those characters regarded as quantitative.

     It should also be remembered in carrying out such an analysis
that most conventional forms of genetic analysis rely on the
assumption that phenotypic variability and thus genetic variation
exists for particular characters and components. This is perhaps a
truism, but nevertheless needs emphasis since generally there is not
much point in choosing a character for modification, (say for example
photosynthetic ability to improve yield) without the demonstration
first of heritable variation which can be analysed and exploited.

Table 1. Major Target Characters in Wheat

Primary Characters	Secondary Characters
Yield per se ("Yield potential" factors)	Fertile tillers per square metre Grain number per spike Grain size Harvest index
Biomass	Maximum photosynthetic rate Quantum efficiency of photosynthesis Respiration rate Photorespiration rate
Adaptation/maturity	Vernalisation response Photoperiod response Earliness per se
Stress Tolerance ("Yield potential limiting" factors)	Disease resistance Pest resistance Herbicide tolerance Drought tolerance Extreme temperature tolerance Salt tolerance Heavy metal tolerance
Quality	Protein quality Protein quantity Grain hardness Sprouting resistance

THE TOOLS FOR LOCATING TARGET GENES

The Development of Comprehensive Genetic Maps

One of the most important current advances in plant genetics
lies in the development of comprehensive genetic maps of plant
chromosomes by identifying and employing 'marker' systems. A genetic
marker can be defined as any characteristic of an individual plant
which is controlled by a single major gene locus segregating in a
mendelian fashion, and where discrete phenotypes are brought about by
differences in the expression or structure of different alleles.

Traditionally, most of the markers available in crop species and the
genetic maps obtained were based on observable morphological
differences, either those occurring naturally or those induced by
mutation treatments. However maps compiled from such markers were
invariably sparse due to the paucity of these types of genes in most
species. In recent years, however, this number has been dramatically
increased by the development of additional systems. Indeed, the
potential now exists to produce saturated maps of plant chromosomes
and maps containing loci equally spaced along the chromosome at
distances of less than 10 cM appear feasible (Tanskley et al., 1988).
Undoubtedly the development of comprehensive genetic maps will be a
crucial factor in targeting both important major gene loci and QTL's
controlling agronomic characters, and their further development is
essential.

The most important types of marker systems now being developed
are biochemical and molecular systems. For example, Table 2 outlines
the major systems being utilised for mapping wheat chromosomes and
these allow the visualisation of naturally occurring cryptic
differences in gene structure or expression through electrophoresis.
Biochemical markers arise from differences in the structure of gene
products, particularly proteins, and this variation is visualised by
separating the proteins on their differential mobility through gels
and using histological staining procedures. First, many species are
polymorphic at loci controlling the amino acid composition of enzymes
(Tanksley and Orton, 1983). Using isoelectric focusing (IEF) with
polyacrylamide gels in particular allows great flexibility and gives
good resolution in detecting and exploiting such variation (Koebner,
pers. comm.). Secondly, storage proteins can be separated into their
component sub-units, and these are often polymorphic. For example,
in wheat there is considerable variation in endosperm protein
sub-unit composition and the proteins can be separated and visualised
by one dimensional sodium dodecyl sulphate polyacrylamide gel
electrophoresis (SDS-PAGE) or by two dimensional separations by
combining SDS-PAGE with IEF (Payne et al., 1985). Using these
systems many of the loci controlling both structural and
non-structural proteins of the wheat endosperm have been identified,
characterised and located. This type of analysis can also be
extended to total proteins extracted from other tissues, such as
leaves. For example, Colas des Francs and Thiellement (1985) working
with wheat have identified 766 separate "spots" corresponding to
different proteins in 2-dimensional separations of proteins extracted
from young leaves. Although the levels of polymorphism for these
proteins have not yet been characterised this technique may have the
potential to identify many more useful biochemical markers in wheat.
Combining these separation techniques with monoclonal antibody
technology by use of "Western" blotting techniques could greatly
increase the precision to detect and locate polymorphisms for such
protein diversity.

Table 2. Biochemical and Molecular Marker Systems in Wheat

Type	Method of Visualisation
Isozymes	One dimensional gel electrophoresis - 'native' electrophoresis. Isoelectric focusing (IEF)
Endosperm protein subunits	One dimensional separation: SDS-PAGE Two dimensional separations (IEF/SDS-PAGE)
Total protein separation (leaves, roots)	Two dimensional separations (IEF/SDS-PAGE)
Total protein separation	Monoclonal antibodies
Restriction fragment length polymorphisms	Separation in agarose gels; use of labelled cDNA and genomic DNA probes: "southern blotting".

Biochemical marker systems exploit variation in gene products. However, with the use of restriction endonucleases it is now possible to exploit variation in DNA structure as genetic markers since regions of chromosomes in different genotypes differ in their base sequence and therefore in sites where these enzymes can cut. Arising from these properties has been the most significant and exciting advance in genetic mapping, namely the development of techniques for revealing restriction fragment length polymorphisms (RFLPs). Many recent reviews have discussed the potential of RFLP analysis in crop improvement (see for example, Beckmann and Soller, 1986; Gale et al., this volume) and their potential as tools in plant genetics. Of all the currently available techniques RFLPs are likely to be the most significant and useful for the location of agronomically important genes.

Although biochemical and molecular marker systems are receiving greatest attention at the present time it is unwise to ignore the traditional source of markers, namely morphological differences. Indeed these will continue to be a source of markers, but may also have greater significance as being themselves desirable target genes for genetic manipulation. Although morphological markers are generally considered to be just visual variation occurring naturally in populations (for example in wheat, the presence or absence of awns, glume or grain colour differences) a much bigger source of such markers is induced morphological changes brought about by applying

karyotypic differences following specific chromosome staining
techniques. Of most importance are trisomics, monosomics and
nullisomics, telocentrics, reciprocal translocations and interstitial
or terminal deletions; and stocks containing these are now available
in many important crop species such as maize, Zea mays, rye, Secale
cereale, barley, Hordeum vulgare, oilseed rape, Brassica napus, potato,
Solanum tuberosum, tomato, Lycopersicum esculentum, and in particular,
wheat. These stocks enable specific marker loci to be readily
located to individual chromosomes, chromosome arms, and when using
banding techniques such as N-banding, C-banding or in situ
hybridisation, to particular chromosome segments. Using cytological
markers augments the genetic map, but more importantly allows the
integration of genetic maps with physical maps of plant chromosomes.
This is achieved by establishing linkage relationships between marker
loci and chromosome landmarks such as centromeres, secondary
constructions or particular bands. This is also an important pathway
to the physical isolation of specific genes.

Overall the formulation of comprehensive genetic maps of plant
chromosomes should aim at the integration of all of these marker
systems and seek to characterise the whole genome genetically and
physically. If this can be achieved in important crop species
virtually all aspects of genetic manipulation would be greatly
simplified.

The Development of Appropriate Genetic Stocks

Precise genetic analysis hinges on the development of
appropriate genetic stocks and the starting point is the choice of
varieties of contrasting phenotype as a basis for developing
segregating generations by hybridization. Usually one of the parents
or populations is chosen because it expresses an extreme or
'interesting' phenotype for the character in question and the other
is taken as being of 'normal' or opposing phenotype. Conventionally,
it is necessary that these parents are uniform and true breeding for
the attribute and hence homozygous at the loci of interest. Such
genotypes are usually readily available in most crop species which
are naturally self-pollinating, but not necessarily available in
outcrossing species, particularly where varieties consist of
heterozygous populations developed by random mating amongst selected
individuals or families. In such populations uniform individuals
have to be produced by inbreeding, by double haploid methods, or
isolated by selection of extreme genotypes. However the movement
towards F_1 hybrid varieties of cross-pollinating species in many
major crops has necessitated the development of inbred lines, and
these can, in many cases, be used as the starting points for
genetical analysis.

The types of families and generations that can be produced from
the initial parents will be the major constraint on the types of

certain crop species this is achieved by using aneuploid lines in cytogenetic manipulations to develop chromosomally defined genetic stocks. The outstanding example of this approach is, of course, in wheat where varietal differences can be analysed chromosome by chromosome through the use of various types of monosomic analysis and by the development and analysis of single chromosome substitution lines (Law et al., 1988).

In wheat by using cytogenetic stocks both inter and intrachromosomal analyses of great precision have been used to locate loci controlling both qualitative and quantitative variation for genes controlling many of the important agronomic characters. It is necessary to develop methods to emulate this approach in other species. There is some progress in this area, for example, in maize by using B-A translocation procedures (Beckett, 1978). Nevertheless to date only in wheat is there a satisfactory level of methodology to routinely produce such genetic stocks. The development of precise genetic maps may offer the opportunity to develop analogous lines in any crop species regardless of the availability of cytogenetical methodologies. In such an approach marker sets from one parent are transferred by backcrossing into the contrasting parent. The integrity of the marker combinations, and hence of particular parental chromosome segments, is maintained by co-dominant marker selection. In this way isogenic lines for different segments of the genome can be developed on a constant genetic background. This can be carried out systematically for all portions of the genome to develop isogenic lines for all individual chromosome arms, or smaller segments, depending on the coverage of the genome available from the genetic map. At the final generation of backcrossing the heterozygous isogenic lines are selfed and the alternative parental homozygotes, and also recombinant lines for the segment, isolated in the progenies for detailed genetical investigation of the individual genome segment. This approach is of course a modification of the classic "isogenic" line development used to study the effects of particular genes in particular genetic backgrounds.

A modification to such a systematic "genome assay" is to develop genotypes which contain random portions of the genome of one parent in the genetic backround of another. Two convenient methodologies, which involve less work than the isogenic lines approach (by omitting genotypic classification and selection at each generation,) can be considered. These methods were originally developed for estimating the numbers of loci segregating for a quantitative character. The first is the Wehrhahn and Allard (1965) "inbred backcross" technique and the second the "genotype assay" technique of Jinks and Towey (1976). The inbred backcross method allows random segments of one parental genome to be isolated in a homozygous state in the genetic background of the contrasting parent. The genotype assay method develops isogenic pairs from advanced selfed generations. By assessing the marker classification of these individual lines it is possible to classify the respective genome contributions of both

parents and to identify those segments of the non-recurrent parent
present in the genetic background of the recurrent parent. Relating
the genome contributions of the alternative parents to the phenotypic
performance of individual lines should allow the localisation of
individual genetic effects.

STRATEGIES FOR TARGETING GENES

Location of Major Genes

     The identification and localisation of a new major gene locus is
a relatively straightforward procedure provided that a comprehensive
genetic map or efficient cytogenetic marker techniques are available.
First it is necessary to hybridize the divergent phenotypes and then
to develop and examine segregating generations to establish that
variation is discrete. Then, the numbers of loci present can be
established from the frequencies of the different genotypes in these
generations. Subsequent localisation of the loci will occur by the
association of the 'new' genes with established genetic markers in
the crosses of interest, and by using maximum likelihood statistical
procedures to evaluate the likely gene order and chromosomal
positions.

     Frequently the identification of the position of a new gene will
provide a new genetic marker for subsequent studies. Once the locus
is "established" the extent of allelic variation in different
varieties can be examined; their effects on the primary character,
and their pleiotropic effects on other characters elucidated. Hence
the consequences of using particular alleles in breeding programmes
can be predicted. In this context marker genes themselves can be
classified into two different groups, namely "functional" markers and
"non-functional" markers. Functional markers can be defined as those
where there is a distinct effect on the phenotypic performance of the
primary character, or a pleiotropic effect on another character. A
non-functional marker locus can be defined as one where there is no
apparent effect of allelic variation on performance and the locus is,
so called, "neutral" in effect. Generally most morphological markers
- dwarfing genes, herbicide resistance genes, pest and disease
reaction loci, etc, have functional significance. For such loci,
specific alleles can be identified which are desirable targets for
genetic manipulation and gene isolation. Hence the development of a
comprehensive genetic map within a species will go hand in hand with
the targeting of major gene loci of agronomic significance. Further,
by studying the underlying physiological and biochemical basis of
specific allelic variation it may also be possible to identify the
biochemical pathways involved and hence determine the molecular basis
of variation.

     Most molecular and biochemical markers are assumed to be neutral
in effect on the phenotype. The usefulness of such markers will be

external stimuli to plants (Table 3).   Thus we can consider
differential responses to disease inoculation in this category where
plants can exhibit different levels of infection and this facilitates
the identification of major gene loci with different alleles
mediating resistant or susceptible responses.   Polymorphisms are also
elicited by treatments with environmental stimuli.   For example, in
wheat low temperature treatments identify loci controlling
vernalisation requirement or cold hardiness; short day lengths
identify photoperiod response loci.   A further group of useful and
important polymorphisms in this category are those exhibited
following treatments with exogenously applied chemicals – chemical
response polymorphisms (CRPs), (Snape and Parker, 1988).   It is well
known that treatments of crop species with a variety of chemicals,
both organic and inorganic, can produce differential changes in plant
morphology, growth rate, variable amounts of leaf damage or even
plant death.   Differential responses to herbicides in particular may
provide a useful source of new such genetic markers (Snape and
Parker, 1988).   These responses also, of course, have obvious
agronomic significance in locating and manipulating herbicide
resistance in crops.

Finally, an important component of genetic mapping strategies in
many crop species is the availability and use of cytological markers.
These arise from natural or induced changes in chromosome number or
structure which are visualised microscopically as

Table 3. Morphological Marker Systems

1.	'Natural' Variation	:	'fixed' morphological differences in plant size, shape or colour
2.	'Induced' Variation	:	Differential changes in morphology induced by applying external stimuli: Disease Inoculation Environmental Stimuli: Cold Temperature Day length Stress Chemical Application

genetic analysis that can be performed and this again, depends on
the breeding system of the species. Most of the major crop species,
even if they are naturally cross-pollinating can, nevertheless, be
artificially self-pollinated. Thus, theoretically there are few
barriers to the types of families and generations that can be
developed, although technical problems may limit the availability of
seed of certain generations. However, in species which have strong
incompatibility mechanisms, or are monoecious or dioecious it may not
be possible to develop true inbred lines or defined segregating
families and the types of genetic analysis that can be performed are
limited.

     The most common generations produced for genetical analysis of
both qualitative and quantitative variation are the F_1 and the
immediate generations derived from these, namely the F_2 and
backcrosses. Subsequently the F_3 and further generations can be
obtained by self-pollinating and crossing the initial generations,
and hence a series of generations with defined genetical
relationships, within and between families, can be developed. Other
types of generations can also be produced by inbreeding and utilising
different mating schemes from these initial generations. Of
particular current significance for identifying target genes are
populations of random homozygous lines. These can be developed by
recurrent inbreeding, particularly by the method of single seed
descent (SSD) (Brim, 1966) or by using doubled haploid procedures
(Snape and Simpson, 1981). This novel procedure allows populations
of completely homozyous lines to be produced from heterozygous
parents in a single generation. Both these methods produce
populations of lines where individual lines are true breeding within
and across generations, and, consequently can be replicated over
experiments and environments (Snape and Simpson, 1981). This allows
sequential analysis to be performed on the same genotypes so that
detailed genetic information on individual crosses can be formulated
over time. Thus linkage maps can be progressively built up, unlike
F_2 or backcross populations where "transient" individual plants have
to be assessed for all attributes at the same time. Destructive
sampling is also facilitated, and quantitative characters can be
measured in different environments and treatments while all analysis
relate to a constant set of genotypes. These populations can thus
simplify the genetic analysis, yet increase its power, particularly
when combined with marker facilitated analysis of target characters.

     These basic generations form the primary populations for linkage
mapping and for the genetical analysis of quantitative characters.
Essentially they are used to focus analysis onto the complete genome,
and to identify all the segregating components. However, where
possible, it can be more efficient and effective to produce genotypes
which can be used to partition the genome into smaller "units" for
analysis. The most convenient subdivision is to attempt to analyse
the effects of individual chromosomes separately by developing
appropriate genotypes to carry out, so called, chromosome assays. In

in "gene-tagging" where a neutral marker, such as an isozyme or an RFLP is associated by close linkage with a locus of functional significance (see Gale et al., this volume). Particular targets for such tagging will be major gene loci where it is difficult to screen allelic variation because of the environmental sensitivity of the variation, or that the relevant test is technically difficult, or time or resource consuming. An excellent current example of such tagging in wheat is the association of an allele at the locus controlling the endopeptidase enzyme with a source of eyespot resistance (Summers et al., 1988).

Electrophoretic markers are generally assumed to be neutral in effect but exceptions can occur where the electrophoretic banding pattern can reveal a functional significance. Of particular note is the association of breadmaking quality in wheat with the sub-unit composition of endosperm proteins, especially the high molecular weight glutenins (Payne et al., 1988). Specific "good" and "poor" banding patterns have been recognised by correlating electrophoretic profiles with breadmaking quality. This also has predictive significance since there is a correlation between the mobility (and hence size) of particular sub-units and quality where higher molecular weight sub-units are associated with better quality. This is thus an aid to electrophoretically screening land races and alien species related to wheat for sub-units of larger size than available within wheat. These then provide candidates for genetic manipulation. This work has in turn led to studies of the biochemical and molecular basis of allelic variation for these proteins and their gene isolation and cloning has subsequently been possible (Flavell et al., 1988).

This analysis of endosperm proteins in wheat provides a model for the way in which major gene variation can be identified and exploited to improve an important agronomic character. However even in species where the genomes are becoming well characterised such examples are rare, consisting of a few disease resistance loci, proteins associated with quality, and dwarfing genes. Much greater efforts in genetic analysis are needed if more than the tip of the iceberg is to be penetrated!

Targeting QTLs

The difficulty of identifying and manipulating genes for quantitative variation relative to major gene variation has meant that very few QTLs have been identified and located in major crop species. The exception is perhaps wheat where the availability of defined, cytogenetic derived, genotypes, particularly chromosome substitution lines and their derivatives, has led to the localisation of QTLs for a variety of characters (Law, 1967; Law et al., 1988; Snape et al., 1985). Most other species are poorly defined, although very recently in certain crops there has been an increasing effort to

locate QTLs, such as in tomato and maize.   This rekindled interest has
been promoted by the development of marker techniques (i.e. RFLPs)
allowing, for the first time, adequate marker coverage of the genome
which can then be applied to analyse quantitative variation.

     The basic approaches to locating QTLs are not new.   The
pioneering work of Thoday and colleagues (Thoday, 1961) with
Drosophila laid the foundations of QTL analysis.   They formulated the
basic methodologies and current approaches offer a refinement rather
than a new or alternative approach (see Weller, 1986; Shrimpton and
Robertson, 1988a, 1988b).   All of these approaches utilise markers to
characterise segments of chromosomes.   Allelic variation for the
markers is then correlated with metrical differences for the
characters being considered.

     The complete marker coverage of a genome allows the total
variation for a quantitative character in a cross to be accounted for
by association with marker polymorphisms.   For example, Paterson et
al (1988) working with a cross in tomato mapped six QTLs controlling
fruit mass, four QTLs controlling the concentration of soluble solids
and five for fruit pH.   In these analyses mean differences in
performance of marker genotypes were analysed by maxiumum likelihood
methods to estimate the recombination frequency between marker loci
and the associated QTL, giving either a definite locus location, or
the most probable position of the QTL relative to surrounding
markers.   However, there are still difficulties with these "whole
genome" approaches that have not yet been resolved.   Primarily there
is the problem of environmental variation affecting the performance
of the quantitative trait.   Hence the number and even the position of
QTLs can change over experiments and so can the measured effects of
alternative alleles.   Thus as Weller (1986) pointed out, his method
was useful only for detecting QTLs with effects greater than one
phenotypic standard deviation.   Nevertheless, since it is only QTLs
with large effects that are of interest these methods should, in most
circumstances, target the few major components of variation, if they
exist, for any particular character.   QTLs of small effect are likely
to get lost as background 'noise' in the analysis.

     Marker mediated location of QTLs is carried out on the
segregating generations between contrasting parents.   As with the
location of major genes, F_2 and backcross populations as well as
homozygous recombinant inbred lines are commonly used.   Although QTLs
can be located to particular chromosome segments, they cannot with
this approach be accurately located to a specific locus as with major
genes, because of the uncertainties of estimation.   A great part of
this uncertainty is the inability to unambiguously classify
individual plants or lines genotypically for alleles at the QTL.
This is because of not only the environmental effects on expression
but also because phenotypic performance at an individual QTL is
obscured by the background segregation of alleles at other QTLs
controlling the character.   In essence classification at a QTL is

only by association and not by the segregational analysis necessary
for genotypic classification and for estimating segregation ratios.
For accurate QTL location, segregational data is required. This can
be obtained by developing isogenic lines for the alternative alleles
on a standard genetic background. The most direct method to do this
is to carry out a backcross programme while selecting for markers
flanking the QTL. Alternatively, lines developed by the inbred
backcross or genotype assay techniques can be developed and
identified for the particular marked segments.

Once appropriate isogenic lines have been developed for
chromosome segments containing alleles of interest, segregational
tests can be used to establish the exact position of the QTL relative
to the marker loci. Subsequent characterisation of the primary and
pleiotropic effects of the alternative alleles can be carried out.
In this way detailed information about targeted QTLs can be compiled
and used to develop strategies for their manipulation and isolation.
To examine the range of allelic variation in other varieties requires
the development of further isogenics for the marked segment. This,
of course, requires that the flanking markers are polymorphic in the
other varieties of interest. Ideally it would be useful to be able
to associate the degrees of polymorphism for flanking or tagging
markers with metrical variation at the QTL. However, at the present
time there is no evidence that such correlations exist and the
predictive value of QTL analysis is, as yet, limited to the crosses
under study.

Targeting Characters Exhibiting No Intraspecific Genetical Variation

A special case of identifying and locating target genes for
important agronomic characters occurs when the characters do not
exhibit phenotypic and hence genotypic variation within a species.
However it is often desirable to locate the genes for such characters
so that the alleles can be isolated and either "improved" or used in
their "native" state for transformation. At the present time two
approaches can be used to locate such loci. The first is to attempt
to create phenotypic variability and hence to generate 'new' alleles
by mutagenesis. Once a mutant is obtained the genetical difference
can be analysed and the relevant loci located as a preliminary to the
further characterisation of the mutant allele. Occasionally such
mutant alleles can be an improvement of the existing allele, although
generally the induced variation will be deleterious. Nevertheless by
creating the allelic differences genetic analysis can proceed.

The second approach is to develop and characterise aneuploid
lines or deletion stocks for whole chromosomes, chromosome arms or
chromosome segments. In such an analysis it is hoped that the
reduction in the allelic dose at particular loci will correlate with
phenotypic changes relative to the euploid. In this way karyotypic
analysis combined with phenotypic analysis can locate relevant loci.

In wheat this has been an invaluable approach in the location of genes for both major gene loci and QTLs (Law et al., 1988). Indeed much of the work on the localisation of biochemical and molecular markers in wheat is based on this approach (Gale et al., 1988).

Overall, the strategies for marker mediated location of target genes are similar for both major genes and QTLs and a generalised strategy is outlined in Table 4.

Table 4. A generalised strategy for targeting genes

STEP 1: Decide on target character and objectives

STEP 2: Identify divergent phenotypes for analysis

STEP 3: Identify and/or develop appropriate 'test' generations by hybridization and selfing

STEP 4: Locate loci controlling the character by genetic analysis – identify marker associated effects

STEP 5: Assess extent of allelic variation in different varieties – "marker polymorphism" prediction

STEP 6: Develop isogenic (or equivalent) lines for assessment of the pleiotropic effects of different alleles

STEP 7: 'Tag' desirable alleles for use in genetic manipulation

STEP 8: Molecular isolation and cloning of specific alleles – understanding of biochemical and molecular basis of allelic variation

STEP 9: Transformation using appropriate alleles in designed "cassettes"

EXPLOITING TARGETED GENES

Following the identification and location of target loci a final consideration is how to employ these for crop improvement by using genetic manipulation techniques. Essentially three options are available in utilising the variation. First, and operationally simplest, is to replace alleles in existing varieties with "better" alleles identified in other varieties or in closely related species and thus to build up desirable allelic combinations at different loci. This is achieved by conventional hybridization techniques and is the standard plant breeding approach. However it could be greatly enhanced by marker mediated selection.

Secondly, it can be useful to increase the dosage of particular existing loci, carrying either the same or different alleles to those already present, in important varieties. This is achieved by the insertion of additional copies, either in tandem with the existing locus or indeed on a separate chromosome segment. This could result in an increased production of a particular gene product and hence an enhanced phenotype. Finally it will be possible to introduce a completely "foreign" locus identified in a separate species into the existing genome. This can be directed to a particular chromosome segment, or more likely with present technology, inserted at random into the genome. This could be in single or multiple copy depending on the transformation events generated.

All of these manipulations will, hopefully, change the existing phenotype into an improved "desired" form. However each of these manipulations requires a different genetic manipulation protocol and their genetic consequences may not be entirely predictable. Indeed the processes of genetic analysis may have to be reapplied to understand the consequences of the genetic manipulations carried out! At the present time little information is available to understand or predict the changes brought about, but hopefully, this will emerge as transformed plants are studied in the field.

This paper has not mentioned strategies for the isolation of identified genes and this has been considered to be outside of the scope of the present discussion. Indeed the technologies to achieve this routinely are not yet available. Presumably they will involve the developing methods of transposon tagging, insertion mutagenesis, and chromosome walking and jumping, and the like. To date very few important identified genes have been isolated. Even for single, well known genes of great agronomic benefit, for example the Norin 10 dwarfing genes Rht1 and Rht2 in wheat, this has not yet been possible. The location and agronomic effects of these genes are well characterised yet the biochemical basis of allelic variation is obscure and attempts to isolate the genes have had no success. This is characteristic of many other such genes.

In conclusion, it should be emphasised, and it is perhaps not generally appreciated, that the genetic analysis of agronomic characters is not easy. Even the location of a few genes for a few characters is a task requiring considerable expenditure of effort in terms of time and resources. Indeed the present paucity of target genes is perhaps proof of the complexities involved and underlines the need for further sustained and determined efforts in important crop species.

References

Baker, R.J., 1984, Quantitative genetic principles in plant breeding, in: "Gene Manipulation in Plant Improvement," J.P. Gustafson, ed., Plenum Press, New York (1984), pp. 147-176.

Beckmann, J.S., and Soller, M., 1986, Restriction fragment length polymorphisms in plant genetic improvement, in: "Oxford Surveys of Plant Molecular and Cell Biology:," Vol. 3, Oxford University Press, Oxford (1986), pp. 197-250.

Beckett, J.B., 1978, B-A translocations in maize. 1. Use in locating genes by chromosome arms, J. Hered., 69:27-36.

Brim, C.A., 1966, A modified pedigree method of selection in soybeans, Crop Sci., 6:220.

Colas des Francs, C., and Thiellement, H., 1985, Chromosomal localisation of structural genes and regulators in wheat by 2D-electrophoresis of ditelosomic lines, Theor. Appl. Genet., 71:31-38.

Falconer, D.S., 1960, Introduction to Quantitative Genetics", Oliver and Boyd, Edinburgh.

Fisher, R.A., 1918, The correlations between relatives on the supposition of Mendelian inheritance, Trans. R. Soc. Edinb., 52:399-433.

Flavell, R.B., Harries, N., O'Dell, M., Sardana, R.K., and Jackson, S, 1988, Transposable elements and the control of ribosomal RNA gene expression in wheat, in: "Proceedings Seventh International Wheat Genetics Symposium", Institute of Plant Science Research, Cambridge (1988), pp. 33-37.

Gale, M.D., and Sharp, P.J., 1988, Genetic markers in wheat-developments and prospects, in: "Proceedings Seventh International Wheat Genetics Symposium," Institute of Plant Science Research, Cambridge (1988), pp. 469-475.

Gale, M.D., and Miller, T.E., 1988, The introduction of alien genetic variation into wheat, in: "Wheat breeding: Its scientific basis," F.G.H., Lupton, ed., Chapman and Hall, London (1988), pp. 173-210.

Jinks, J.L., and Towey, P., 1976, Estimating the number of genes in a polygenic system by genotype assay, Heredity, 37:69-81.

Law, C.N., 1967, The location of genetic factors controlling a number of quantitative characters in wheat, Genetics, 56:445-461.

Law, C.N., Snape, J.W., and Worland, A.J., 1988, Aneuploidy in wheat
    and its uses in genetic analysis, in: "Wheat breeding: Its
    scientific basis," F.G.H. Lupton, ed. Chapman and Hall, London
    (1988), pp. 71-108.
Mather, K., and Jinks, J.L., 1971, Biochemical Genetics (2nd
    Edition), Chapman and Hall, London.
Micke, A., Donini, B., and Maluszynski, M., 1987, Induced mutations
    for crop improvement - a review, Trop. Agric. (Trindad), 64:259-
    278.
Paterson, A.H., Lander, E.S., Hewitt, J.D., Peterson, S., Lincoln,
    S.E., and Tanksley, S.D., 1988, Resolution of quantitative traits
    into mendelain factors by using a complete linkage map of
    restriction fragment length polymorphisms, Nature, 335:721-726.
Payne, P.I., Holt, L.M., Jarvis, M.G., and Jackson, E.A., 1985, Two-
    dimensional fractionalisation of the endosperm proteins of bread
    wheat (Triticum aestivum): Biochemical and genetic studies,
    Cereal Chem., 62:319-326.
Payne, P.I., Holt, L.M., Krattiger, F., and Carrillo, J.M., 1988.
    Relationship between seed quality charactersitics and HMW
    glutenin subunit composition determined using wheats grown in
    Spain, J. Cereal Sci., 7:229-235.
Rasmussen, D.C., 1984, Ideotype research and plant breeding, in:
    "Gene Manipulation in Plant Improvement," J.P. Gustafson, ed.,
    Plenum Press, New York (1984) pp. 95-120.
Shrimpton, A.E., and Robertson, A., 1988a, The isolation of polygenic
    factors controlling bristle score in Drosophila melanogaster (a).
    I. Isolation of third chromosome sternopleural bristle effects to
    chromosome segments, Genetics, 118:437-443.
Shrimpton, A.E., and Robertson, A., 1988b, The isolation of polygenic
    factors controlling bristle score in Drosphila melanogaster. (b).
    II. Distribution of third chromosome bristle effects within
    chromosome segments, Genetics, 48:455-459.
Snape, J.W., 1981, The use of doubled haploids in plant breeding, in:
    "Induced variability in plant breeding," Centre for Agricultural
    Publishing and Documentation, Wageningen (1981), pp. 52-58.
Snape, J.W., Law, C.N., Parker, B.B., and Worland, A.J., 1985,
    Genetical analysis of chromosome 5A of wheat and its influence on
    important agronomic characters, Theor. Appl. Genet., 71:518-526.
Snape, J.W., and Parker, B.B., 1988, Chemical response polymorphisms:
    An additional source of genetic markers in wheat, in:
    :Proceedings Seventh International Wheat Genetics Symposium,"
    Institute of Plant Science Research, Cambridge (1988), pp.
    651-656.
Snape, J.W., and Simpson, E., 1981, The genetical expectations of
    doubled haploid lines derived from different filial generations,
    Theor. Appl. Genet., 60:123-128.
Snape, J.W., Wright, A.J., and Simpson, E., 1984, Methods for
    estimating gene numbers for quantitative charcters using doubled
    haploid lines, Theor. Appl. Genet., 67:143-148.

Summers, R.W., Koebner, R.M.D., Hollins, T.W., Forster, J., and
    MaCartney, D.P., 1988, The use of an isozyme marker in breeding
    wheat (Triticum aestivum) restricted to eyespot pathogen
    (Pseudocercosporella herpotrichoides), in: "Proceedings of the
    Seventh International Wheat Genetics Symposium," Institute of
    Plant Science Research, Cambridge (1988), pp. 1195-1197.
Tanksley, S.D., and Orton, T.J., 1983, "Isozymes in Plant Genetics
    and Breeding", Elsevier, Amsterdam.
Tanksley, S.D., Miller, J., Paterson, A., and Bernatzky, R., 1988,
    in: "Chromosome Structure and Function: Impact of new concepts",
    J.P. Gustafson and R. Appels, eds., Plenum Press, New York
    (1988), pp. 157-174.
Thoday, J.M., 1961, Location of polygenes, Nature 191:368-370.
Weir, B.S., Eisen, E.J., Goodman, M.M., and Mamkoong, G., 1988,
    "Proceedings of the Second International Conference on
    Quantitative Genetics," Sinamer Associates, Inc., Sunderland,
    Massachusetts.
Wehrhahn, C., and Allard, R.W., 1965, The detection and measurement
    of the effects of individual genes involved in the inheritance of
    a quantitative character in wheat, Genetics, 51:109-119.
Weller, J.I., 1986, Maximum likelihood techniques for the mapping and
    analysis of quantitative trait loci with the aid of genetic
    markers, Biometrics, 42:627-640.
Worland, A.J., 1986, Gibberellic acid insensitive dwarfing genes in
    southern European wheats, Euphyticia, 35:857-866.
Worland, A.J., Gale, M.D., and Law, C.N., 1988a, Wheat genetics, in:
    "Wheat breeding: Its scientific basis," F.G.H. Lupton, ed.,
    Chapman and Hall, London (1988), pp. 129-172.
Worland, A.J., Petrovic, S., and Law, C.N., 1988b, Genetic analysis
    of chromosome 2D of wheat. II. The importance of this chromosome
    in Yugoslavia varieties, Plant Breeding, 100:247-259.
Worland, A.J., Law, C.N., and Petrovic, S., 1988c. Pleiotropic
    effects of the chromosome 2D genes Ppdl, Rht8 and Yr16, in:
    "Proceedings of the Seventh International Wheat Genetics
    Symposium," Institute of Plant Science Research, Cambridge
    (1988), pp. 669-674.

INCOMPATIBILITY BARRIERS OPERATING IN CROSSES OF *Oryza sativa* WITH

RELATED SPECIES AND GENERA

Lesley A. Sitch

Department of Plant Breeding
International Rice Research Institute
P. O. Box 933, Manila, Philippines

INTRODUCTION

The genus *Oryza* includes 20 wild species and 2 cultigens, *O. sativa* and *O. glaberrima*. Most *Oryza* species, including the cultigens, are diploid (2n=24); eight are tetraploid (2n=48). Six genomes have been identified, found either as diploids or allotetraploids; AA, BB, BBCC, CC, CCDD, EE, and FF. *O. meridionalis* is assumed to have the AA genome. Five additional species, whose genomic composition is unknown, have been identified (Table 1).

The wild *Oryza* species and related genera are a rich source of genes for rice improvement (Table 2). Although considerable work has been done to produce interspecific hybrids for taxonomic and phylogenic analysis (Bouharmont, 1962; Chu et al., 1969a; Nezu et al., 1960; Wuu et al., 1963), little effort has been made to transfer desirable traits from wild species into cultivated rice.

Among the closely related species *O. sativa* f. *spontanea* was used as the source of the wild abortive (WA) male sterile cytoplasm (Virmani and Shinjyo, 1988). More than 95% of the hybrid rice grown in China is derived from WA cytosterile lines (Yuan and Virmani, 1988). In 1977, Khush transferred the dominant gene *Gs* conferring grassy stunt virus resistance from *O. nivara* to several rice cultivars. *O. longistaminata* possesses very long and well-exerted stigmas, a trait that has been exploited by many hybrid rice breeding programs (Taillebois and Guimaraes, 1988). The more distantly related species are a rich source of insect resistance (Heinrichs et al., 1985). Recently, brown planthopper resistance was transferred from *O. officinalis* into cultivated rice (Jena and Khush, 1986).

CROSSABILITY OF WILD *Oryza* SPECIES WITH CULTIVATED RICE

The percentage seed set reported in the literature for crosses between *O. sativa* and the wild *Oryza* species or related genera varies considerably between species and between reports (Table 3).

Gene Manipulation in Plant Improvement II
Edited by J. P. Gustafson
Plenum Press, New York, 1990

Table 1. Species of *Oryza*, chromosome numbers, genome symbols and geographical distributions
(from Chang, 1985)

Species name (synonym)	X=12 2n=	Genome group	Distribution
O. sativa L.	24	AA	Asia
O. glaberrima Steud.	24	A^gA^g	West Africa
O. nivara Sharma & Shastry (*O. fatua*, *O. sativa* f. *spontanea*)	24	AA	South and Southeast Asia and southern China
O. rufipogon W. Griffith (*O. perennis*, *O. fatua*, *O. perennis*, subsp. *balunga*)	24	AA	South and Southeast Asia and southern China
O. barthii A. Chev. (*O. breviligulata*)	24	A^gA^g	West Africa
O. longistaminata A. Chev. & Roehr. (*O. barthii*)	24	A^lA^l	Africa
O. glumaepatula Steud. (*O. perennis* subsp. *cubensis*)	24	$A^{cu}A^{cu}$	South America and West Indies
O. meridionalis N. Q. Ng	24	—	Australia
O. punctata Kotschy ex Steud.	24,48	BBCC BB(?)	Africa
O. eichingeri A. Peter	24,48	CC BBCC	East and central Africa
O. officinalis Wall. ex Watt	24	CC	South and Southeast Asia, southern China, New Guinea
O. minuta J.S. Presl ex C.B. Presl	48	BBCC	Southeast Asia
O. alta Swallen	48	CCDD	Central and South America
O. grandiglumis (Doell) Prod.	48	CCDD	South America
O. latifolia Desv.	48	CCDD	Central and South America
O. australiensis Domin	24	EE	Australia
O. brachyantha A. Chev. & Roehr.	24	FF	West and central Africa
O. meyeriana (Zoll. & Morrill ex Steud.) Baill.	24	—	Southeast Asia, southern China
O. granulata Nees & Arn. ex Hook f.	24	—	South and Southeast Asia
O. ridleyi Hook f.	48	—	Southeast Asia
O. longiglumis Jansen	48	—	New Guinea
O. schlechteri Pilger	—	—	New Guinea

Table 2. Some wild species of *Oryza* and related genera with traits of economic importance

Wild species	Genome	Useful trait[1]
O. glaberrima	A^gA^g	Resistance to GLH; early vegetative vigor
O. nivara	AA	Resistance to grassy stunt virus and blast
O. rufipogon	AA	Tolerance of acid sulphate soils and stagnant flooding; source of CMS
O. barthii	A^gA^g	Resistance to bacterial blight
O. longistaminata	A^lA^l	Floral characteristics for outcrossing
O. eichingeri	CC BBCC	Resistance to BPH, GLH, and WBPH
O. officinalis	CC	Resistance to BPH, GLH, and WBPH
O. minuta	BBCC	Resistance to BPH, GLH, WBPH, blast, and bacterial blight
O. australiensis	EE	Resistance to BPH and drought
O. brachyantha	FF	Resistance to stemborer and rice whorl maggot
O. ridleyi	—	Resistance to rice whorl maggot, bacterial blight, and blast
O. longiglumis	—	Resistance to bacterial blight and blast
Porteresia coarctatum	—	Tolerance of salinity

[1]   BPH : brown planthopper; GLH : green leafhopper; WBPH : white-backed planthopper, CMS : cytoplasmic male sterility

As expected, the highest seed sets were obtained from crosses with the AA genome species; seed sets ranging from 9.1-76.9%. The variation between reports probably reflects differences in the *O. sativa* cultivar and wild species accession used. For example, crosses of *O. sativa* cultivar IR36 with *O. nivara* Acc. 101973 gave 9.1% seed set; crosses with *O. nivara* Acc. 103826 gave 62.2% seed set (Sitch, unpublished).

Crosses with species possessing the BB, BBCC, CC, CCDD, EE, and FF genomes yielded 0-30.0% seed set. Among these species, the lowest seed sets were obtained from crosses with *O. australiensis* (EE) and *O. brachyantha* (FF).

Crosses attempted with *O. granulata, O. ridleyi* and the related genus, *Rhynchoryza subulata*, were unsuccessful.

These results suggest that strong incompatibility mechanisms do not operate in crosses of *O. sativa* with the AA genome species, although seed sets may be low in some combinations. However, the low seed sets obtained from intergenomic crosses are presumably caused by one or more incompatibility barriers.

The barriers encountered in crosses between distantly related species have been discussed by a number of workers (Brar and Khush, 1986; Stebbins, 1958). Two main types of barriers limit hybrid production and survival: pre-fertilization and post-fertilization barriers.

Table 3. The percentage seed set obtained from crosses of *O. sativa* with wild *Oryza* species and
related genera

Pollinator	Genome	Seed set (%)	Reference
O. glaberrima	A[g]A[g]	56.7	Sitch (unpublished)
		47.0	Oka (1964)
		42.0	Chu et al. (1969a)
		34.4	Nezu et al. (1960)
		29.5	Morinaga and Kuriyama(1957)
		+[1]	Bouharmont (1962), Bouharmont et al. (1985), Kihara (1959), Morinaga (1959), Sano et al. (1979), Sano (1983), Seetharaman (1962)
O. nivara	AA	76.9	Nezu et al. (1960)
		9.1-62.2	Sitch (unpublished)
		+	Bouharmont (1962), Kihara (1959), Li et al. (1962)
O. rufipogon	AA	18.5-73.0	Sitch (unpublished)
		35-54	Kihara (1959)
		35-49	Chu et al. (1969a)
		39.3	Nezu et al. (1960)
		+	Bouharmont (1962), Li et al. (1962)
O. barthii	A[g]A[g]	53.8	Chu and Oka (1965)
		38.0	Chu et al. (1969a)
		36.4	Nezu et al. (1960)
		33.0-54.0	Oka (1964)
		+	Kihara (1959), Li et al. (1962), Morinaga (1959), Yeh and Henderson (1961)
O. longistaminata	A[l]A[l]	+	Bouharmont (1962)
O. glumaepatula	A[cu]A[cu]	39.0	Oka (1964)
		+	Gotoh and Okura (1933)
O. punctata	BB(?), BBCC	11.4	Ogawa and Katayama (1971)
		+	Katayama (1966a)
O. eichingeri	CC,BBCC	20.0	Nezu et al. (1960)
		+	Jena (pers. comm.), Kihara (1959), Morinaga (1959)
O. officinalis	CC	8.8-17.3	Jena and Khush (1986)
		5.9	Nezu et al. (1960)
		+	Kihara (1959), Li et al. (1962), Morinaga (1959), Nandi (1938), Shin and Katayama (1979)

Table 3 contd.

Pollinator	Genome	Seed set (%)	Reference
O. minuta	BBCC	21.3	Nezu et al. (1960)
		0-4.2	Sitch (unpublished)
		1.7	Capinpin and Magnaye (1951)
		+	Katayama (1966a), Kihara (1959), Li et al. (1962), Morinaga (1959)
O. alta	CCDD	30.0	Nezu et al. (1960)
		+	Katayama (1966b), Kihara (1959)
O. grandiglumis	CCDD	6.7	Sitch (unpublished)
		+	Katayama (1966b)
O. latifolia	CCDD	25.0	Nezu et al. (1960)
		0-19.7	Sitch (unpublished)
		+	Bouharmont (1962), Gotoh and Okura (1933), Jena (pers. comm.), Katayama (1966b), Kihara (1959), Li et al. (1962), Morinaga (1959)
O. australiensis	EE	3.8	Nezu et al. (1960)
		0.5-3.2	IRRI (1985)
		1.9	Li et al. (1963)
		+	Kihara (1959), Morinaga (1959)
O. brachyantha	FF	0.2	Sitch (unpublished)
		+	Wuu et al. (1963), Yang et al. (1965)
		0	Kihara (1959), Nezu et al.(1960)
O. granulata	—	0	Kihara (1959), Nezu et al. (1960), Sitch (unpublished)
O. ridleyi	—	0	Nezu et al. (1960), Sitch (unpublished)
Rhynchoryza subulata	—	0	Kihara (1959), Nezu et al. (1960), Sitch (unpublished)

[1] seed set data not indicated

PRE-FERTILIZATION BARRIERS LIMITING PRODUCTION OF INTERSPECIFIC
AND INTERGENERIC HYBRIDS

Pre-fertilization barriers include all factors that hinder fertilization and may result
from the failure of pollen grain germination, slow pollen tube growth, and incomplete
fertilization.

Failure of pollen grain germination has been observed in crosses between dicotyle-
donous species, such as Poplar (Knox et al., 1972), *Nicotiana* and *Lycopersicum* (Sree
Ramulu et al., 1979) but not in cereals. Indeed, pollen tubes were detected in the ovaries
of wide crosses such as *Secale cereale/Zea mays, S. cereale/Hordeum bulbosum*, and
Triticum aestivum/Z. mays (Zenkteler and Nitzsche, 1984); *T. aestivum/H. bulbosum*
(Sitch and Snape, 1987); and *Z. mays*/Sorghum and pearl millet/Sorghum (Reger and
James, 1982).

Fertilization may be limited by the failure of pollen tubes to reach the micropyle,
resulting from slow pollen tube growth or the premature inhibition of growth. In cereals,
this phenomenon has been observed in maize/sorghum crosses (Mock and Loeschner,
1973); maize/*Tripsacum* crosses (James-Sprague, 1982); *T. aestivum/S. cereale* crosses
(Jalani and Moss, 1980); and *T. aestivum/H. bulbosum* crosses (Sitch and Snape, 1987).
Inhibition of pollen tube growth is often, but not exclusively, associated with the formation of callose plugs, bursting, and swelling of pollen tube tips (Jalani and Moss, 1980;
Pickering, 1981).

Incomplete fertilization, fertilization of either the egg nucleus or the polar nuclei
only, has been observed in *T. aestivum/H. bulbosum* crosses (Mól and Zenkteler, 1982;
Sitch, 1985). Mól and Zenkteler (1982) showed that fertilization of wheat by *Brachypo-
dium pinnatum* failed, although micropylar penetration was successful. In barley/rye
crosses, the sperm nuclei remained in the synergid or in the space between the egg and the
central cell (Bannikova and Khvedynich, 1974).

In rice, reports on pre-fertilization incompatibility are limited to an investigation of
the mechanism of self-incompatibility in *O. barthii* (Chu et al., 1969b). Following self-
pollination, *O. barthii* pollen grains germinated, but the majority of pollen tubes failed to
penetrate the stigma. A study of the extent to which pre-fertilization incompatibility
mechanisms operate in interspecific and intergeneric combinations involving *O. sativa* is
described here.

Materials and methods

Crosses were made between *O. sativa* (AA) cultivar IR31917-45-3-2 and six wild
species, *O. officinalis* (CC), *O. eichingeri* (CC), *O. minuta* (BBCC), *O. alta* (CCDD), *O.
brachyantha* (FF), and *Rhynchoryza subulata*. Ovaries were fixed from two panicles per
cross combination, from eight replications. Pollen grain germination and pollen tube
growth was examined in 30 ovaries per cross combination per replicate, using the technique of Kho and Baër (1968), modified as follows:

1. Fix ovaries 24 hours after pollination in 3 parts absolute alcohol : 1 part glacial acetic
   acid.
2. Macerate ovaries in 8N NaOH for 2 hours.

3. Stain ovaries with 0.05% toluidine blue for 10 minutes.
4. Stain overnight with 0.1% aniline blue (water soluble) in 0.1N K_3PO_4.
5. Mount ovaries in glycerol and examine under ultraviolet illumination, using a 320-400 nm exciter filter and a 470 nm barrier filter.

Enlarged ovaries were considered fertilized. Pollen tubes detectable in the wall of fertilized ovaries were broken as a result of ovary swelling.

Percentage of pollen grain germination

The mean percentage of pollen grain germination for each pollinator is shown in Table 4.

A highly significant difference (p<0.001) was detected in the percentage of pollen grain germination between pollinators; ranging from 90.3% in crosses with *O. minuta* to 48.2% in crosses with *R. subulata*. A comparison of the means (data transformed to angles) showed that pollen grain germination in *R. subulata* was significantly lower than that in the remaining species. The differences between the remaining species were not significant.

Percentage of pollen tubes reaching the style

A comparison of the number of germinated pollen grains and the number of pollen tubes detected within the style is shown in Table 5.

A highly significant difference (p<0.001) was detected between pollinators, ranging from 62.7% of *O. officinalis* pollen tubes reaching the style to 0% in *R. subulata*. Two groups of pollinators could be distinguished. A high frequency of pollen tube penetration of the stigma and style was observed in pollinations with *O. eichingeri, O. alta, O. minuta,* and *O. officinalis.* A significantly lower percentage of pollen tubes reached the style in pollinations with *O. brachyantha* and *R. subulata.*

Variation in the extent of pollen tube growth

A comparison of the percentage of ovaries in which pollen tube growth ceased at four positions within the ovary: P1 on the stigma surface, P2 within the stigma and style, P3 within the ovary, and P4 at the micropyle is shown in Table 6. The data were analyzed using a contingency X^2 analysis.

A significant difference (X^2=616.51, 15 d.f., p<0.001) in the extent of pollen tube growth was detected between pollinators. Two groups of pollinators could be differenti-ated (X^2=470.54, 3 d.f., p<0.001): *R. subulata* and *O. brachyantha* in which pollen tube penetration of the stigma was observed in only 1% and 11% of the ovaries examined, and the remaining pollinators in which pollen tube penetration occurred in more than 80% of the ovaries examined.

Significantly more (X^2=6.90, 2 d.f., p=0.05-0.01) pollen tubes entered the stigma in crosses with *O. brachyantha* than with *R. subulata.* Pollen tube growth persisted into the ovary wall in 1% of *O. brachyantha*-pollinated ovaries. Among the remaining four pollinators, *O. eichingeri* showed significantly fewer (X^2=101.70, 3 d.f., p<0.001) pollen

Table 4. The mean percentage pollen grain germination of six wild species on stigmas of cultivar
IR31917-45-3-2

Pollinator	Pollen grains		Mean percentage pollen grain germination[1]
	examined (No.)	germinated (No.)	
O. minuta	2751	2465	90.3 (72.4 a)
O. officinalis	1064	940	88.6 (71.4 a)
O. eichingeri	956	809	83.3 (66.6 a)
O. brachyantha	1060	896	75.1 (64.4 a)
O. alta	1436	1191	83.3 (62.8 a)
R. subulata	796	443	48.2 (43.1)

[1] Average of 8 replications; transformed means in parentheses; any two means having a common
letter are not significantly different at the 5% level of significance.

Table 5. The number of germinated pollen grains, the number of pollen tubes observed in the style and
mean percentage of pollen tubes reaching the style 24 hours after pollination of cultivar
IR31917-45-3-2 with six wild species

Pollinator	Pollen grains germinated (No.)	Pollen tubes in style (No.)	Mean percentage of pollen tubes reaching the style[1]
O. officinalis	940	529	62.7 (53.7 a)
O. minuta	2465	694	49.3 (44.5 a,b)
O. alta	1191	369	45.5 (41.5 a,b,c)
O. eichingeri	809	275	40.1 (38.9 b,c)
O. brachyantha	896	9	0.8 (3.0 d)
R. subulata	443	0	0 (0.0 d)

[1] Average of 8 replications; transformed means in parentheses; any two means having a common
letter are not significantly different at the 5% level of significance.

Table 6. Variation in the extent of pollen tube growth, 24 hours after pollination, in ovaries of
cultivar IR31917-45-3-2 pollinated with six wild species

Pollinator	Ovaries examined (No.)	Perentage of ovaries[1]			
		P1	P2	P3	P4
R. subulata	84	99	1	0	0
O. brachyantha	114	89	10	1	0
O. eichingeri	126	18	39	36	7
O. officinalis	124	7	8	20	65
O. alta	109	6	7	32	55
O. minuta	125	3	16	37	44

[1] expressed as the percentage of ovaries in which pollen tube growth ceased at: P1 = the stigmal
surface, P2 = stigma/style, P3 = ovary, P4 = micropyle

tubes penetrating the stigma and ovary than the other species; 82% rather than 93% for *O. officinalis*, 94% for *O. alta*, and 97% for *O. minuta*. In *O. eichingeri*-pollinated ovaries, pollen tubes were observed at the micropyle in only 7% of the ovaries examined.

In crosses with *O. brachyantha* and *R. subulata*, the pollen tubes formed a spiral before growth ceased on the stigma surface. A swelling was often seen at the tip. In the remaining species, cessation of pollen tube growth was occasionally associated with the formation of callose plugs.

Percentage of micropylar penetration versus percentage of fertilization

A comparison of the percentage of ovaries in which pollen tubes were observed at the micropyle and the percentage of fertilized ovaries for each combination is shown in Table 7.

A significant difference (X^2=112.47, 5 d.f., p<0.001) in the frequency of fertilization was detected between pollinators. Two groups of pollinators could be differentiated (X^2=79.96, 1 d.f., p<0.001). Low fertilization frequencies were obtained from pollinations with *R. subulata*, *O. brachyantha*, and *O. eichingeri*. Higher fertilization frequencies were obtained from *O. minuta, O. officinalis*, and *O. alta* pollinations. The success of fertilization differed significantly among these three pollinators (X^2=19.97, 2 d.f., p<0.001).

For four of the six cross combinations examined, the percentage of ovaries with pollen tubes at the micropyle exceeded the percentage of fertilized ovaries. Fertilization was most successful in *O. alta*-pollinated ovaries; 64% of the ovaries with pollen tubes at the micropyle were fertilized. Fertilization was less successful in *O. officinalis*-, *O. eichingeri*-, and *O. minuta*-pollinated ovaries, where only 37%, 29%, and 25% of the ovaries were fertilized, respectively.

Table 7. The percentage of ovaries with pollen tubes at the micropyle and the percentage of fertilized ovaries obtained from crosses of cultivar IR31917-45-3-2 with six wild species.

Pollinator	Ovaries examined (No.)	Percentage of ovaries[1] with pollen tubes at micropyle	fertilized
O. officinalis	124	65	24
O. alta	109	55	35
O. minuta	125	44	11
O. eichingeri	126	7	2
O. brachyantha	114	0	0
R. subulata	84	0	0

[1] Average of 8 replicates

Implications

Inhibition of pollen grain germination was observed only in the intergeneric cross with *R. subulata.* Germination was not, however, prevented, implying that this incompatibility mechanism is not solely responsible for the absence of seed set obtained from *O. sativa/R. subulata* crosses (Table 3).

Hybrid production in crosses of *O. sativa* with *O. brachyantha* and *R. subulata* is clearly limited by the failure of the pollen tubes to penetrate the stigma. This mechanism is presumably the most important barrier to fertilization in *O. sativa/O. brachyantha* and *O. sativa/R. subulata* crosses.

Cessation of pollen tube growth within the stigma, style, and ovary was observed in crosses with *O. officinalis, O. eichingeri, O. minuta, O. alta,* and *O. brachyantha,* being most prevalent in pollinations with *O. eichingeri.* A comparison of the frequency of micropylar penetration and fertilization in these crosses also suggests that fertilization may not necessarily follow pollen tube penetration of the micropyle. These mechanisms may be responsible for the low frequency of fertilization (Table 7) and seed set (Table 3) obtained from crosses with *O. officinalis, O. eichingeri,* and *O. minuta,* and further reduce fertilization and seed set in crosses with *O. brachyantha.* Microscopic studies on fertilization and early zygote development are necessary to complement these results and provide a greater understanding of the mechanisms limiting seed set in *Oryza* interspecific and intergeneric crosses.

POST-FERTILIZATION BARRIERS LIMITING PRODUCTION AND SURVIVAL OF INTERSPECIFIC AND INTERGENERIC HYBRIDS

Post-fertilization barriers cause hybrid inviability or weakness. They may affect many stages, from development of the zygote and seed through growth of the hybrid plant to the formation of gametes. Four phenomena will be discussed: the role of parental ploidy level, embryo and endosperm degeneration, seedling mortality, and hybrid sterility.

Role of parental ploidy level

Differences in ploidy level of the parents often lead to incompatibility. In 1953, Crowder crossed *Lolium perenne* L. (2n=14) with *Festuca arundinacea* Schreb. (2n=42) and obtained higher seed sets from crosses with *L. perenne* as the female parent. Incompatibility resulting from differences in ploidy level can be overcome by doubling the chromosome number of one parent and making crosses with the resultant autopolyploid. Wernesman et al. (1965) obtained no seed from the cross *Lotus tenuis* Wald. et Kit. (2n=2x=14) by *L. corniculatus* L. (2n=4x=24). However, seed set from reciprocal crosses between an advanced generation tetraploid *L. tenuis* (2n=4x=24) and *L. corniculatus* averaged more than 8 seeds per pod.

A similar phenomenon was observed in rice by Nezu et al. (1960). Reciprocal crosses between nine diploid and three tetraploid *Oryza* species and showed that crosses using the diploid species as the female parent gave higher seed sets and numbers of viable F_1 hybrids than crosses using the tetraploid as the female parent.

Embryo and endosperm degeneration

The failure/arrest of embryo differentiation or endosperm development has been observed in interspecific and intergeneric crosses in many crops (Hu and Wang, 1984).

Embryo and endosperm degeneration is common in interspecific crosses involving *O. sativa*, particularly in intergenomic crosses. Before embryo rescue techniques were available, interspecific hybrid production was limited to intragenomic crosses. Very few intergenomic hybrids were reported, a notable exception being the *O. sativa/O. minuta* hybrid reported by Capinpin and Magnaye (1951). Many new intergenomic hybrids have been produced using the embryo rescue technique (Jena and Khush, 1986; Kihara, 1959; Wuu et al., 1963). Embryos are cultured 10-14 days after pollination on 1/4-strength Murashige and Skoog medium devoid of auxin and allowed to germinate in the dark. Upon germination the plantlets are transferred to the light, from there into nutrient solution, and then into the soil.

Chu et al. (1969a)[1] examined the germinability of F_1 hybrid seeds between *O. sativa* and four *O. perennis* accessions (Asia, Africa, America, Oceania), *O. glaberrima*, and *O. breviligulata*. Although the germination rate was high (94-100%) for most crosses, poor germinability (4%) was observed in crosses with *O. perennis* (Africa). In the *O. sativa/O. perennis* (Africa) cross, poor germinability was due to embryo and endosperm deterioration, starting 6 days after fertilization. In the reciprocal cross, degeneration began 3 days after fertilization. This barrier was controlled by a set of complementary dominant lethal genes which disturbed tissue differentiation in young F_1 zygotes, primarily affecting endosperm development.

Seedling mortality

Seedling mortality is a common cause of hybrid inviability. In interspecific crosses of *Crepis*, this is caused by a single lethal gene *l* (Hollingshead, 1930). A similar gene that affects the viability of *T. monococcum/Ae. umbellulata* hybrids was identified (Sears, 1944).

In rice, Chu et al. (1969a)[1] observed a low frequency of F_1 weakness in crosses of *O. sativa* with *O. perennis* (Asia) and *O. perennis* (Africa), at a rate of 0.6% and 3.2% of the progeny respectively. The F_1 plants ceased growing approximately 30 days after germination, showed poor root development, premature leaf senescence, and produced few or no tillers. Oka (1964), however, did not detect F_1 weakness in comparable interspecific hybrids.

Hybrid sterility

Hybrid sterility may be due to disharmonious nuclear-cytoplasmic interactions, genic, or chromosomal differences between parents.

---

[1]Recent reclassification of the *Oryza* species means that *O. perennis* (Africa) refers to *O. bathii; O. perennis* (Asia) refers to both *O. nivara* and *O. rufipogon*; *O. perennis* (America) is assumed to represent *O. glumaepatula*; and *O. perennis* (Oceania) is assumed to represent *O. meridionalis*.

Disharmonious interaction between the nucleus of one species and the cytoplasm of the other often leads to reciprocal differences in the growth and development of hybrids, particularly male sterility. Yabuno (1977) examined the fertility of reciprocal interspecific cytoplasm substitution lines of *O. sativa* (japonica) and *O. glaberrima*. All substitution lines possessing *glaberrima* cytoplasm were completely fertile with characteristics of the sativa parent, suggesting that the *glaberrima* cytoplasm did not influence the expression of the *sativa* genome. However, all substitution lines possessing *sativa* cytoplasm were male sterile, indicating that the control of pollen fertility of these substitution lines is of a cytoplasmon-genic nature. Similar results were obtained by Sano (1985) in reciprocal *O. sativa* (indica)/*O. glaberrima* cytoplasm substitutions.

The wild species are an important source of cytoplasmic male sterility (CMS) in rice. Of the 35 sources of CMS, 18 are derived from the wild species *O. sativa* f. *spontanea* and *O. rufipogon* (Virmani and Shinjo, 1988). Rutger and Shinjyo (1980)[1] demonstrated that male sterile cytoplasm exists in Asian (64% frequency) and American (4%) strains of *O. perennis*. No male sterile cytoplasm was found in the African and Oceanian strains. More recently, Virmani and Wan (1988) identified two accessions of *O. breviligulata* and one accession of *O. glaberrima* as prospective CMS sources. *O. nivara* (Acc. 101508) was also found to induce CMS in indica and japonica rices (Hybrid Rice Newsletter, 1989).

Male sterility can also result from genetic differences between parents. Chu et al. (1969a)[1] observed that crosses involving the AA genome species varied in pollen fertility. Crosses between *O. sativa* and *O. perennis* (Asia) gave 84% pollen fertility and those with *O. perennis* (Africa) gave 35% pollen fertility. Crosses with *O. perennis* (America, Oceania) were completely sterile. However, such hybrids showed normal pairing at diakinesis and metaphase I with few meiotic abnormalities. The authors assumed that the hybrid sterility exhibited by these crosses is under genetic control.

Three genes which control male sterility in *O. sativa*/*O. glaberrima* hybrids have been identified. Sano et al. (1979) demonstrated that strains of *O. sativa* and *O. glaberrima* possess $S_1{}^a/S_1{}^a$ S_2/S_2 and S_1/S_1 $S_2{}^a/S_2{}^a$ genotypes, respectively. The two loci were shown to be independent. Neither locus adversely affected microspore development in the homozygote, but each induced abortion of pollen not carrying the locus in the heterozygote S/S^a. In 1983, Sano identified an additional locus $(S_3/S_3{}^a)$ that influences sterility in this hybrid combination.

Lack of chromosomal homology due to differences in the structure and number of chromosomes of the parental species results in sterility in numerous interspecific and intergenomic hybrids. In intergenomic crosses involving the *Oryza* species, lack of chromosome homology gives rise to a low frequency of chiasma. The consequent formation of hypoploid gametes results in complete pollen sterility (Katayama, 1966a; 1966b; Nezu et al., 1960; Yang et al., 1965).

It is often possible to restore fertility in such hybrids by colchicine treatment to produce the amphiploid. Watanabe (1975) produced amphiploids of four F_1 hybrids, *O.*

---

[1]Recent reclassification of the *Oryza* species means that *O. perennis* (Africa) refers to *O. bathii; O. perennis* (Asia) refers to both *O. nivara* and *O. rufipogon; O. perennis* (America) is assumed to represent *O. glumaepatula*; and *O. perennis* (Oceania) is assumed to represent *O. meridionalis*.

sativa/O. officinalis (AACC), *O. sativa/O. minuta* (AABBCC), *O. sativa/O. latifolia* (AACCDD), and *O. sativa/O. australiensis* (AAEE). Unfortunately all were male sterile. In comparison, Ogawa and Katayama (1971) produced a fertile *O. sativa/O. punctata* (AABB) amphiploid by colchicine treatment. Although amphiploids involving *O. sativa* are frequently male sterile an increased frequency of fertile female gametes and hence greater crossability can be expected. At IRRI, colchicine treatment of completely sterile *O. sativa/O. minuta* hybrids produced partially fertile plants that yielded a greater frequency of BC_1 progeny than was obtained by backcrossing the F_1 hybrids directly (Sitch, unpublished). Backcrossing of diploid *O. sativa/O. australiensis* (AE) hybrids with *O. sativa* was unsuccessful due to hybrid sterility (Jena, pers. comm.). By doubling the chromosome number of the *O. sativa* parent and crossing the resultant autotetraploid with *O. australiensis*, Jena produced triploid *O. sativa/O. australiensis* (AAE) hybrids. These hybrids were more fertile and backcrossing was successful.

PROSPECTS FOR WIDE HYBRIDIZATION AS AN APPROACH TO RICE IMPROVEMENT

Both pre- and post-fertilization incompatibility barriers operate in crosses of *O. sativa* with the wild *Oryza* species. The majority of *Oryza* species, particularly those species whose genome composition have been identified, are crossable with cultivated rice although seed sets are often low. Degeneration of the hybrid embryo and endosperm occurs in all intergenomic crosses and even in crosses with certain accessions of the AA genome species. In such crosses, embryo rescue must be used to ensure hybrid survival. Once the hybrid has been produced, transfer of desirable traits from the wild species into *O. sativa* should be possible using procedures described elsewhere (Brar and Khush, 1986).

Gene transfer from species whose genomic composition is unknown, such as *O. granulata, O. meyeriana*, and *O. ridleyi*, and from related genera, such as *Rhynchoryza subulata*, and *Porteresia coarctatum* is much more difficult. Fertilization is prevented by strong pre-fertilization incompatibility barriers. In *O. ridleyi, P. coarctatum* (Sitch, unpubl.ished), and *R. subulata*, incompatibility is caused primarily by the inhibition of pollen tube penetration of the stigma. A number of approaches toward overcoming such incompatibility have been described:

- pollination with a mixture of inviable compatible pollen (killed with ethanol) and incompatible pollen (mentor pollen technique). Pollen walls are known to contain extracellular proteins (recognition factors) which are released on the stigma after pollination. The recognition factors released from the wall of the dead compatible pollen mask the rejection reaction of the recipient stigma, thus allowing the alien pollen to germinate and penetrate the stigma (Sree Ramulu et al., 1979);

- removal of antagonistic exudate from the incompatible stigma with alcohol and replacement with a suitable exudate by direct physical contact of the incompatible and compatible stigma (Tilton and Russell, 1984);

- mechanical removal of the pistil and pollination of the exposed end of the style (Collins et al., 1984; Tilton and Russell, 1984);

- grafting a compatible style onto the incompatible ovary in lieu of the incompatible style (Collins et al., 1984; Tilton and Russell, 1984);

- application of growth regulators such as gibberellic acid, indole acetic acid (Larter and Chaubey, 1965), amino acids and sucrose solutions (Matsubara, 1984) to promote pollen tube growth;

- manipulation of temperature and relative humidity to maximize pollen tube penetration of the stigma and style (Carter and McNeilly, 1975; Franken et al., 1988; Okazaki and Hinata, 1987); and

- intraovarian pollination by injecting a pollen suspension into the ovary (Zenkteler, 1980).

These methods should be explored as approaches to overcoming the pre-fertilization incompatibility mechanisms operating in distant crosses involving *O. sativa.* If such methods prove unsuccessful, somatic hybridization represents an alternative route to hybrid production (Collins et al., 1984). Somatic hybrids can then be manipulated using conventional methods to facilitate gene transfer (Brar and Khush, 1986).

REFERENCES

Bannikova, V. P., and Khvedynich, O. A., 1974, Features of fertilization in the course of remote hybridization in plants, *in*: "Fertilization in Higher Plants", H. F. Linskens, ed., North-Holland Publishing Company, Amsterdam (1974), pp 301-307.
Brar, D. S., and Khush, G. S., 1986, Wide hybridization and chromosome manipulation in cereals, *in*: "Handbook of Plant Cell Culture, Volume 4, Techniques and Applications," D. A. Evans, W. R. Sharp and P. V. Ammirato, ed., Macmillan Publishing Company, New York and Collier Macmillan Publishers, London (1986), pp 221-263.
Bouharmont, J., 1962, Recherches cytogénétiques chez quelques hybrides interspecifiques d'*Oryza, La Cellule,* 63:53-132 (English summary).
Bouharmont, J., Olivier, M., and Dumont du Chassart, M., 1985, Cytological observations in some hybrids between the rice species *Oryza sativa* L. and *O. glaberrima* Steud., *Euphytica,* 34:75-81.
Capinpin, J. M., and Magnaye, A. B., 1951, Hybridization of cultivated and wild rice, *Philippine Agric.,* 34:219-229.
Carter, A. L., and McNeilly, T., 1975, Effects of increased humidity on pollen tube growth and seed set following self pollination in brussels sprout (*Brassica oleracea* var. *gemmifera), Euphytica,* 24:805-813.
Chang, T. T., 1985, Crop history and genetic conservation: Rice - A Case Study, *Iowa State Journal of Research,* 59:425-455.
Chu, Y. E., and Oka, H. I., 1965, Deterioration of F$_1$ embryos in hybrids between *Oryza sativa* and *O. perennis* subsp. *barthii, Ann. Rep. Nat. Inst. Genet., Japan.,* 16:71.
Chu, Y. E., Morishima, H., and Oka, H. I., 1969a, Reproductive barriers distributed in cultivated rice species and their wild relatives, *Japan. J. Genet.,* 44:207-233.
Chu, Y. E., Morishima, H., and Oka, H. I., 1969b, Partial self-incompatibility found in *Oryza perennis* subsp. *barthii, Japan. J. Genet.,* 44:225-229.

Collins, G. B., Taylor, N. L., and DeVerna, J. W., 1984, *In vitro* approaches to interspecific hybridization and chromosome manipulation in crop plants, *in*: "Gene Manipulation in Plant Improvement," J. P. Gustafson, ed., Plenum Press, New York and London (1984), pp 323-383.

Crowder, L. V., 1953, Interspecific and intergeneric hybrids of *Festuca* and *Lolium*, *J. Heredity*, 44:195-203.

Franken, J., Custers, J. B., and Bino, R. J., 1988, Effects of temperature on pollen tube growth and fruit set in reciprocal crosses between *Cucumis sativus* and *C. metuliferus*, *Plant Breed.*, 100:150-153.

Gotoh, K., and Okura, E., 1933, A preliminary note of cytological studies of *Oryza*, *J. Soc. Trop. Ag.*, 5:363-364.

Heinrichs, E. A., Medrano, F. G., and Rapusas, H. R., 1985, "Genetic Evaluation For Insect Resistance In Rice," International Rice Research Institute, Manila.

Hollingshead, L., 1930, A lethal factor in *Crepis* effective only in an interspecific hybrid, *Genetics*, 15:114-140.

Hu, C., and Wang, P., 1984, Embryo culture: technique and applications, *in*: "Handbook of Plant Cell Culture, Volume 4, Techniques and Applications," D. A. Evans, W. R. Sharp and P. V. Ammirato, ed., Macmillan Publishing Company, New York and Collier Macmillan Publishers, London (1986), pp 43-96.

Hybrid Rice Newsletter, 1989, *Oryza nivara* cytoplasm found to induce male sterility in rice, *Hybrid Rice Newsletter*, 3(1):4.

IRRI, 1985, Wide hybridization for rice improvement, *in*: "Annual Report for 1984," International Rice Research Institute, Manila (1985), pp 133-135.

Jalani, B. S., and Moss, J. P., 1980, The site of action of the crossability genes (*Kr1, Kr2*) between *Triticum* and *Secale*. 1. Pollen germination, pollen tube growth and number of pollen tubes, *Euphytica*, 29:571-579.

James-Sprague, J., 1982, Combining genomes by conventional means, *in*: "Plant Improvement and Somatic Cell Genetics," I. K. Vasil, W. R. Scowcroft, and K. J. Frey, ed., Academic Press, New York (1982), pp 99-118.

Jena, K. K., and Khush, G. S., 1986, Production of monosomic alien addition lines of *O. sativa* having a single chromosome of *O. officinalis*, *in*: "Rice Genetics," International Rice Research Institute, Manila (1986), pp 199-208.

Katayama, T., 1966a, Cytogenetical studies on the genus *Oryza*. 2. Chromosome pairing in the interspecific hybrid with the ABC genomes, *Japan. J. Genet.*, 41: 309-316.

Katayama, T., 1966b, Cytogenetical studies on the genus *Oryza*. 3. Chromosome pairing in the interspecific hybrid with the ACD genomes, *Japan. J. Genet.*, 41:317-324.

Kho, Y. O., and Baër, J., 1968, Observing pollen tubes by means of fluorescence, *Euphytica*, 17:298-302.

Khush, G. S., 1977, Disease and insect resistance in rice, *Adv. Agron.*, 29:265-341.

Kihara, H., 1959, Considerations on the origin of cultivated rice, *Seiken Zihô*, 10:68-83.

Knox, R. B., Willing, R. R., and Ashford, A. E., 1972, Pollen wall proteins: role as recognition substances in interspecific incompatibility in poplars, *Nature*, 237:381-383.

Larter, E., and Chaubey, C., 1965, Use of exogenous growth substances in promoting pollen tube growth and fertilization in barley-rye crosses, *Can. J. Genet. Cytol.*, 7:511-518.

Li, H. W., Weng, T. S., Chen, C. C., and Wang, W. H., 1962, Cytogenetical studies of *Oryza sativa* L. and its related species. 2. A preliminary note on the interspecific hybrids within the section Sativa Roschev., *Bot. Bull. Sinica*, 3:209-219.

Li, H. W., Chen, C. C., Weng, T. S., and Wuu, K. D., 1963, Cytogenetical studies of *Oryza sativa* L. and its related species. 4. Interspecific cross involving *O. australiensis* with *O. sativa* and *O. minuta*, *Bot. Bull. Sinica*, 4:65-74.

Matsubara, S., 1984, Overcoming the self-incompatibility of *Raphanus sativus* by application of plant hormones, amino acids and vitamines, and by temperature treatment of pollen, *Euphytica*, 33:113-121.

Mock, J. J., and Loescher, W. H., 1973, Incompatibility of maize and sorghum manifest in failure of pollen growth, *Egypt. J. Genet. Cytol.*, 2: 338-344.

Mól, R., and Zenkteler, M., 1982, Cytological studies of some wild species as pollinators of wheat and barley, *Z. Pflanzenzücht.*, 89:31-38.

Morinaga, T., 1959, Note on genome analysis in Oryza species, *Int. Rice Commun. Newsletter*, 8:10-11.

Morinaga, T., and Kuriyama, H., 1957, Cytogenetical studies on *Oryza sativa* L.. IX. The F_1 hybrid of *O. sativa* L. and *O. glaberrima* Steud., *Japan. J. Breed.*, 7:57-65.

Nandi, H. K., 1938, Interspecific hybridization in *Oryza*. Cytological evidence of the hybrid origin of *Oryza minuta*, *Bose Res. Inst. Trans., Calcutta*, (1935-1936), 11:99-121.

Nezu, M., Katayama, T. C., and Kihara, H., 1960, Genetic study of the genus *Oryza*. 1. Crossability and chromosomal affinity among 17 species, *Seiken Zihô*, 11:1-11.

Ogawa, T., and Katayama, T., 1971, Cytogenetical studies on the genus *Oryza*. 5. Chromosome pairing in the interspecific hybrid between genomes A and B (*O. punctata*), *Japan. J. Breed.*, 21:151-154.

Oka, H. I., 1964, Pattern of interspecific relationships and evolutionary dynamics in *Oryza, in*: "Proc. First Int. Symp. on Rice Genetics and Cytogenetics," International Rice Research Institute, Manila (1964).

Okazaki, K., and Hinata, K., 1987, Repressing the expression of self-incompatibility in crucifers by short-term high temperature treatment, *Theor. Appl. Genet.*, 73:496-500.

Pickering, R. A., 1981, Pollen tube-stylodium interaction in *Hordeum vulgare* L. x *H. bulbosum* L., *in*: "Barley Genetics IV," Proc. 4th Int. Genet. Sym., Edinburgh University Press, Edinburgh (1981), pp 666-676.

Reger, B. J., and James, J., 1982, Pollen germination and pollen tube growth of sorghum when crossed to maize and pearl millet, *Crop Sci.*, 22:140-144.

Rutger, J. N., and Shinjo, C., 1980, Male sterility in rice and its potential use in breeding, *in*: "Innovative Approaches to Rice Breeding," International Rice Research Institute, Manila (1980), pp 53-66.

Sano, Y., 1983, A new gene controlling sterility in F_1 hybrids of two cultivated rice species, *J. Heredity*, 74:435-439.

Sano, Y., 1985, Interspecific substitutions of an indica strain of *Oryza sativa* L. and *O. glaberrima* Steud., *Euphytica*, 34:587-592.

Sano, Y., Chu, Y. E., and Oka, H. I., 1979, Genetic studies of speciation in cultivated rice. 1. Genic analysis for the F_1 sterility between *O. sativa* L. and *O. glaberrima* Steud. *Japan. J. Genet.*, 54:121-132.

Sears, E. R., 1944, Inviability of intergeneric hybrids involving *Triticum monococcum* and *T. aegilopoides*, *Genetics*, 29:113-127.

Seetharaman, R., 1962, Studies on hybridization between Asian and African species of cultivated rices and their significance, *Science and Culture*, 28:286-289.

Shin, Y. B., and Katayama, T., 1979, Cytogenetical studies on the genus *Oryza*. XI. Alien addition lines of *O. sativa* with single chromosomes of *O. officinalis*, *Japan. J. Genet.*, 54:1-10.

Sitch, L. A., 1985, The production and utilization of wheat doubled haploids from the interspecific cross with *Hordeum bulbosum*, Ph.D. thesis, University of Cambridge, United Kingdom.

Sitch, L. A., and Snape, J. W., 1987, Factors affecting haploid production in wheat using the *Hordeum bulbosum* system. 1. Genotypic and environmental effects on pollen grain germination, pollen tube growth and the frequency of fertilization, *Euphytica*, 36:483-496.

Sree Ramulu, K., Bredemeijer, G. M. M., and Dijkhuis, P., 1979, Mentor pollen effects on gametophytic incompatibility in *Nicotiana, Oenothera* and *Lycopersicum, Theor. Appl. Genetics*, 54:215-218.

Stebbins, G. L., 1958, The inviability, weakness and sterility of interspecific hybrids, *Adv. Genet.*, 9:147-215.

Taillebois, J., and Guimaraes, E. P., 1988, Improving outcrossing rate in rice (*Oryza sativa* L.), *in*: "Hybrid Rice," International Rice Research Institute, Manila (1988), pp 175-180.

Tilton, V. R., and Russell, S. H., 1984, Applications of *in vitro* pollination/fertilization technology, *BioScience*, 34:239-242.

Virmani, S. S., and Shinjyo, C., 1988, Current status of analysis and symbols for male sterile cytoplasms and fertility restoring genes, *Rice Genetics Newsletter*, 5:9-15.

Virmani, S. S., and Wan, B. H., 1988, Development of CMS lines in hybrid rice breeding, *in*: "Hybrid Rice," International Rice Research Institute, Manila (1988), pp 103-114.

Watanabe, Y., 1975, Cytogenetic studies on rice and its wild relatives. 1. Cytogenetic investigations of induced polyploids, with special reference to the sterility of synthesized amphiploids, *Bull. Nat. Inst. Ag. Sci., Ser. D*, (1975), pp 1-90.

Wernesman, E. A., Davis, R. L., and Keim, W. F., 1965, Interspecific fertility of two Lotus species and their F_1 hybrids, *Crop Sci.*, 5:452-454.

Wuu, K. D., Jui,Y., Lu, K. C. L., Chou, C., and Li, H. W., 1963, Cytogenetical studies of *Oryza sativa* L. and its related species. 3. Two intersectional hybrids, *O. sativa* Linn. x *O. brachyantha* A. Chev. et Roehr. and *O. minuta* Presl x *O. brachyantha* A. Chev. et Roehr., *Bot. Bull. Sinica*, 4:51-59.

Yabuno, T., 1977, Genetic studies on the interspecific cytoplasm substitution lines of japonica varieties of *Oryza sativa* L. and *O. glaberrima* Steud., *Euphytica*, 26:451-463.

Yang, K. K. S., Ho, K. C., and Li, H. W., 1965, Cytogenetical studies of *Oryza sativa* L. and its related species. 8. Study on meiotic division of F_1 hybrid of *O. sativa* L. x *O. brachyantha* A. Chev. et Roehr., *Bot. Bull. Sinica*, 6:32-38.

Yeh, B., and Henderson, M. T., 1961, Cytogenetic relationship between cultivated rice, *Oryza sativa* L., and five wild diploid forms of *Oryza, Crop Sci.*, 1:445-450.

Yuan, L. P., and Virmani, S. S., 1988, Status of hybrid rice research and development, *in*: "Hybrid Rice," International Rice Research Institute, Manila (1988), pp 7-24.

Zenkteler, M., 1980, Intraovarian and *in vitro* pollination, *in*: "Perspectives in Plant Cell and Tissue Culture," I. K. Vasil, ed., *Inter. Rev. Cytol. Suppl.* 11B, Academic Press, New York (1980), pp 137-156.

Zenkteler, M., and Nitzsche, W., 1984, Wide hybridization experiments in cereals, *Theor. Appl. Genet.*, 68:311-315.

WHEAT X MAIZE AND OTHER WIDE SEXUAL HYBRIDS: THEIR POTENTIAL FOR GENETIC MANIPULATION AND CROP IMPROVEMENT

David A. Laurie, Louise S. O'Donoughue
and Michael D. Bennett[1]

Institute of Plant Science Research, Maris Lane, Trumpington,
Cambridge, CB2 2JB, U.K.
[1]Jodrell Laboratory, Royal Botanic Gardens, Kew, Richmond,
Surrey, TW9 5DS, U.K.

1. INTRODUCTION

"The view generally entertained by naturalists is that species, when inter-crossed, have been specially endowed with the quality of sterility, in order to prevent the confusion of all organic forms." With these words Charles Darwin opens his chapter on hybridism in The Origin of Species. Darwin, however, goes on to argue that the degree of sterility in crosses is *not* a specially endowed quality but is highly variable, arising through the processes of natural selection by the accumulation of what we would now call genetic differences.

The lack of absolute barriers to interspecific hybridization has proved to be of great value to plant geneticists and plant breeders and has given rise to a discipline generally referred to as wide-crossing. In the context of this paper, wide-crosses can be defined as interspecific or intergeneric hybrids either between two crop species or between a crop species and one of its wild relatives.

The history of wide-crossing in cereals, the group of plants with which we will principally be concerned, is one of a progressive increase in the

Gene Manipulation in Plant Improvement II
Edited by J. P. Gustafson
Plenum Press, New York, 1990

number of species combinations in which hybrids have been recognized or synthesized and an increase in the ease with which mature hybrid plants are produced and made fertile. For example, the first reported hybrids of wheat (*Triticum aestivum*) and rye (*Secale cereale*) were sterile (Wilson, 1876) and 15 years elapsed before fertile hybrids were reported (Rimpau, 1891). Furthermore, the hybrid nature of the plants was not confirmed cytologically for another 45 years (Müntzing, 1936). Subsequent technical developments such as colchicine treatment for doubling chromosome numbers and improved embryo culture methods have made the production of amphiploids such as primary triticales much simpler. The first wheat x rye hybrids were scientific curiosities, but commercial triticale varieties are now grown on at least 750 000 ha worldwide (Gregory, 1987).

The ease with which a wide hybrid is produced depends on many bio-logical parameters, including mechanical factors such as the size and shape of the reproductive cells, and genetic factors such as incompatibility and karyotypic stability in the hybrids. However, another important aspect which is not sufficiently recognized is the perceived difficulty of producing the hybrid plant. Wheat x barley (*Hordeum vulgare*) crosses illustrate this point.

Efforts to hybridize wheat and barley began over 80 years ago with Farrer (1904), but no well-documented success was achieved for nearly 70 years until Kruse (1973) used barley as the female parent to produce a few hybrid seeds. Despite his report this combination was still regarded as very difficult or even in doubt by many workers. Success using barley as the male, reported by Kruse (1976) and Islam *et al.* (1978, 1981), put the matter beyond question. Once the cross was perceived to be possible several other groups were able to produce hybrid plants by standard methods (Fedak, 1985a; Sethi *et al.*, 1986). The history of sexual hybridization suggests that many wide-hybrids are wrongly thought to be difficult or impossible to obtain and that this view, in itself, limits the chances of success.

Of course, the desire to make a particular hybrid, even when coupled with a report of its production, does not inevitably presage its successful development as the considerable efforts to obtain maize (*Zea mays*) x sorghum (*Sorghum bicolor*) hybrids show. James (1978, 1979) claimed that "non-classical" hybrids (i.e. plants with 20 maize and 1 to several sorghum chromosomes) could be produced at low frequency. However, this combina-tion is still widely viewed as very difficult or impossible. Thus, while optimism is often in order, no one can predict which species combinations will be successful in yielding hybrids and which will prove recalcitrant. It must be concluded that wide-hybridization remains very much an art as well as a rapidly advancing science.

The relative ease with which hybrids are formed by sexual means has long been regarded as a measure of the relatedness of the species concerned. However, recent work on crosses which link two subfamilies of the Gramineae, namely the Pooidea and the Panicoidea (according to the classification of Hutchinson (1959)) suggests that this is not necessarily so. In 1984 Zenkteler and Nitzsche reported fertilization in a number of novel and remarkable cereal wide-crosses. They claimed, for example, that globular embryos were formed when hexaploid wheat was pollinated with maize. These results excited our interest, but we confess to being sceptical since their crosses had been analyzed by light microscopy of thick (about 10μm) sections. This method gives little information on the chromosome complement of embryos and we therefore felt that the rigorous proof required to substantiate such an extraordinary claim was lacking. However, we soon found that they were right.

2. WHAT HAPPENS WHEN CHINESE SPRING WHEAT IS POLLINATED WITH SENECA 60 MAIZE?

When the hexaploid wheat variety Chinese Spring was crossed with the maize variety Seneca 60, pollen readily germinated on the stigma and one or more pollen tubes reached the embryo sac in about 80% of florets. In about 2% of florets the maize pollen tube severely damaged the wheat gametes, in about 34% the pollen tube entered the embryo sac but failed to release the sperm nuclei and in about 16% the sperm nuclei were released but failed to fuse with the wheat gametes (Laurie and Bennett, unpublished). In the remaining 28% of florets fertilization occurred (Laurie and Bennett, 1988a), a remarkably high frequency in view of the taxonomic distance separating the two species.

In florets where fertilization had occurred it was relatively rare to find both an embryo and an endosperm. In a sample of 343 florets fixed 48h after pollination, 80 (23.3%) had only an embryo, 8 (2.3%) had only an endosperm and 12 (3.5%) had both. Thus, the frequency of embryo development was 4.6-fold higher than the frequency of endosperm development.

Evidence for the hybrid origin of the embryos came from cytological studies of zygotes at metaphase. Firstly, we calculated the expected sizes of chromosomes in a hybrid. By using the 4C nuclear DNA content of Chinese Spring (69.3pg, Bennett and Smith, 1976), the relative DNA content of the individual chromosome arms (Furuta et al., 1984) and the chromosome arm ratios (Sears, 1954), the amount of DNA in each chromosome arm of wheat could be estimated. The same calculation was made for Seneca 60 maize using the 4C DNA content (9.8pg, Laurie and Bennett, 1985) and relative chromosome sizes and arm ratios from measurements of electron micro-graphs of ten serially thin sectioned root-tip metaphase nuclei (Bennett and Smith, unpublished). A diagram of expected chromosome sizes in a hybrid was then drawn assuming direct proportionality of arm DNA content and arm length. These calculations showed that a hybrid zygote should contain 21 wheat chromosomes and 10 smaller maize chromosomes, the largest of which should be about half the size of the smallest wheat chromosome (Fig. 1a).

Zygotes at metaphase had this predicted combination (Fig. 1b), confirming the hybrid origin of the embryos. However, the hybrids were karyotypically unstable and eliminated the maize chromosomes early in development. These observations, which are described in detail elsewhere (Laurie and Bennett, 1989), can be summarized as follows:
(1) In zygotes at anaphase the maize chromosomes did not appear to be attached to the spindle and tended to lag behind the wheat chromosomes as the latter moved towards the spindle poles (Fig. 1c).
(2) About 70% of embryos consisting of two cells at interphase had one or more micronuclei, showing elimination of one or more maize chromosomes at the first cell division.
(3) Metaphase preparations from 2-celled embryos had four to ten maize chromosomes and anaphase or telophase preparations always showed lagging maize chromosomes.
(4) Metaphase preparations from 4-celled embryos had none to five maize chromosomes and it appeared that most of these were eliminated during the third cell division since no maize chromosomes were observed in metaphase preparations from embryos with eight or more cells. Thus, elimination of the maize chromosomes was effectively complete within three cell division cycles in this material. All embryos with four or more cells had micronuclei (Fig. 1d), indicating that embryo development was dependent on karyogamy.
(5) Chromosome behaviour in the endosperm was much more difficult to quantify since fewer were formed. However, the expected F_1 combination of 42 wheat and 10 maize chromosomes was found in the single primary endosperm mitosis observed and the presence of micronuclei in 85% of endo-

---→

Fig. 1. Embryology of wheat (var. Chinese Spring) x maize (var. Seneca 60) hybrids.
(a) Idiogram of expected chromosome sizes in a hybrid.
(b) Zygote metaphase showing the expected combination of 21 wheat and 10 maize chromosomes (solid arrows). (c) Zygote anaphase showing 21 wheat daughter chromosomes moving to each pole and 20 maize daughter chromosomes (solid arrows), most of which are lagging behind the wheat chromosomes.
(d) Embryo showing micronuclei (examples arrowed). Open arrows in 1b and 1c indicate the wheat satellites. Bars represent 10μm.

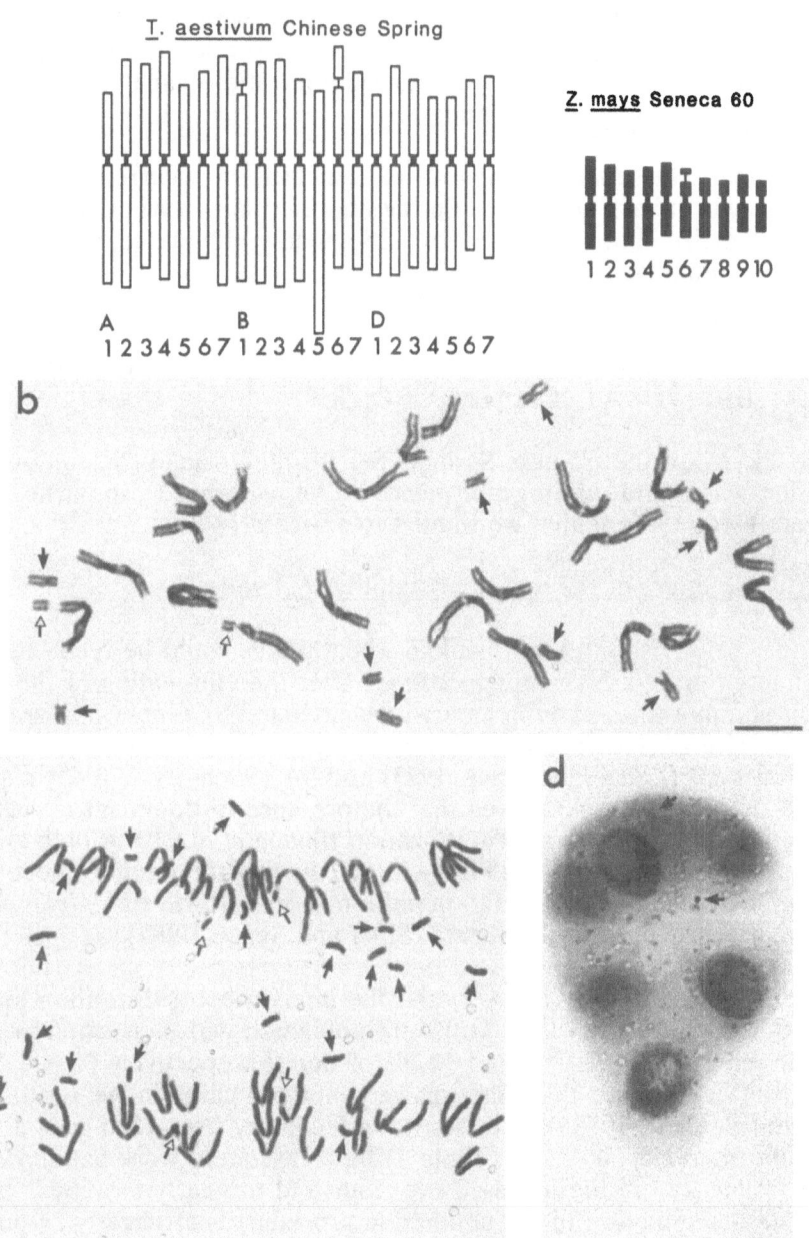

Figure 1 (full page plate)

sperms with four or more nuclei showed that most arose from hybrid nuclei. The endosperm was almost always highly abnormal, with few nuclei in comparison to self-pollinated Chinese Spring caryopses produced under similar conditions (Bennett *et al.*, 1975), and showed aberrations such as micronuclei, misshapen nuclei and chromatin bridges.

In summary, these studies showed that fertilization occurred in crosses between the wheat variety Chinese Spring and the maize variety Seneca 60 and that the maize chromosomes were subsequently eliminated, giving rise to haploid wheat embryos.

3. HOW DO CHINESE SPRING X SENECA 60 CROSSES COMPARE WITH OTHER WHEAT X MAIZE CROSSES?

The success of the Chinese Spring x Seneca 60 crosses prompted us to ask if other wheat and maize genotypes could be hybridized. In particular, there were two questions that we wanted to answer;

a) What is the effect of the wheat crossability (*Kr*) genes?

Barclay (1975) showed that haploid wheat plants could be recovered from Chinese Spring x *H. bulbosum* crosses after the elimination of the *H. bulbosum* chromosomes early in embryo development. Chinese Spring carries recessive alleles (*kr1*, *kr2*) at crossability loci on the long arms of chromosomes 5B (Lange and Riley, 1973) and 5A (Sitch *et al.*, 1985) respectively, and subsequent work showed that the presence of dominant alleles greatly reduced the frequency of fertilization (Snape *et al.*, 1979; Falk and Kasha, 1981, 1983; Sitch *et al.*, 1985), with *Kr1* having the greater effect. The reduction in crossability was due to inhibition of the growth of *H. bulbosum* pollen tubes at the base of the stigma (Sitch and Snape, 1987).

When Chinese Spring (*kr1*, *kr2*) and the intervarietal substitution line Chinese Spring (Hope 5B) (*Kr1*, *kr2*) were pollinated with Seneca 60 maize, fertilization occurred in 29.7% and 30.7% of florets respectively (Table 1) and further work showed that other maize genotypes gave similar results (Laurie and Bennett, 1988a). Crosses using Highbury (*Kr1*, *Kr2*) gave a fertilization frequency of 14.4% (Table 1) but subsequent work has shown that this frequency can be increased over four-fold to nearly 60% by appropriate manipulation of the pollination procedure, particularly by pollinating when the florets are at a developmental age corresponding to the day of anthesis (Laurie, 1989a). The highest frequency obtained in Highbury x Seneca 60 crosses was more than 100-fold higher than the frequency of fertilization in crosses between Highbury and three *H. bulbosum* clones (Sitch and Snape, 1987).

These results are important because they show that maize pollen tubes appear to be unaffected by the action of dominant *Kr* alleles, which are present in the material of most wheat breeding programs and which have greatly restricted the use of the *H. bulbosum* system for wheat haploid production.

b) <u>Does the ploidy level of the parents affect chromosome stability?</u>

This question is of interest since the proportion of hybrids to haploids recovered from barley x *H. bulbosum* crosses is known to be affected by the ploidy levels of the parents. Crosses between diploids tend to produce haploid barley, although the ratio of haploids to hybrids is strongly influenced by the parental genotypes (Simpson *et al.*, 1980; Pickering, 1983). Crosses between diploid barley and tetraploid *H. bulbosum* usually give triploid hybrid progeny (Subrahmanyam and Kasha, 1973; Kasha, 1974).

Hybrid embryos were formed in crosses between hexaploid wheats and three tetraploid maize genotypes (Table 1) but there was no evidence of increased karyotypic stability and all maize chromosomes were rapidly lost (Laurie and Bennett, 1988a). The effect of the ploidy level of the female parent was investigated by O'Donoughue and Bennett (1988), using the diploid maize variety Seneca 60 as the male parent. Fertilization occurred in the tetraploid macaroni wheat (*T. durum*) cultivar Kubanka, and in 12 out of 14 other crosses involving wild relatives of wheat from the genera *Triticum* and *Aegilops* including *T. urartu* and *Ae. squarrosa*, the respective donors of the A and D genomes of hexaploid wheat. Fertilization frequencies were significantly different between ploidy levels, with diploids giving the lowest values, and some tetraploids and hexaploids giving higher frequencies than the hexaploid bread wheats (Table 1). Only *Ae. speltoides* and *T. thaoudar* failed to give fertilization in these experiments, but the latter was subsequently found to hybridize at low frequency.

An interesting feature of the above crosses was the variation in the frequency of single and double fertilization. For example, the tetraploid macaroni wheat cultivar Kubanka gave a much higher frequency of double fertilization than the hexaploid wheats (Table 1), and this difference has been confirmed in subsequent experiments.

All the hybrids were unstable, losing the maize chromosomes within a few cell cycles in all cases. There was no clear correlation between ploidy level and stability, although there was evidence that some crosses retained maize chromosomes for slightly longer than the hexaploid wheats. For example, four maize chromosomes were observed in a metaphase cell from a six-day old *Ae. squarrosa* x maize embryo in which there were at least 58 cells.

Table 1. Fertilization frequencies in crosses between various Triticeae species and maize, sorghum or pearl millet.

	n	em	en	em + en	Total fertilized n	%
Diploids						
Ae. comosa x S60	106	0	1	2	3	2.8[1]
Ae. mutica x S60	80	1	0	1	2	2.5[1]
Ae. umbellulata x S60	50	0	1	5	6	12.0[1]
Ae. speltoides x S60	46	0	0	0	0	0[1]
Ae. squarrosa x S60	100	0	13	3	16	16.0[1]
T. thaoudar x S60	72	0	0	0	0	0[1]
T. urartu x S60	34	0	1	0	1	2.9[1]
H. vulgare Sultan x S60	100	4	8	16	28	28.0[2]
S. cereale PS x S60	150	4	10	14	28	18.7
Tetraploids						
Ae. crassa x S60	36	1	2	5	8	22.2[1]
Ae. triaristata x S60	30	1	0	1	2	6.7[1]
T. dicoccoides x S60	75	4	2	5	11	14.7[1]
T. dicoccum x S60	100	0	0	2	2	2.0[1]
T. timopheevi x S60	22	0	5	0	5	22.7[1]
T. durum Kubanka x S60	104	8	3	17	28	26.9[1]
Kubanka x 23BE	100	2	6	40	48	48.0[3]
Hexaploids						
Ae. crassa x S60	82	11	12	9	32	39.0[1]
Ae. triaristata x S60	15	4	2	2	8	53.3[1]
T. aestivum						
CS (*kr1, kr2*) x S60	343	80	8	12	100	29.7[4]
CS x *Ac*	40	10	1	0	11	27.5[2]
CS x S9B	100	57	2	10	69	69.0[5]
CS x 23BE	220	17	4	42	63	28.6[3]
CS(H5A) (*kr1, Kr2*) x S60	22	4	2	3	9	40.9[2]
CS(H5B) (*Kr1, kr2*) x S60	189	36	7	15	58	30.7[6]
Hope (*Kr1, Kr2*) x S60	32	2	3	2	7	21.9[2]
Highbury (*Kr1, Kr2*) x S60	194	19	0	9	28	14.4[6]
Highbury (*Kr1, Kr2*) x 23BE	100	2	3	27	32	32.0[3]

Notes to table 1 . n is the number of ovaries dissected. e is the number which had only an embryo, en is the number which had only an endosperm and e+en is the number

The results from crosses using Kubanka, and several other tetraploid cultivars (O'Donoughue unpublished), suggest that wheat x maize crosses could be a valuable new method for producing haploids in commercial macaroni wheats.

4. CAN OTHER WIDE-CROSSES BE MADE WHICH SPAN COMPARABLE TAXONOMIC DISTANCES?

The ease with which wheat and maize could be hybridized prompted us to investigate other cereal wide-crosses which spanned a comparable taxonomic distance. Four of these are described below.

a) Hexaploid wheat x sorghum crosses

When Chinese Spring was pollinated with the diploid grain sorghum genotype S9B the fertilization frequency was more than double that found in Chinese Spring x Seneca 60 maize (69% and 29.7% respectively, Table 1). However, the tendency towards fertilization of only the egg-cell remained, and 57 out of 69 fertilized florets had only an embryo. Of the eight zygotes at metaphase that were obtained, seven had the expected F_1 combination of 21 wheat and 10 very small sorghum chromosomes (Figs. 2a and 2b). The eighth zygote had 20 sorghum chromosomes, either from fusion of two sperm nuclei with the egg-cell or because of fertilization by an unreduced gamete. The hybrid embryos were karyotypically highly unstable and in at least one case all the sorghum chromosomes were lost at the zygotic division. Endosperms were again highly abnormal.

---

with both an embryo and an endosperm. The total number showing fertilization is also given.
Abbreviations: CS is Chinese Spring. CS(H5A) and CS(H5B) are single chromosome substitution lines in which chromosomes 5A and 5B were replaced by the corresponding chromosome from the variety Hope. PS is Petkus Spring rye. S60 is Seneca 60 maize. S9B is S9B sorghum. 23BE is Tift 23BE pearl millet. *Ac* is a maize line carrying the transposable element *Activator*. All male parents were diploids.
Superscripts refer to the original sources: 1 O'Donoughue and Bennett (1988). 2 Laurie and Bennett (1988a). 3 Laurie (1989b). 4 Laurie and Bennett (1986, 1987). 5 Laurie and Bennett (1988b). 6 Laurie and Bennett (1987).

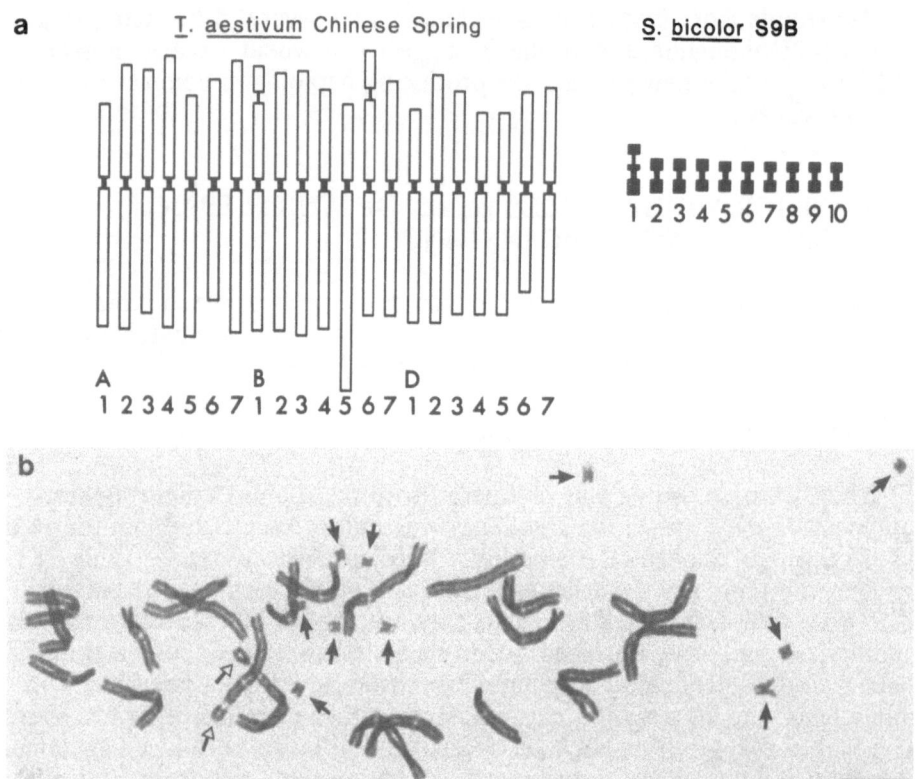

Fig. 2. Embryology of wheat (var. Chinese Spring) x sorghum (var. S9B)
hybrids.
(a) Idiogram of expected chromosome sizes in a hybrid.
(b) Zygote metaphase showing the expected combination of
21 wheat and 10 sorghum chromosomes (solid arrows). Open
arrows indicate the wheat satellites. Bars represent 10μm.

b) Hexaploid wheat x pearl millet crosses

When Chinese Spring was pollinated with the pearl millet (*Pennisetum
glaucum* syn. *P. americanum*) variety Tift 23BE the fertilization frequency was
similar to that found in Chinese Spring x Seneca 60 maize crosses (28.6%
and 29.7% respectively, Table 1). However, the results differed in that
fertilized florets from crosses with pearl millet usually had both an embryo
and an endosperm. Evidence for the hybrid origin of the embryos again
came from cytological studies of zygotes at metaphase. Three had the
expected combination of 21 wheat and 7 pearl millet chromosomes while a

fourth had 14 pearl millet chromosomes. Loss of pearl millet chromosomes at the first cell division was rare but it commenced at the second and appeared to be complete by the fourth. The expected combination of 42 wheat and 7 pearl millet chromosomes was also found in the only primary endosperm mitosis in which chromosomes could be counted (Fig. 3d).

When the hexaploid wheat variety Highbury (*Kr1*, *Kr2*) was used as the female, pearl millet gave fertilization in 32% of florets (Table 1). This frequency was not significantly different from the frequency in Chinese Spring (*kr1*, *kr2*) indicating that pearl millet, like maize, is insensitive to the action of dominant alleles at the *Kr* loci. Fertilization was also found in 48% of florets from crosses using the tetraploid wheat variety Kubanka as the female parent, suggesting that pearl millet, like maize, has potential for haploid production in both tetraploid and hexaploid wheats.

c) Barley x maize crosses

Successful hybridization was also achieved in crosses between the barley (*H. vulgare*) variety Sultan and the maize variety Seneca 60. The overall frequency of fertilization was similar to that found in Chinese Spring wheat x Seneca 60 crosses (28.0% and 29.7% respectively) but the proportion of double fertilization was higher (Table 1). Once again proof of the hybrid origin of the embryos was obtained from preparations of zygotes at metaphase where the expected combination of 7 barley and 10 small maize chromosomes was observed (Figs. 4a and 4b). In marked contrast to the situation in wheat x maize crosses, the centromeres of maize chromosomes and the maize nucleolus organizer region (NOR) (Fig. 4c) were clearly visible in zygotes and there was evidence that the hybrids retained at least some maize chromosomes for several cell cycles.

Karyotypically unstable interspecific *Hordeum* crosses show evidence of a hierarchy of dominance which determines which parental chromosome complement will be lost (Subrahmanyam, 1982; Subrahmanyam and von Bothmer, 1987). The slight increase in stability in barley x maize crosses in comparison to wheat x maize crosses suggests that other *Hordeum* x maize combinations would be of interest. We are therefore currently investigating crosses using diploid or tetraploid *H. bulbosum* as the female parent.

d) Rye x maize crosses

When the rye variety Petkus Spring was pollinated with Seneca 60, fertilization occurred in 18.7% of the florets studied (Table 1). Most fertilized florets had both an embryo and an endosperm and preliminary studies indicated that elimination of the maize chromosomes occurred early in embryo and endosperm development.

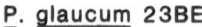

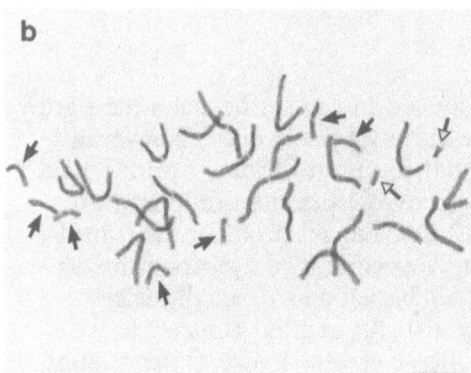

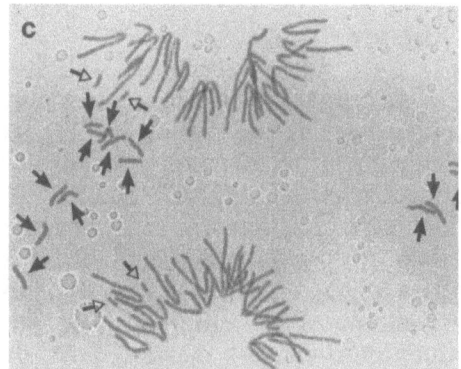

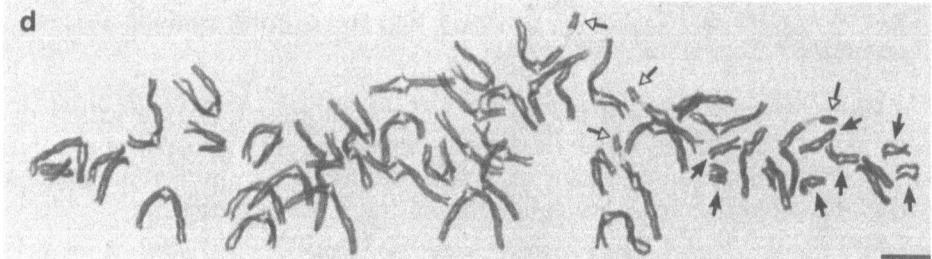

Fig. 3. Embryology of wheat (var. Chinese Spring) x pearl millet (var. Tift 23BE) hybrids.
(a) Idiogram of expected chromosome sizes in a hybrid.
(b) Zygote metaphase with the expected combination of 21 wheat and 7 pearl millet chromosomes (solid arrows).
(c) Anaphase from a 2-celled embryo showing elimination of pearl millet chromosomes (arrowed). (d) Primary endosperm division just entering anaphase, showing 42 wheat and 7 pearl millet chromosomes (solid arrows). Open arrows indicate the wheat satellites. Bars represent 10μm.

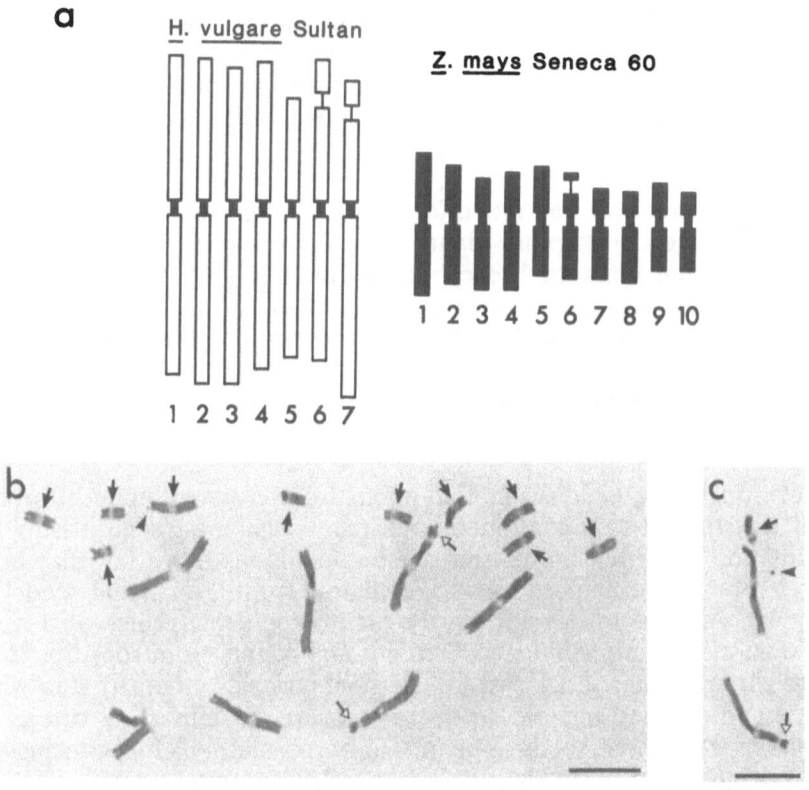

Fig. 4. Embryology of barley (var. Sultan) x maize (var. Seneca 60) hybrids.
(a) Idiogram of expected chromosome sizes in a hybrid.
(b) Zygote metaphase showing the expected combination of 7
barley and 10 smaller maize chromosome (solid arrows). The
arrowhead indicates a possible maize satellite, although it is
not clear which is the rest of the maize chromosome 6. Open
arrows indicate the barley satellites. (c) Part of another
zygote metaphase showing the maize chromosome 6 with
satellite (arrowhead). Bars represent 10μm.

5. CAN PLANTS BE PRODUCED FROM THESE CROSSES?

The above results clearly show that sexual hybridization between taxonomically diverse cereal crops can easily be achieved, but such observations would be of limited interest if plants could not be produced. In all of the crosses discussed above it has proved very difficult to recover plants using conventional embryo rescue techniques (i.e. by leaving the seeds to develop on plants for two to three weeks before attempting embryo culture). The reasons for this are not known, but seed abortion is probably caused by the absence, or poor development, of the endosperm. Clearly, an alternative embryo rescue method was necessary.

Initially we concentrated our efforts on hexaploid wheat x maize crosses, where it was found that haploid plants could be recovered using spikelet culture (Laurie and Bennett, 1988c). In this method (Fig. 5) spikelets were removed from plants two days after cross-pollination, surface sterilized and transferred to Murashige and Skoog medium containing 0.1mg l^{-1} 2,4-D and 60g l^{-1} sucrose. After three weeks at 20°C in continuous light the ovaries were dissected and embryos were transferred to conventional embryo rescue medium.

Interestingly, the yield of haploid plants from cross-pollinated florets was higher if the spikelet also contained a normally developing seed arising from self pollination. The reason for this nurse effect is unknown but may be due to the release of one or more growth regulators from the normal seed. However, leaving embryos to develop on the plant in spikelets which also had a normal seed did not aid survival. Chromosome counts from root-tips showed 21 wheat chromosomes in all cases. The seedlings were then treated with colchicine and doubled haploid seed was recovered in almost all cases. Plants grown from these seeds appeared to be normal, but this conclusion is currently being tested more rigorously in a field experiment comparing parent lines with wheat doubled haploids derived from crosses with maize or *H. bulbosum*. Fertile doubled haploid wheat plants have also been obtained from wheat x maize crosses by Comeau *et al.* (1988), who used ovary culture to overcome the problem of seed abortion.

Recently we have found that spikelet culture can be used to produce haploid barley from crosses with Seneca 60 maize, but the yield of plants was low, only three being recovered from 365 florets. This is only 4.6% of the expected number, assuming that an embryo would have been formed in 20% of florets (Table 1), suggesting that barley is less suited than wheat to this culture system.

It is clearly important to continue the development of embryo rescue methods, preferably so that embryos can be induced to develop on plants to

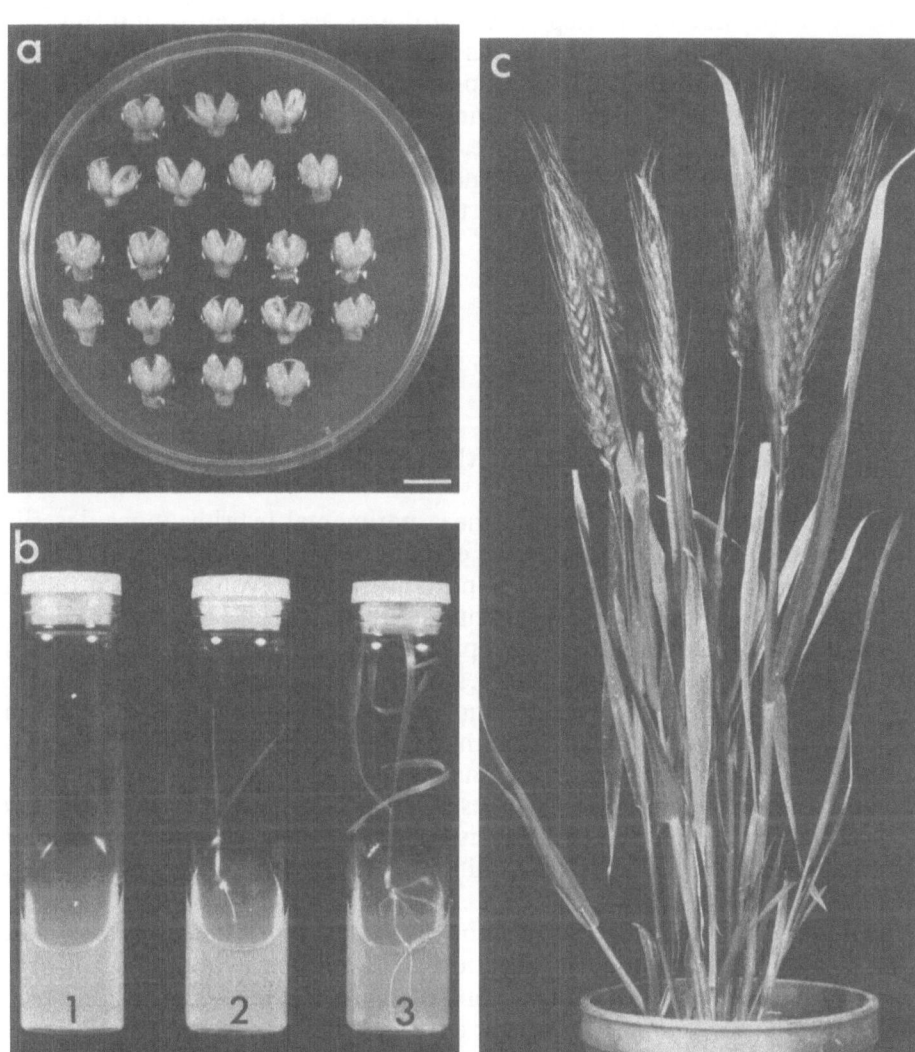

Fig. 5. Spikelet culture and embryo rescue.
(a) Spikelets of Chinese Spring wheat transferred to culture medium two days after pollination. (b) Embryo culture tubes showing - (1) A freshly excised embryo, (2) A germinating embryo and (3) A haploid seedling ready to be transplanted to soil. (c) A haploid plant of Highbury (*Kr1, Kr2*) wheat from a cross with Seneca 60 maize. Bars represent 10mm.

a point where they can be excised onto conventional embryo culture medium. We have recently found that haploid embryos from hexaploid wheat x maize crosses will continue to develop *in vivo* if florets are treated with 0.5mg l⁻¹ 2,4-D each day for two to three weeks after pollination. This method, which also enabled three haploid Chinese Spring wheat plants to be produced from crosses with pearl millet, has the advantage of simplicity and promises to be a useful alternative to spikelet culture. In summary, there are already three successful embryo rescue protocols, and we can expect that future technical developments will increase the proportion of haploid embryos that can be recovered as plants.

DISCUSSION

1. WIDE-HYBRIDIZATION IN PLANT EVOLUTION

In the light of the results from the experimental hybridizations described above it is of interest to consider the evolutionary role of wide-hybridization. Wide-hybrids have obviously been important in plant evolution since they are the origin of allopolyploid species including the tetraploid and hexaploid wheats discussed above. Other examples are crops such as *Brassica* species, oats (*Avena sativa*), plums and cherries (*Prunus* spp.), strawberries (*Fragaria* spp.), tobacco (*Nicotiana tabacum*) and cotton (*Gossypium hirsutum*) and the wild species *Spartina townsendii* (Simmonds, 1979; Hawkes, 1983; Swanson *et al.*, 1980). The diploid ancestors of many of these are known, and in several cases the polyploids have been synthesized experimentally. However, many areas of uncertainty, such as the precise origin of the B genome in polyploid wheats, still remain (Dvořák, 1988; Kimber, 1988).

There is, however, no reason to believe that the formation of polyploids is the only rôle which wide-hybridization plays in plant evolution. It is also possible that segments of genomes (one or more chromosomes, or one or more chromosome segments) may be transferred. It is even possible that the nuclear genome of the maternal parent of the original hybrid could be completely replaced, producing a natural alloplasmic line. As an example of the latter, Doebley and Sisco (1989) have recently suggested that the S type of cytoplasmic male sterility in maize may have originated from hybridization with teosinte (*Z. mays* ssp. *mexicana*) such as that presently found in the Copandiro region of Mexico.

Evolution by chromosome introgression might be expected to be especially prevalent in polyploids such as wheat which readily tolerate aneuploidy under experimental conditions. This might even occur in crosses which are usually karyotypically unstable, since there are reports of the occasional retention of an alien chromosome in hexaploid wheat x *H. bulbosum* crosses (Miller and Chapman, 1976; Miller *et al.*, 1983). Introgression might account for the odd behaviour of the group 4[1] chromosomes in polyploid wheats. Hybrids between hexaploids and the proposed A genome donor show six bivalents instead of the expected seven in the great majority of meiocytes (Chapman *et al.*, 1976; Dvorák, 1976, 1983; Miller *et al.*, 1981; Kimber, 1988). This has led to speculation that the unpaired chromosome may have entered the polyploid wheats from another source (Driscoll, 1981; Miller, 1983).

Of particular interest in the context of introgression are chromosomes from several *Aegilops* species which are preferentially transmitted when introduced into hexaploid wheat (Miller, 1983; Endo, 1988a). Plants monosomic for the alien chromosome produce offspring that are frequently or exclusively disomic upon self-pollination and the alien chromosome is also transmitted to most or all of the progeny upon outcrossing. Finch *et al.* (1984) showed that in the case of the *Ae. sharonensis* group 4 chromosome, and probably for others (Endo, 1988b), chromosome breakage occurs in about half the meiospores at the first post-meiotic mitosis. Thus, it appears that cells lacking the alien chromosome are rendered inviable by chromosome breakage and only those gametes containing the alien chromosome remain functional.

When crossed onto euploid wheats, preferentially transmitted chromosomes may also induce post-zygotic chromosome aberrations, including translocations and deletions and this tendency is markedly increased if the wheat parent is monosomic for chromosome 4B (Endo, 1988a). This property makes these chromosomes of interest for the transfer of alien genes into wheat and for genetic studies using deletion mapping (Endo, 1988b). Furthermore, if introgression between wheat and these *Aegilops* species has occurred naturally, chromosome rearrangements may serve to distinguish evolutionary phylogenies.

Interspecific hybridization might also be one of the mechanisms by which B-chromosomes, which often differ markedly from the remainder of the chromosome complement, arise. For example, Sapre and Deshpande (1987) reported that one or two chromosomes of *Coix gigantea* showed B-chromosome like behaviour when incorporated in the genome of *C. aquatica*

Footnote.

[1] Nomenclature for the group 4 chromosomes of wheat follows the recommendation of the Proc. 7th Int. Wheat Genet. Symp. (eds. Miller, T.E., and Koebner, R.M.D.), p. 1219. i.e. that the chromosome previously designated 4A should become chromosome 4B and *vice versa*.

via sexual hybridization. In this example the B-chromosomes are derived from the A-chromosomes of a different genome, but it is also possible that B-chromosomes which had become established in one species might be transmitted to, or "infect", another.

In addition, wide-hybridization should also enable the introgression of smaller chromosome segments, including single genes. Indeed, many authors consider cycles of hybridization between wild progenitor species and cultivated plants to have been of great importance in crop evolution (Zohary, 1970; Harlan *et al.*, 1973). Molecular techniques are likely to be particularly useful in identifying instances of this type of evolution. To take a recent example from polyploid wheats, Gill and Appels (1988) found that the *Nor* locus on chromosome 6G in AAGG tetraploids was not closely related to any of the other *Nor* loci studied and they suggested that this reflects introgression from a third as yet uncharacterized species.

Transposable elements are a varied class of sequences which are of particular importance in considerations of genome evolution (Flavell, 1986; Brookfield, 1986; Flavell *et al.*, 1988; Lambowitz, 1989) and introgression by wide-hybridization. In some cases, such as the *Wis-2* element in wheat (Harberd *et al.*, 1987) and the *Bs1* and *Cin4* elements in maize (Johns *et al.*, 1985; Schwarz-Sommer *et al.*, 1987 respectively) there are similarities to retrotransposons, suggesting that these elements have spread via RNA intermediates by utilizing reverse transcriptase. Consequently, they may have considerable potential for increasing in copy number. Another interesting example, which concerns a transposon of a different type, is the *Mutator* (*Mu*) element of maize which can be transmitted in a highly non-Mendelian fashion in appropriate situations, with up to 90% of progeny having *Mutator* activity despite repeated outcrossing (Robertson, 1978, 1981). Thus, transposable elements may be capable of rapid spread between the genomes of a novel wide-hybrid, and of introgression between species.

In addition, most higher plant genomes contain considerable amounts of highly repeated sequences (it is estimated that single copy coding sequences account for no more than 0.5% of the genome of hexaploid wheat, while highly repeated sequences account for some 75% (Flavell *et al.*, 1987)), and it has become increasingly clear that there is a considerable turnover of such sequences by amplification, deletion and rearrangement within the genome. and such mechanisms may enable sequences to move between genomes. The process by which the characteristics of a genome are changed as a result of the behaviour of DNA itself has been referred to as "molecular drive" (Dover, 1982; Dover and Tautz, 1986).

The above discussion focuses on the potential for genomes, chromosomes or individual DNA sequences to be transferred between species during evolution. However, spontaneous wide-hybridization also has implications for the immediate future in regard to the use of biotechnology in agriculture. For instance, resistance to the herbicide glyphosate encoded by the *aroA* gene of *Salmonella typhimurium* has already been transferred to several dicotyledons, including tobacco and tomato (*Lycopersicon esculentum*), using the *Agrobacterium* Ti plasmid system (Botterman and Leemans, 1988). Outcrossing of genetically engineered plants to other species or wild populations could result in the emergence of resistant weeds, a point discussed for this and other examples by Botterman and Leemans (1988) and Woolhouse (1987). Such problems could arise once transgenic cereals are produced, especially in crops such as maize, sorghum and barley where fertile hybrids are known to occur with wild relatives (i.e. where wild relatives are included in what Harlan *et al.* (1973) describe as the "primary gene pool"). The likelihood of transfer is hard to assess at present but the possibility should not be ignored.

Finally spontaneous wide-hybridization may impose selection pressures even when gene transfer does not occur. It has been suggested that the crossability (*Kr*) genes of wheat, which are described in section 3a, evolved to prevent reduction in fertility caused by pollination by alien species (Riley and Chapman, 1967).

2. CAN WIDE-HYBRIDIZATION INDUCE CHANGES IN DNA CONTENT?

The previous section discussed the exchange of DNA sequences via sexual hybridization, but there are some clear examples of genomic change which appear to be *induced* by hybridization. Price *et al.* (1983) reported that hybrids between *Microseris douglasii* and *M. bigelovii* had F_1 DNA contents which spanned almost the entire range between the parental DNA contents (a 10% difference) rather than having an intermediate value. F_2 progeny had mean DNA contents similar to the F_1 plants from which they were derived suggesting that the changes could be fixed and inherited in a regular manner. A different situation is apparent in the data of Furuta *et al.* (1975) where a tetraploid wheat cultivar was crossed with a series of *Ae. squarrosa* accessions with different DNA contents. The synthetic hexaploids had indistinguishable DNA contents instead of the expected range, suggesting that in this case all were adjusted to the same level.

Another example concerns the bizarre "megachromosomes" found in certain interspecific *Nicotiana* hybrids, where some cells have chromosomes of up to ten times their normal length. The amplified chromosome segments appeared to be heterochromatic in *N. tabacum* x *N. otophora* crosses (Gerstel and Burns, 1966) but were euchromatic in *N. tabacum* x *N. plumbaginifolia*

(Gerstel and Burns, 1976). The enlarged chromosomes were too big to separate properly on the spindle and were not inherited. Moav *et al.* (1968) found a less extreme example in *N. tabacum* x *N. plumbaginifolia* crosses. Chromosomes of *N. plumpaginifolia* origin became enlarged, but to a more modest extent, and were transmitted to the next generation. The mechanism of amplification is not known but it is interesting to speculate that it may be similar to that observed in cultured mammalian cells selected for resistance to drugs such as methotrexate (Bostock, 1986).

The above examples provide clear evidence that wide-hybridization may, in itself, induce alterations in the genome. These are cases where the changes are substantial enough to be detected cytologically, and it is reasonable to suppose that other hybrids may invoke less dramatic variation. A interesting example might be the activation of quiescent transposable elements. It is clear that more needs to be known about the effects of hybridity on genome composition, particularly with respect to the nature of the induced changes and to their potential for modifying the phenotype in ways which are useful for plant improvement.

3. WHAT ARE THE POTENTIAL USES OF NOVEL WIDE-HYBRIDS FOR GENETIC MANIPULATION?

The likelihood of evolution by polyploidization, chromosome introgression or gene transfer is dependent on the ability of the hybrid seed to develop into a viable plant. In an experimental situation, failure of seed development can often be overcome by transferring immature embryos to artificial media, which makes the potential of for experimental wide-hybridization much greater than the potential for spontaneous wide-hybridization. How can wide-crosses of the kind described earlier in this paper be utilized for genetic manipulation?

a) Haploid production

The crosses we have studied in which wheat or barley have been pollinated with maize, sorghum or pearl millet have all been karyotypically highly unstable, and they therefore offer new opportunities for exploiting chromosome elimination for haploid production. The most obvious application is for wheats carrying dominant alleles of the *Kr1* and *Kr2* genes, which give very low fertilization frequencies when crossed with *H. bulbosum*, but which give much higher frequencies when pollinated with maize or pearl millet. Further work is needed to increase the efficiency of the embryo rescue techniques, but it has already been possible to produce fertile doubled haploid plants of Hope (*Kr1, Kr2*) and Highbury (*Kr1, Kr2*) from crosses with the maize variety Seneca 60.

Haploids are of importance to plant breeding programs because they offer a means of rapidly advancing selected breeding lines to complete homozygosity and of increasing the efficiency of selection (Snape and Simpson, 1981). The barley x *H. bulbosum* system has been used in many barley breeding programs (Anderson and Reinbergs, 1985) and has led to the release of commercial varieties, the first being Mingo in 1979 (Ho and Jones, 1980).

In addition, haploids are of great value to geneticists because they allow the fixation of recombinant gametes as completely homozygous material. Doubled haploid lines, usually produced from F_1 or F_2 material, allow the experimental replication of individual recombinants and hence are particularly attractive for the genetic analysis of characters which cannot be assessed easily on a single plant basis (Choo *et al.*, 1979; Snape and Simpson, 1981, 1986; Snape *et al.*, 1984). Such homozygous recombinant lines are potentially powerful tools for localizing agronomically important quantitative trait loci (QTL) using RFLP maps, an area of research which is currently receiving great attention (Burr *et al.*, 1988; Paterson *et al.*, 1988; Lander and Botstein, 1989; Tanksley *et al.*, 1989).

In a broader context, the results described in this paper suggest that chromosome elimination in wide-crosses may be more common than has previously been believed and that the instability which has been extensively studied in barley or wheat x *H. bulbosum* crosses is by no means an exceptional phenomenon. For instance, haploid plants can be produced from hexaploid oat x maize crosses (H. Rines, pers. comm). Thus, it appears likely that wide-crosses can be utilized for haploid production via chromosome elimination in a much wider range of species, perhaps including noncereals, than had previously been thought.

b) Gene transfer

The transfer of agronomically important characters into crop species by wide-crossing has recently been considered in several reviews (Fedak, 1985b, 1989; Mujeeb-Kazi and Kimber, 1985; Goodman *et al.*, 1987; Gale and Miller, 1987; Lange and Balkema-Boomstra, 1988). Historically, wide-crossing has concentrated on the transfer of disease resistance genes and, in wheat, the success of the technique has relied on the ability to induce homoeologous pairing at meiosis by suppression of the action of the *Ph1* gene. Homoeologous recombination allows agronomically unfavourable characters carried by the alien chromosome to be separated from the desired trait. Chromosome segments from very distantly related species are likely to be much more difficult to manipulate by such methods and the time and effort required may make it uneconomic to use such sources of novel germplasm in plant breeding programs.

In spite of this potential problem, the novel wide-crosses we have discussed in this paper offer exciting areas for experiments in alien gene transfer. For instance, it is of interest to determine if maize DNA can be incorporated into wheat or barley, if it is transcriptionally active, and whether the phenotype of the resulting plant is modified in any way which may be of potential advantage to plant breeders. The question of gene activity is obviously crucial since there is little point in introducing genes from alien species if they do not act in the right place at the right time. However, there are encouraging examples from transformation experiments which show that regions controlling transcription may operate correctly in very distantly related hosts. For example, Colot *et al.* (1987) showed that chimaeric constructs consisting of upstream regions of low- or high-molecular weight wheat glutenin genes linked to the coding region of a bacterial chloramphenicol acetyl transferase gene were expressed only in the endosperm of transgenic tobacco plants.

The transfer of a complex trait such as C_4 photosynthesis from maize to wheat might be too ambitious, although there is evidence that wheat already contains genes coding for enzymes of the C_4 pathway (Chao *et al.*, 1989), but several other goals should be attainable such as the transfer of zein genes and, in particular, the transfer of active maize transposable elements.

Transposable elements acting as insertional mutagens are powerful tools for the isolation of genes not easily accessibly by other methods such as the construction of cDNA libraries. Wheat has endogenous sequences which have the characteristics of retrotransposon type mobile elements, and several examples of these have been investigated in detail (Harberd *et al.*, 1987; Flavell *et al.*, 1988). In addition Mitchell *et al.* (1989) have recently described an element which has similarities to the short terminal repeat class of transposons (e.g. the *Ac/Ds* elements of maize). However, no mobility of the endogenous wheat elements has been observed, and it is of great interest to try to introduce active transposable elements from other sources.

Maize transposable elements have been used to clone several maize genes involved in anthocyanin biosynthesis, starting with the *bronze* locus (Fedoroff *et al.*, 1984). More recently the *o2-m5* allele of the *opaque-2* locus, a trans-acting regulator of zein gene expression in the endosperm (Schmidt *et al.*, 1987; Motto *et al.*, 1988) and the *Knotted* locus which affects leaf development (Hake *et al.*, 1988) have been isolated. Furthermore, several schemes have been proposed for cloning other genes, such as those conferring disease resistance (Bennetzen *et al.*, 1988; Ellis *et al.*, 1988) which are of direct interest to plant breeders and which are attractive targets for sequence analysis, *in vitro* manipulation and transformation. Another obvious target in wheat would be the gibberellin insensitive *Rht* dwarfing genes.

We are therefore attempting to transfer active maize transposable elements to wheat and barley by sexual hybridization with the aim of developing stocks from which genes can be isolated by "transposon tagging". For this to be successful the elements must retain mobility in their new host, but this is a reasonable expectation since, for example, the maize *Ac* element has already been shown to transpose in tobacco (Baker *et al.*, 1986), *Arabidopsis thaliana* and carrot (*Daucus carota*) (Van Sluys *et al.*, 1987), potato (*Solanum tuberosum*) (Knapp *et al.* 1988) and tomato (Yoder *et al.*, 1988) when introduced by the *Agrobacterium* Ti-plasmid method. Likewise the *En/Spm* element has been shown to transpose in tobacco (Masson and Fedoroff, 1989; Pereira and Saedler, 1989).

Hybrid embryos are formed when wheat or barley are pollinated with maize lines carrying *Ac* or *Mu* (Laurie and Bennett, 1988a; Laurie, unpublished) but all hybrids that we have examined cytologically have been karyotypically highly unstable, and have lost the maize chromosomes within the first few cell division cycles. This means that a transposable element would have little opportunity to move into the wheat or barley genome before the maize chromosomes were eliminated, even assuming that an appropriate transposase had been produced. Furthermore, one of the best characterized maize elements (*Ac*) has been shown to have a strong tendency to move to genetically linked sites (Greenblatt, 1984; Schwartz, 1988). This implies that in most cases movement of the element would involve re-insertion in the same maize chromosome, which would subsequently be lost.

We are therefore trying to stably transfer maize DNA to wheat and barley by inducing intergenomic translocations before elimination of the maize chromosomes by exposing zygotes or two-celled embryos to ionizing radiation. The timing of the treatment is important since our cytological studies indicate that in the genotype combinations we are using the number of maize chromosomes present after the second cell division is very low. Several doubled haploid wheat plants have been recovered from irradiated wheat x maize crosses, but it is not yet known if any transfer of maize DNA has been achieved.

4. FUTURE PROSPECTS FOR SEXUAL WIDE-HYBRIDIZATION

It is certain that interspecific and intergeneric crosses of the kind traditionally employed in wide-hybridization will continue to be used, and will continue to be of great benefit to plant breeding programs. Furthermore, the controlled introgression of alien genes is likely to become more efficient with the development of more sophisticated marker systems, particularly RFLPs (Young and Tanksley, 1989).

It is also clear that the range of species which can be hybridized by conventional sexual means is much greater than has previously been thought. To date our particular research interests have led us to concentrate on cereal wide-hybrids which involve major crop species, but it seems highly unlikely that the crosses we have studied here represent the taxonomic limits of plant sexual hybridization. It will be of value to know what other wide-hybrids can be produced and what potential they have for haploid production or gene transfer.

Furthermore, the results from detailed studies of wheat x maize crosses indicate that fertilization occurs with high efficiency if a maize sperm nucleus reaches the wheat egg-cell. This may either mean that cell recognition signals between gametes are unimportant in cereals, or that such signals are conserved. In either case this result, together with recent progress in producing fertile plants from cereal protoplasts (Abdullah *et al.*, 1986; Shillito *et al.*, 1989; Prioli and Söndahl, 1989), suggests that it may eventually be feasible to produce novel wide-hybrids by *in vitro* fertilization using isolated gametes. This may be thought of as an extension of sexual hybridization.

ACKNOWLEDGEMENTS

David Laurie was funded by the United Kingdom Overseas Development Administration, project R3797. Louise O'Donoughue was funded by a Commonwealth Scholarship. We thank Dr. Howard Rines, USDA-ARS research geneticist at the University of Minnesota, for permission to cite unpublished results. We also thank George Coupland, Robert Koebner, Terry Miller and John Snape for comments on the manuscript.

REFERENCES

Abdullah, R., Cocking, E.C., and Thompson, J.A., 1986, Efficient plant regeneration from rice protoplasts through somatic embryogenesis, Bio/Technology, 4: 1087-1090.

Anderson, M.K., and Reinbergs, E., 1985, Barley breeding, in: "Barley", D.C. Rasmusson, ed., American Society of Agronomy, Crop Science Society of America, Soil Science Society of America, Publishers Madison, Wisconsin, pp. 231-268.

Baker, B., Schell, J., Lörz, H., and Fedoroff, N., 1986, Transposition of the maize controlling element "Activator" in tobacco, Proc. Natl. Acad. Sci. USA, 83: 4844-4848.

Barclay, I.R., 1975, High frequencies of haploid production in wheat (*Triticum aestivum*) by chromosome elimination, Nature, 256: 410-411.

Bennett, M.D., and Smith, J.B., 1976, Nuclear DNA amounts in angiosperms, Philos. Trans. Roy. Soc. Lond. Ser. B, 274: 227-274.

Bennett, M.D., Smith, J.B., and Barclay, I.R., 1975, Early seed development in the Triticeae, Phil. Trans. R. Soc. Lond. Ser. B, 272: 199-227.

Bennetzen, J.L., Qin, M.M., Ingels, S., and Ellinghoe, A.H., 1988, Allele-specific and mutator-associated instability at the *Rp1* disease resistance locus of maize, Nature, 332: 369-370.

Bostock, C.J., 1986, Mechanisms of DNA sequence amplification and their evolutionary consequences, Phil. Trans. R. Soc. Lond. Ser. B, 312: 261-273.

Botterman, J., and Leemans, J., 1988, Engineering herbicide resistance in plants, Trends in Genetics, 4: 219-222.

Brookfield, J.F.Y., 1986, The population biology of transposable elements, Phil. Trans. R. Soc. Lond. Ser. B, 312: 217-226.

Burr, B., Burr, F.A., Thompson, K.H., Albertson, M.C., and Stuber, C.W., 1988, Gene mapping with recombinant inbreds in maize, Genetics, 118: 519-526.

Chao, S., Raines, C.A., Longstaff, M., Sharp, P.J., Gale, M.D., and Dyer, T.A., 1989, Copy number and chromosomal location in wheat and some of its close relatives of the genes for enzymes involved in photosynthetic CO_2 fixation, Mol. Gen. Genet., (in press).

Chapman, V., Miller, T.E., and Riley, R., 1976, Equivalence of the A genome of bread wheat and that of *Triticum urartu*, Genet. Res. Camb., 27: 69-76.

Choo, T.M., Christie, B.R., and Reinbergs, E., 1979, Doubled haploids for estimating genetic variances and a scheme for population improvement in self-pollinating crops, Theor. Appl. Genet., 54: 267-271.

Colot, V., Robert, L.S., Kavanagh, T.A., Bevan, M.W., and Thompson, R.D., 1987, Localization of sequences in wheat endosperm protein genes which confer tissue-specific expression in tobacco, The EMBO J., 6: 3559-3564.

Comeau, A., Plourde, A., St. Pierre, C.A., and Nadeau, P., 1988, Production of doubled haploid wheat lines by wheat x maize hybridization (Abstract), Genome, 30: Supplement 1, p 482.

Doebley, J., and Sisco, P.H., 1989, On the origin of the maize male sterile cytoplasms: its completely unimportant, that's why its so interesting, Maize Genetics Cooperation News Letter, 63: 108-109.

Dover, G.A., 1982, Molecular drive: a cohesive mode of species evolution, Nature, 299: 111-117.

Dover, G.A., and Tautz, D., 1986, Conservation and divergence in multigene families: alternatives to selection and drift, Phil. Trans. R. Soc. Lond. Ser. B, 312: 275-289.

Driscoll, C.J., 1981, Perspectives in chromosome manipulation, Phil. Trans. R. Soc. Lond. Ser. B, 292: 535-546.

Dvorák, J., 1976, The relationship between the genome of *Triticum urartu* and the A and B genomes of *Triticum aestivum*. Can. J. Genet. Cytol. 18: 371-377.

Dvořák, J., 1983, The origin of wheat chromosomes 4A and 4B and their genome reallocation, Can. J. Genet. Cytol., 25: 210-214.

Dvořák, J., 1988, Cytogenetical and molecular inferences about the evolution of wheat, in: "Proc. 7th Int. Wheat Genet. Symp., Vol. I", T.E. Miller and R.M.D. Koebner, eds., Institute of Plant Science Research, Cambridge, U.K., pp. 47-51.

Ellis, J.G., Lawrence, G.J., Peacock, W.J., and Pryor, A.J., 1988, Approaches to cloning plant genes conferring resistance to fungal pathogens, Annu. Rev. Phytopathol., 26: 245-263.

Endo, T.R., 1988a, Chromosome mutations induced by gametocidal chromosomes in common wheat, in: "Proc. 7th Int. Wheat Genet. Symp., Vol. I", T.E. Miller and R.M.D. Koebner, eds., Institute of Plant Science Research, Cambridge, U.K., pp. 259-265.

Endo, T.R., 1988b, Induction of chromosomal structural changes by a chromosome of *Aegilops cylindrica* L. in common wheat, J. Hered., 79: 366-370.

Falk, D.E., and Kasha, K.J., 1981, Comparison of the crossability of rye (*Secale cereale*) and *Hordeum bulbosum* onto wheat (*Triticum aestivum*), Can. J. Genet. Cytol., 23: 81-88.

Falk, D.E., and Kasha, K.J., 1983, Genetic studies of the crossability of hexaploid wheat with rye and *Hordeum bulbosum*, Theor. Appl. Genet., 64: 303-307.

Farrer, W., 1904, Some notes on the wheat "Bobs", its peculiarities, economic value and origin, Agric. Gazette of N.S.W., 15: 849-854.

Fedak, G., 1985a, Wide crosses in *Hordeum*, in: "Barley", D.C. Rasmusson, ed., American Society of Agronomy, Crop Science Society of America, Soil Science Society of America, Publishers Madison, Wisconsin, pp. 155-186.

Fedak, G., 1985b, Alien species as sources of physiological traits for wheat improvement, Euphytica, 34: 673-680.

Fedak, G., 1989, Wide hybridization for cereal improvement, in: "Current options for cereal development", M. Malusznyski, ed., Kluwer Academic Publishers, Dortrecht, Boston, London, pp. 39-48.

Fedoroff, N., Furtek, D.B., and Nelson, O.E., 1984, Cloning of the *bronze* locus in maize by a simple and generalizable procedure using the transposable controlling element *Activator* (*Ac*), Proc. Natl. Acad. Sci. USA, 81: 3825-3829.

Finch, R.A., Miller, T.E., and Bennett, M.D., 1984, "Cuckoo" *Aegilops* addition chromosome in wheat ensures its transmission by causing chromosome breaks in meiospores lacking it, Chromosoma, 90: 84-88.

Flavell, R.B., 1986, Repetitive DNA and chromosome evolution in plants, Phil. Trans. R. Soc. Lond. Ser. B, 312: 227-242.

Flavell, R.B., Bennett, M.D., Seal, A.G., and Hutchinson, J., 1987, Chromosome structure and organization, in: "Wheat breeding: its scientific basis", F.G.H. Lupton, ed., Chapman and Hall, London, New York, pp. 211-268.

Flavell, R.B., Harris, N., O'Dell, M., Sardana, R.K., and Jackson, S., 1988, Transposable elements and the control of ribosomal RNA gene expression in wheat, in: "Proc. 7th Int. Wheat Genet. Symp., Vol. I", T.E. Miller and R.M.D. Koebner, eds., Institute of Plant Science Research, Cambridge, U.K., pp. 33-37.

Furuta, Y., Nishikawa, K., and Makino, T., 1975, Intraspecific variation of nuclear DNA content in *Aegilops squarrosa*, Jap. J. Genet., 50: 257-263.

Furuta, Y., Nishikawa, K., Makino, T., and Sawai, Y., 1984, Variation in DNA content of 21 individual chromosomes among six subspecies in common wheat, Jpn. J. Genet., 59: 83-90.

Gale, M.D., and Miller, T.E., 1987, The introduction of alien genetic variation in wheat, in: "Wheat Breeding: Its scientific basis", F.G.H. Lupton, ed., Chapman and Hall, London, New York, pp. 173-210.

Gerstel, D.U., and Burns, J.A., 1966, Chromosomes of unusual length in hybrids between two species of *Nicotiana*, Chromosomes Today, 1: 41-56.

Gerstel, D.U., and Burns, J.A., 1976, Enlarged euchromatic chromosomes ("megachromosomes") in hybrids between *Nicotiana tabacum* and *N. plumbaginifolia*, Genetica, 46: 139-153.

Gill, B.S., and Appels, R., 1988, Relationships between *Nor*-loci from different Triticea species, Plant Syst. Evol., 160: 77-89.

Goodman, R.M., Hauptli, H., Crossway, A., and Knauf, V.C., 1987, Gene transfer in crop improvement, Science, 236: 48-54.

Greenblatt, I.R., 1984, A chromosome replication pattern deduced from pericarp phenotypes resulting from movements of the transposable element, Modulator, in maize, Genetics, 108: 471-485.

Gregory, R.S., 1987, Triticale breeding, in: "Wheat breeding: its scientific basis", F.G.H. Lupton, ed., Chapman and Hall, London, New York, pp. 269-286.

Hake, S., Vollbrecht, E., and Freeling, M., 1989, Cloning *Knotted*, the dominant morphological mutant in maize using *Ds2* as a transposon tag, The EMBO J., 8: 15-22.

Harberd, N.P., Flavell, R.B., and Thompson, R.D., 1987, Identification of a transposon-like insertion in a *Glu-1* allele of wheat, Mol. Gen. Genet., 209: 326-332.

Harlan, J.R., De Wet, J.M.J., and Price, E.G., 1973, Comparative evolution of cereals, Evolution, 27: 311-325.

Hawkes, J.G., 1983, "The diversity of crop plants", President and Fellows of Harvard College, U.S.A., Cambridge (Massachusetts), London.

Ho, K.M., and Jones, G.E., 1980, Mingo barley, Can. J. Plant Sci., 60: 279-280.

Hutchinson, J., 1959, "The families of flowering plants. Vol. II. Monocotyledons", Clarendon Press, Oxford.

Islam, A.K.M., Shepherd, K.W., and Sparrow, D.H.B., 1978, Production and characterization of wheat-barley addition lines, in: "Proc. 5th Int. Wheat Genet. Symp.", New Delhi, pp. 365-371.

Islam, A.K.M.R., Shepherd, K.W., and Sparrow, D.H.B., 1981, Isolation and characterization of euplasmic wheat-barley chromosome addition lines, Heredity, 46: 161-174.

James, J., 1978, New maize x *Tripsacum* hybrids for maize improvement, Euphytica, 28: 239-247.

James, J., 1979, New types of maize x *Tripsacum* and maize x sorghum hybrids - their use in maize improvement, in: "Proc. 10th meeting of the maize and sorghum section of EUCARPIA", Varna, pp. 120-125.

Johns, M.A., Mottinger, J., and Freeling, M., 1985, A low copy number, *copia*-like transposon in maize, The EMBO J., 4: 1093-1102.

Kasha, K.J., 1974, Haploids from somatic cells, in: "Haploids in higher plants, Proc. 1st Int. Symp., Guelph, K.J., Kasha, ed., University of Guelph, Ontario, pp. 67-87.

Kimber, G., 1988, Evolutionary patterns in the wheat group, in: "Proc. 7th Int. Wheat Genet. Symp., Vol. I", T.E. Miller and R.M.D. Koebner, eds., Institute of Plant Science Research, Cambridge, U.K., pp. 47-51.

Knapp, S., Coupland, G., Uhrig, H., Starlinger, P., and Salamini, F., 1988, Transposition of the maize transposable element *Ac* in *Solanum tuberosum*, Mol. Gen. Genet., 213: 285-290.

Kruse, A., 1973, *Hordeum* x *Triticum* hybrids, Hereditas, 73: 157-161.

Kruse, A., 1976, Reciprocal hybrids between the genera *Hordeum*, *Secale* and *Triticum*, Hereditas, 84: 244.

Lambowitz, A.M., 1989, Infectious introns, Cell, 56: 323-326.

Lander, E.S., and Botstein, D., 1989, Mapping Mendelian factors underlying quantitative traits using RFLP linkage maps, Genetics, 121: 185-199.

Lange, W., and Balkema-Boomstra, A.G., 1988, The use of wild species in breeding barley and wheat, with special reference to the progenitors of the cultivated species, in: "Cereal breeding related to integrated cereal production", M.L. Jorna and L.A.J. Slootmaker, eds., Pudoc, Wageningen, pp. 157-178.

Lange, W., and Riley, R., 1973, The position on chromosome 5B of wheat of the locus determining crossability with rye, Genet. Res. Camb., 22: 143-153.

Laurie, D.A., 1989a, Factors affecting the frequency of fertilization in *Triticum aestivum* cv. Highbury x *Zea mays* cv. Seneca 60 crosses, Plant Breeding, 103: 133-140.

Laurie, D.A., 1989b, The frequency of fertilization in wheat x pearl millet crosses, Genome, (in press).

Laurie, D.A., and Bennett, M.D., 1985, Nuclear DNA content in the genera *Zea* and *Sorghum*. Intergeneric, interspecific and intraspecific variation, Heredity, 55: 307-313.

Laurie, D.A., and Bennett, M.D., 1986, Wheat x maize hybridization, Can. J. Genet. Cytol., 28: 313-316.

Laurie, D.A., and Bennett, M.D., 1987, The effect of the crossability loci *Kr1* and *Kr2* on fertilization frequency in hexaploid wheat x maize crosses, Theor. Appl. Genet., 73: 403-409.

Laurie, D.A., and Bennett, M.D., 1988a, Chromosome behaviour in wheat x maize, wheat x sorghum and barley x maize crosses, in: "Kew Chromosome Conference III", P.E. Brandham, ed., Her Majesty's Stationary Office, London, pp. 167-177.

Laurie, D.A., and Bennett, M.D., 1988b, Cytological evidence for fertilization in hexaploid wheat x sorghum crosses, Plant Breeding, 100: 73-82.

Laurie, D.A., and Bennett, M.D., 1988c, The production of haploid wheat plants from wheat x maize crosses, Theor. Appl. Genet., 76: 393-397.

Laurie, D.A., and Bennett, M.D., 1989, The timing of chromosome elimination in hexaploid wheat x maize crosses, Genome, 32: 953-961.

Masson, P., and Fedoroff, N.V., 1989, Mobility of the maize *Suppressor-mutator* element in transgenic tobacco cells, Proc. Natl. Acad. Sci. USA, 86: 2219-2223.

Miller, T.E., 1983, Preferential transmission of alien chromosomes in wheat, in: "Kew Chromosome Conference II", P.E. Brandham and M.D. Bennett, eds., George Allen and Unwin, London, pp. 173-182.

Miller, T.E., and Chapman, V., 1976, Aneuhaploids in bread wheat, Genet. Res. Camb., 28: 37-45.

Miller, T.E., Reader, S.M., and Gale, M.D., 1983, The effect of homoeologous group 3 chromosomes on chromosome pairing and crossability in *Triticum aestivum*, Can. J. Genet. Cytol., 25: 634-641.

Miller, T.E., Shepherd, K.W., and Riley, R., 1981, The relationship of chromosome 4A of diploid wheat to that of hexaploid wheat: a clarification of an earlier study, Cer. Res. Commun., 9: 327-329.

Mitchell, L.E., Dennis, E.S., and Peacock, W.J., 1989, Molecular analysis of an alcohol dehydrogenase (*Adh*) gene from chromosome 1 of wheat, Genome, 32: 349-358.

Moav, J., Moav, R., and Zohary, D., 1968, Spontaneous morphological alterations in *Nicotiana* hybrids, Genetics, 59: 57-63.

Motto, M., Maddaloni, M., Ponziani, G., Brembilla, M., Marotta, R., Di Fonzo, N., Soave, C., Thompson, R., and Salamini, F., 1988, Molecular cloning of the *o2-m5* allele of *Zea mays* using transposon tagging, Mol. Gen. Genet., 212: 488-494.

Mujeeb-Kazi, A., and Kimber, G., 1985, The production, cytology and practicality of wide hybrids in the Triticeae, Cereal Res. Commun., 13: 111-124.

Müntzing, A., 1936, Über die Entstehungsweise 56-Chromosomiger Weizen-Roggen Bastarde, Der Zuchter, 8: 188-191.

O'Donoughue, L.S., and Bennett, M.D., 1988, Wide hybridization between relatives of bread wheat and maize, in: "Proc. 7th Int. Wheat Genet. Symp., Vol. I", T.E. Miller and R.M.D. Koebner, eds., Institute of Plant Science Research, Cambridge, U.K., pp. 397-402.

Paterson, A.H., Lander, E.S., Hewitt, J.D., Peterson, S., Lincoln, S., and Tanksley, S., 1988, Resolution of quantitative traits into Mendelian factors by using a complete linkage map of restriction fragment length polymorphisms, Nature, 335: 721-726.

Pereira, A., and Saedler, H., 1989, Transpositional behavior of the maize En/Spm element in transgenic tobacco, The EMBO J., 8: 1315-1321.

Pickering, R.A., 1983, The influence of genotype on doubled haploid barley production, Euphytica, 32: 863-876.

Price, H.J., Chambers, K.L., Bachmann, K., and Riggs, J., 1983, Inheritance of nuclear 2C DNA content variation in intraspecific and interspecific hybrids of Microseris (Asteraceae), Am. J. Bot., 70: 1133-1138.

Prioli, L.M., and Söndahl, M.R., 1989, Plant regeneration and recovery of fertile plants from protoplasts of maize (Zea mays L.), Bio/Technology, 7: 589-594.

Riley, R., and Chapman, V., 1967, The inheritance in wheat of crossability with rye, Genet. Res. Camb., 9: 259-267.

Rimpau, W., 1891, Kreuzungsprodukte landwirtschaftlicher Kulturplanzen, Landwirtschaftl. Jahrb., 20: 335-371.

Robertson, D.S., 1978, Characterization of a mutator system in maize, Mut. Res., 51: 21-28.

Robertson, D.S., 1981, Tests of two models for the transmission of the Mu mutator in maize, Mol. Gen. Genet., 183: 51-53.

Sapre, A.B., and Deshpande, D.S., 1987, Origin of B chromosomes in Coix L. through spontaneous interspecific hybridization, J. Hered., 78: 191-196.

Schmidt, R.J., Burr, F.A., and Burr, B., 1987: Transposon tagging and molecular analysis of the maize regulatory locus opaque-2, Science, 238: 960-963.

Schwartz, D., 1989, Pattern of Ac transposition in maize, Genetics, 121: 125-128.

Schwarz-Sommer, Z., Leclercq, L., Göbel, E., and Saedler, H., 1987, Cin4, an insert altering the structure of the A1 gene in Zea mays, exhibits properties of nonviral transposons, The EMBO J., 6: 3878-3880.

Sears E.R., 1954, The aneuploids of common wheat, Univ. Missouri Agr. Expt. Stat. Res. Bull., 572.

Sethi, G.S., Finch, R.A., and Miller, T.E., 1986. A bread wheat (Triticum aestivum) x cultivated barley (Hordeum vulgare) hybrid with homoeologous pairing, Can. J. Genet. Cytol., 28: 777-782.

Shillito, R.D., Carswell, G.K., Johnson, C.M., DiMaio, J.J., and Harms, C.T., 1989, Regeneration of fertile plants from protoplasts of elite inbred maize, Bio/Technology, 7: 581-587.

Simmonds, N.W., 1979, Principles of crop improvement, Longman Group Limited, London, New York.

Simpson, E., Snape, J.W., and Finch, R.A., 1980, Variation between *Hordeum bulbosum* genotypes in their ability to produce haploids in barley, *Hordeum vulgare*, Z. Pflanzenzüchtg., 85: 205-211.

Sitch, L.A., and Snape, J.W., 1987, Factors affecting haploid production in wheat using the *Hordeum bulbosum* system. 1. Genotypic and environmental effects on pollen grain germination, pollen tube growth and the frequency of fertilization, Euphytica, 36: 483-496.

Sitch, L.A., Snape, J.W., and Firman, S.J., 1985, Intrachromosomal mapping of crossability genes in wheat (*Triticum aestivum*), Theor. Appl. Genet., 70: 309-314.

Snape, J.W., and Simpson, E., 1981, Uses of doubled haploid lines for genetical analysis in barley, in: "Barley Genetics IV", Proc. 4th Int. Barley Genet. Symp., pp. 704-709.

Snape, J.W., and Simpson, E., 1986, The utilisation of doubled haploid lines in quantitative genetics, Bull. Soc. bot. Fr. Actualités bot., 133: 59-66.

Snape, J.W., Wright, A.J., and Simpson, E., 1984, Methods for estimating gene numbers for quantitative characters using doubled haploid lines, Theor. Appl. Genet., 67: 143-148.

Snape, J.W., Chapman, V., Moss, J., Blanchard, C.E., and Miller, T.E., 1979, The crossabilities of wheat varieties with *Hordeum bulbosum*, Heredity, 42: 291-298.

Subrahmanyam, N.C., 1982, Species dominance in chromosome elimination in barley hybrids, Curr. Sci., 51: 28-31.

Subrahmanyam, N.C., and Kasha, K.J., 1973, Selective chromosomal elimination during haploid formation in barley following interspecific hybridization, Chromosoma, 42: 111-125.

Subrahmanyam, N.C., and von Bothmer, R., 1987, Interspecific hybridization with *Hordeum bulbosum* and development of hybrids and haploids, Hereditas, 106: 119-127.

Swanson, C.P., Merz, T., and Young, W.J., 1980, "Cytogenetics. The chromosome in division, inheritance and evolution", (2nd edition), Prentics Hall, Inc., London.

Tanksley, S.D., Young, N.D., Paterson, A.H., and Bonierbale, M.W., 1989, RFLP mapping in plant breeding: New tools for an old science, Bio/Technology, 7: 257-264.

Van Sluys, M.A., Tempé, J., and Fedoroff, N., 1987, Studies on the introduction and mobility of the maize *Activator* element in *Arabidopsis thaliana* and *Daucus carota*, The EMBO J., 6: 3881-3889.

Wilson, A.S., 1876, On wheat and rye hybrids, Trans. Proc. Bot. Soc. Edinburgh, 12: 826-828.

Woolhouse, H.W., 1987, New plants and old problems, Ann. Bot., 60: Suppl., 4, 189-198.

Yoder, J.I., Palys, J., Alpert, K., and Lassner, M., 1988, *Ac* transposition intransgenic tomato plants, Mol. Gen. Genet., 213: 291-296.
Young, N.D., and Tanksley, S.D., 1989, RFLP analysis of the size of chromosomal segments retained around the *Tm-2* locus of tomato during backcross breeding, Theor. Appl. Genet., 77: 353-359.
Zenkteler, M., and Nitzsche, W., 1984, Wide hybridization experiments in cereals, Theor. Appl. Genet., 68: 311-315.
Zohary, D., 1970, Centres of diversity and centres of origin, in: "Genetic resources in plants - their exploitation and conservation", O.H.Frankel and E. Bennett, eds., Blackwell, Oxford, pp. 33-42.

INDUCED MUTATIONS - AN INTEGRATING TOOL

IN GENETICS AND PLANT BREEDING

Miroslaw Maluszynski

Plant Breeding and Genetics Section
Joint FAO/IAEA Division
and
Department of Genetics
Silesian University
Katowice, Poland

INTRODUCTION

During the last few years breeders and plant geneticists have shown rekindled interest in mutation techniques as a simple tool to generate desired variability for breeding or basic research. The wide use of model plant mutants, in studies of molecular organization of genomes or of metabolic pathways, has stimulated interest to utilize these techniques in crop improvement by conventional or molecular genetic methods. During the approximately 60 years history of the use of induced mutations in plants a wide range of techniques and approaches have been developed to increase the probability of finding desirable mutants. For example, the traditional methods employing mutagenic chemicals and radiation have been supplemented by the use of in-vitro culture and transposable elements.

Achievements and methods of mutation induction have been described and summarized in reviews published during the last few years (Sigurbjörnsson, 1983; Konzak, 1984; Bird and Neuffer, 1987; Micke et al., 1987; Maluszynski et al., 1988; Neuffer and Chang, 1989). About sixty years ago, Stadler (1930), on the basis of his own experiments, pointed out: "The practical value of induced mutations in the improvement of crop plants has been much overrated, at least as regards immediate application...There are, however, certain special cases in which induced mutation, even in the present state of knowledge, offers a fair possibility of successful application". The present paper, following Stadler's idea,

Gene Manipulation in Plant Improvement II
Edited by J. P. Gustafson
Plenum Press, New York, 1990

will review only some of the more recent developments in the application of mutation techniques in genetics and plant breeding.

GENERAL SCHEME OF INDUCED MUTATIONS

The general scheme of induced mutations and mutant selection in seed propagated crops has remained practically unchanged for 50 or 60 years (Stadler, 1930). This scheme has a strong basis in genetic rules and is very easy to implement in every breeding programme involving seed propagated, autogamous species (Fig. 1). Seeds from a selfpollinated line, called the M_o generation, are treated with mutagens. During mutagenic treatment genes can be changed to another allelic form. Each cell where this process occurs becomes heterozygotic. This means that M_o seeds become, during treatment, the M_1 generation - and also of course, the whole plant obtained therefrom, which then develops seeds of the M_2 generation. Plants of the M_2 generation will segregate, but very often a significant deviation from the expected segregation ratio will be observed, which is mainly due to chimerism of the M_1 plants.

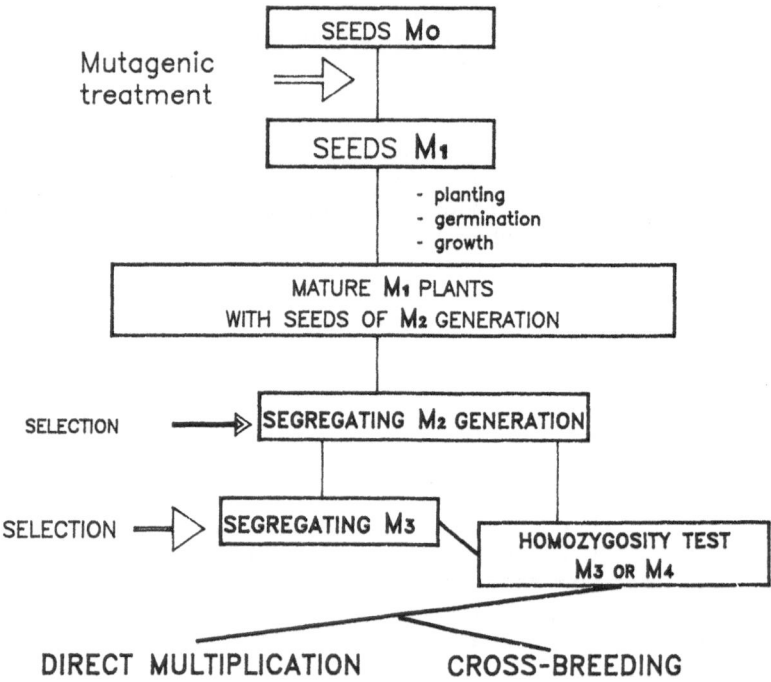

Fig. 1.   "Classical" scheme of induced mutation techniques in
          selfpollinated plants.

Chimerism

Embryos are multi-cellular structures with very peculiar
organization. The success of the whole mutagenic procedure will
depend on induced inherited changes in DNA of initial cells from
which generative tissues develop. The frequency of these changes,
modified by DNA repair processes, will determine the frequency of
mutants observed in the M_2 and following generations. Unfortunately, all of the mutagenic agents known todate, both chemical or
physical in nature, can mutagenize DNA not only of initial cells
but also any other cells in the embryo or seed.

The shoot apex is well developed in the mature seed of
Angiosperms, with two to three morphogenetic cell layers usually
existing. It is generally accepted that a cell germ line is initiated from the second layer. The number of initial cells in each
layer varies significantly, not only among species but even among
varieties within a species. Each initial cell is responsible for
development of a particular part of the plant and thus partici-
pates in the process of plant organogenesis. In barley (Hordeum
vulgare L.), Mullenax and Osborne (1966) found that most cultivars
had in the dormant embryo three primordial leaves and two axillary
buds in addition to the apical meristem. Kirby and Appleyard
(1987) postulate that three or four leaf primordia are already
present in the shoot apex of the wheat (Triticum aestivum L.)
embryo. In the axils of the leaves, tiller buds are initiated.
Primary tillers born in the axils of the main shoot can form a
complex of tillers of a secondary, tertiary or even higher order.

Jacobsen (1966) on the basis of combined genetical and anatom-
ical analysis, postulated that at least 6 separate shoot meristems
or prospective shoot meristems were present in a barley embryo.
Each of them had 1 or 2 functional initial cells for their sporo-
genous tissue.

Initial cells of the main shoot and primary tillers may mutate
independently during mutagenic treatment of the embryo and tissues
developed from these initials may thus carry different mutations.
This process is the source of the chimeric structure of M_1
plants, which develops from any multicellular meristem after mutation induction.

If one considers the number of cells in an embryo, the number
of genes per nucleus and the more or less random distribution of
mutagenic events in the nuclei, the complexity of the genetic constitution induced in embryos by mutagenic treatment becomes
obvious. Generally, the level of genetic changes (both their
frequency and genomic distribution) in each initial cell will
decide the expression of chimeric structure of M_1 plants developed from treated embryos. Of course, effects of mutagens on other
cell structures and on cell metabolism will modify cellular and

developmental processes in initial as well as non-initial cells.
Such dose dependent somatic effects will influence the vigour of
M_1 plants, particularly in the early stages of development.

INDUCED MUTATIONS IN PLANT BREEDING PROGRAMMES

        Understanding the mutagenic processes in multicellular meri-
stems, together with understanding the biology of reproduction of
the species being considered is necessary to understand both the
limitations and the expectations of mutation techniques in plant
breeding. Of course their possible use and prospects of success
also depend on:
-    genetic description of species
-    availability of gene sources in existing germplasm collections
-    possibility of application of other breeding techniques
-    plant character(s) to be improved
-    availability of mass screening procedures
-    expected frequency of mutations for desirable character(s)
-    time required and other economic inputs related to the use of
     different breeding methods.

The Choice of Method

        As with all specific plant breeding methods, the mutation
approach also has some limitations. Each plant, organ, tissue or
cell can be treated mutagenically and a wide spectrum of genetic
changes may be induced. But in consequence two important problems
can arise:
-    how to detect mutant(s) with desirable change(s)
-    how to obtain the necessary change only in a specific locus
     without undesirable or even deleterious mutations in other
     loci important for breeding and agronomic value of a particu-
     lar variety.

        Detailed knowledge of the biology of reproduction has funda-
mental importance for the implementation of mutation techniques.
Different schemes should be applied for mutagen treatment and
mutant selection for example in barley, rye (Secale cereale L.),
wheat or maize (Zea mays Mill.). This is not only because of their
self- or cross-pollination, but also because of ploidy level and
the nature of flower development.

        There is little chance of success if several characters are
hoped to be improved by mutagenic treatment in one and the same
mutant. A simple calculation of the probability in parallel occurrence of three, four or more desired mutations in one nucleus
should prohibit such an unrealistic approach utilizing mutagenic
agents.

This problem is closely related to the estimation of the necessary size of populations, in M_1 and M_2 and following generations, after mutagenic treatment. It is impossible to suggest an optimum size of M_2 population for all species and all objectives of mutagenic treatment. However, on the basis of experience (Maluszynski, 1982), it is easy to predict that a semi-dwarf form(s) will be found among a few thousand M_2 barley plants obtained from a treatment with an optimal dose of MNH (N-methyl-N--nitroso-urea). But it will be necessary to screen many thousands of M_2 or M_3 seedlings to select a powdery mildew resistant mutant in the same material.

Generally, a very large population (between 30-100,000) of M_2 plants should be planted to increase the probability for selection of rare and desirable mutation(s). Classical cross-breeding methods usually employ much smaller numbers of progeny of a single cross and when it is assumed that a simply inherited source for the needed character is really available.

The necessary large size of mutated generations requires a simple, clear and easy mass-screening procedure. This is an absolute requirement before attempting any breeding programme using mutation techniques. In spring barley for example, following MNH treatment, only two to three months (usually in the winter season) are required to screen, in a greenhouse about 4-500,000 M_2 seedlings against powdery mildew in order to realise about 30-40 putative mutants. Selected seedlings mature in a greenhouse. In the spring the M_3 progenies are planted in the field and resistance is carefully checked under natural conditions. As a final result, after reconfirmation of their resistance in the M_4 generation, 2 or 3 resistant mutants may later on be used in a cross breeding programme (R. Madajewski, pers. comm.).

Morphological characters like plant height, leaf size, flower or fruit colour, early or late maturity, grain colour and size are examples of characters which can easily be found in large populations of mutagenically treated plant material. A breeding programme using mutation techniques for such characters as seed protein content or oil quality can be established only when a well equipped service laboratory is available, and when for various reasons a cross-breeding programme cannot be applied or was unsuccessful.

In addition to these logistical difficulties, the most important limitation in using mutation techniques, especially as a method supporting cross-breeding programmes, is the lack of conviction that they can give a real economic profit. Usually this derives from a poor understanding of the proper use of induced mutations in practice.

Objectives and Parent Material for Mutaton Induction

The proper choice of parent material is crucial for the entire
strategy and very often determines future results. The choice
should give a breeder the chance to realise his main objective.
Instead of a long list of cases where mutation techniques should
not be applied we would like to present a few examples of their
proper implementation.

Improvement of Local (Native) Well Adapted Varieties

"Upland" areas are often characterised by unfavourable parameters such as altitude, topography, drastically changeable daylight,
temperature, irregular and often insufficient rainfall and soil
type. Only well adapted varieties with a long history of cultivation in a particular area, can ensure a stable yield in such ad-
verse environmental conditions.

Four upland rice (Oryza sativa L.) varieties native to the
drought-prone plateau of the Bihar region in India were chosen by
Singh and Sinha (1986) for mutation induction. All these varieties
are drought tolerant. Earliness and nonlodging were the most desirable characters. Dry seeds were irradiated with a wide range of
doses from 10 to 60 kR gamma rays. From 12,000 irradiated seeds 98
early maturing mutants with long panicles were selected in the M_3
generation. The frequency of early maturing mutants was much
higher but very often this mutation occurred parallel to others resulting in undesirable changes. Detailed examination of agronomic
performance of mutants in the M_4 and M_5 populations led to the
selection of 7 early maturing, drought resistant, high yielding
mutants. All of these seven mutants were obtained from one variety, Brown Gora.

Finally, after evaluation of three generations (M_6 - M_8)
in replicated yield trials, two mutants were chosen for cultivation
tests in farmers' field conditions. This evaluation was made on
about 700 test plots in six locations. Both mutants were always
significantly superior to the check variety. They matured in 85
days (15 days earlier than the parent variety), but nevertheless,
their yield was higher by 51.7% and 61.8%, respectively (Fig. 2).

It is worth noting that all seven early maturing mutants in
this experiment were shorter in comparison to the parent variety by
20.4 to 31.9%. It is generally accepted that both these traits can
negatively affect characters as drought tolerance and yield.
Results obtained from Singh and Sinha (1986) demonstrated that this
principle cannot be applied in the case of improvement of this well
adapted local variety (Table 1).

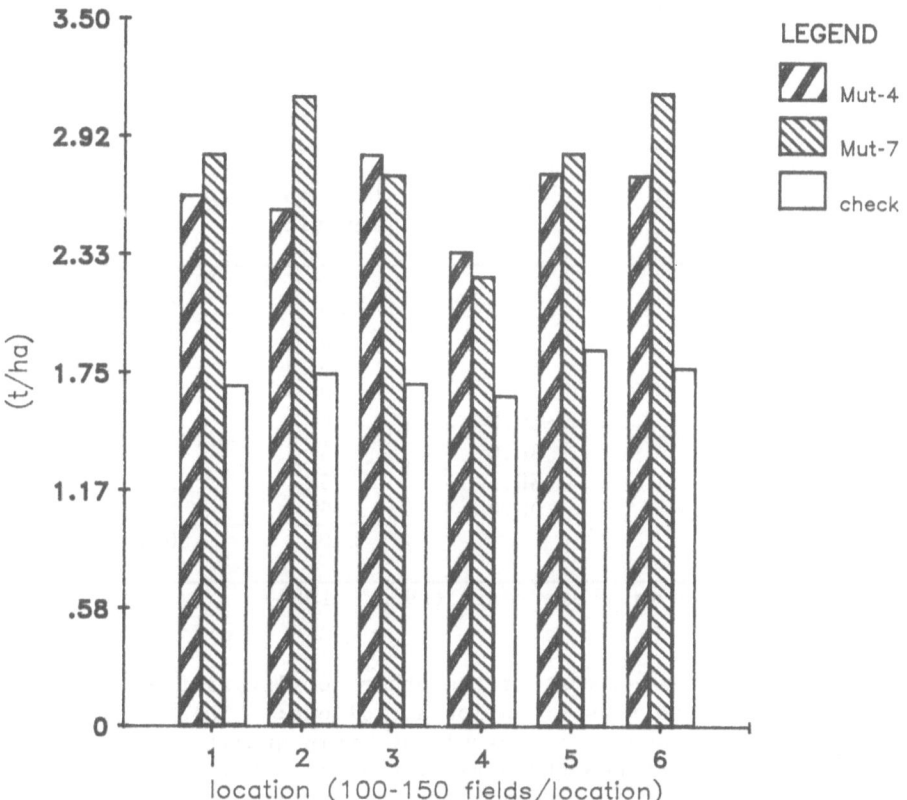

1-Tangarbashli, 2-Girmi, 3-Pungi, 4-Banshjari,5-Katchacho, 6-Burhakhukhra.

Fig. 2. Yield performance of rice cv. Brown Gora and its mutants
in 700 farmers' fields, 1984.
(from data presented by Singh & Sinha, 1986)

From the point of view of using mutation techniques it was
possible to obtain such successful results because a wide range of
doses were applied and four varieties were used. The occurrence of
promising mutants in only one variety can be explained by its
higher susceptibility to gamma rays as compared to the other
varieties. The lack of data regarding somatic effects in M_1
generation and frequency of mutations in M_2 make it impossible to
draw further conclusions.

Several local varieties are not always available, and very
often there is a need to improve a particular variety to meet
market requests. The tall indica rice variety, Basmati 370, is
traditionally cultivated on a large area of Pakistan, India and
Bangladesh, countries where IRRI semi-dwarf varieties have

Table 1. Agronomic performance of short and early
          mutants from cv. Brown Gora in drought-
          prone upland area of Bihar
          (from data presented by Singh & Sinha, 1986)

Variety/Mutant	Height (cm)	Days to maturity	Yield * (t/ha)
Brown Gora	125.3	100	2.17
Mutant 1	90.5	95	2.67
Mutant 2	85.4	95	2.50
Mutant 3	86.8	90	2.71
Mutant 4	99.7	85	3.16
Mutant 5	86.0	85	2.74
Mutant 6	88.3	85	2.78
Mutant 7	100.0	85	3.06
		LSD	0.11

* Mean yield over three years (1981 - 1983)

overwhelmingly replaced most of the other local varieties. Bari et
al. (1984) described this situation as follows: "This variety has
survived primarily due to its excellent grain quality and its
pleasant aroma. Although the variety Basmati 370 yields far less
than variety IR6, its foreign markets are so firmly established,
its consumer preferences are so strongly set, and the growers of
this variety get a premium for their produce to such an extent that
it has remained a major fine-grain variety in Pakistan".

A grain yield improvement of this particular variety could be
obtained by shortening culm length to make it suitable for a high
level of nitrogen cultivation. The DGWG gene, responsible for
short stature and stiffness, is present in practically all high-
yielding semi-dwarf varieties of rice spread around Asia. A simple
cross of semi-dwarf variety with Basmati 370 would be sufficient to
transfer the DGWG gene to a Basmati background. Unfortunately,
this conventional method has not been successful (Cheema and Awan,
1985). Together with the gene for semi-dwarfness, other genes were
transferred which drastically decreased grain quality and the
unique aroma which dictates the value of Basmati 370 on local and
foreign markets. Induced mutations have become an alternative
approach. A mutation programme was initiated and from 83,888 M_2
plants Awan (1986) selected 18 semi-dwarf mutants (Table 2). A
large number of early flowering mutants were found in the same
population. Finally, three semi-dwarf mutants were evaluated in an
M_4 yield trial. All investigated mutants significantly out-

Table 2. Semi-dwarf and early mutants in M_2 population after gamma-ray treatment of rice
cv. Basmati 370
(from data presented by Awan, 1986)

Dose (kR)	No. of M_2 plants	Mutants		
		semi-dwarf	early*	total
15	28,258	9	28	37
20	28,530	2	25	27
25	27,100	7	26	33
Total	83,888	18	79	97

* More than a week earlier than parent

yielded the parent variety Basmati 370 (Fig. 3). Earliness was
also a desirable character in this area. A cross of a short stature mutant with an early mutant should produce a high yielding recombinant with both desirable characters, in the Basmati 370 genetic
background, making it suitable for short season planting in rotation with wheat.

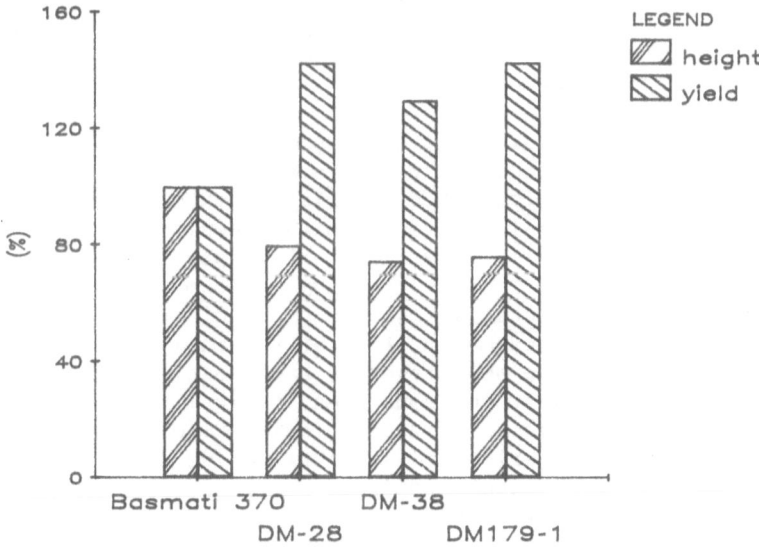

Fig. 3. Agronomic evaluation of gamma-ray induced semidwarf
mutants from cv. Basmati 370.
(after data presented by Awan, 1986)

Introduction of New Species

   In Peru and Bolivia, there are large areas of flat land, on
altitudes above 3,200 meters. In this environment the sources of
nutrition are mainly natural pasture plants, potato (<u>Solanum</u>
<u>tuberosum</u> L.), faba bean (<u>Vicia faba</u> L.), fodder barley and a few
vegetables. Romero-Loli from La Molina University, Lima, has initiated a plant breeding programme with the objective to introduce
bread wheat and high-yielding barley to these highland areas. More
than 10,000 wheat accessions from around the world were collected,
grown and selected for drought, ground frost, hail and high altitude. After several years a few accessions have been chosen as
suitable for farmers' field production and, in replicated small
trials they have yielded over 6 t/ha. During the next two years a
demonstration programme for local farmers will be implemented on a
large scale. Nevertheless, their unique suitability for cultivation on the "Altiplano" can still be improved by decreasing height
and adding disease resistance. The very narrow genetic base among
selected accessions practically excluded prospects of a major
crossing programme, because of the problem of losing specific ad-
aptability. Therefore, it was suggested that mutation techniques
are suitable to generate new variability in selected accessions
currently suitable for cultivation on the "Altiplano". Such a programme has already been initiated at the La Molina University. We
hope that with the use of the doubled haploid method it will be
possible to obtain desired homozygous mutant lines during the next
few years.

Improvement of "Neglected Crops"

   Discussions concerning the development of new approaches or
techniques in plant breeding usually focus on economically most
relevant crops, like maize, wheat, barley, tomato (<u>Lycopersicon</u>
<u>esculentum</u> Mill.). The list of crop plants neglected by re-
searchers but economically very important in particular areas of
the world is very long. Sesame (<u>Sesamum indicum</u> L.) can be
considered as a representative of this group of crops. Ashri
(1988) concludes that sesame is still at the early stage of its
improvement. Forty-nine new varieties have been released during
the last 40-50 years in three countries, typical for different
climatic areas of sesame cultivation (India, R. of Korea and
Venezuela). Nineteen varieties were obtained as a result of hybridization, two by induced mutations, but 28 resulted from selection out of local or introduced plant material. Tallness, profuse
branching, undeterminate growth and seed shattering, are the main
characters which together with others, like pest and disease susceptibility, limit sesame improvement. In other words, a general
change of so-called "plant architecture" is necessary to obtain a
new, more productive type of plant.

The use of induced mutations was considered, to obtain characters such as nonshattering, improved plant architecture, modified
growing period and resistance to pests and diseases, especially
since certain of these desired characters are not available in the
natural germplasm pools. As a result of gamma ray treatment of the
local Israelian variety No. 45, a determinate habit mutant was induced (Ashri, 1981). Genetic analysis indicated that it was mono-
genic and recessive. The author gave the following description of
the mutant line: "is much shorter than the source variety. The
main stem and the branches terminate with a cluster of 5-7 cap-
sules, and the upper internodes are telescoped ... The seeds are
large and plump and seed set is good". A large scale cross-breeding programme with the mutant was immediately initiated with the
objective to obtain well adapted, high yielding, determinate cultivars by transferring the determinate mutation into many different
sesame cultivars. Results obtained so far are very promising
(Ashri, 1988). Our data on the number of mutant cultivars contains
3 sesame varieties obtained by treatment with X-rays, gamma rays
and EMS. In the FAO/IAEA List of Mutant Cultivars it is noted that
very often mutation techniques are successfully used for improvement of "neglected crops". The term "neglected crops" applies to
about 70 crop species.

Restoring Genetic Diversity of Crop Plants

The process of replacing thousands of local varieties or
land-races with a few, very uniform, high yielding ones has accel-
erated the genetic erosion of important food and cash crops around
the world. Economic superiority of modern varieties as well as
agrotechnical requirements related to their introduction has stimulated structural changes in agricultural practices favouring large-
area monocultures. This strategy has greatly improved world food
production but has also generated some problems related to environment and agricultural production per se. It has become obvious
that genetic uniformity, which arose with cultivation of only a few
varieties with a similar genetic background, can make a particular
crop vulnerable to epidemics of pests, diseases and to climatic
stress. Some examples of dramatic situations resulting from
genetic simplification are listed in Table 3. Corn hybrids with
the Texas type cytoplasmic male sterility have become vulnerable to
southern corn leaf blight (Helminthosporium maydis) due to attacks
of pest in the USA in 1970. A similar epidemic was observed in
India in 1970, in pearl millet hybrid plantations, and in China
where rice hybrids with the same cyto- plasmic background were
affected by blast (Swaminathan, 1986). The problem of a decline of
crop genetic diversity is widely discussed by Plucknett et al.
(1987).

Table 3. Examples of genetic vulnerability to pests, diseases
          or severe weather

Crop	Factor	Year	Country	Source
potato	late blight	1846	Ireland	Plucknett et al., 1987
citrus trees	canker	1984	USA/Florida	Plucknett et al., 1987
rice	brown spot	1943	India	Hoyt, 1988
wheat	stem rust	1953-54	USA	Hoyt, 1988
corn *	southern leaf blight	1970	USA	Plucknett et al., 1987
rice	tungro virus	1970	the Philippines	Hoyt, 1988
pearl millet	downy mildew	1970	India	Swaminathan, 1986
wheat	cold weather	1972	USSR	Plucknett et al., 1987
tobacco	cold weather blue mold	1979-80	USA/Canada Cuba	Plucknett et al., 1987
rice	cold weather blast	1980	R. of Korea	Plucknett et al., 1987
citrus trees	canker	1984	USA/Florida	Plucknett et al., 1987
rice	blast	1985	China	Swaminathan, 1986
apple trees	frost	1987	Poland	

* with Texas cytoplasmic male sterility (T-type)

     The cultivation of semi-dwarf, high-yielding rice and wheat
varieties has dominated the growing area in both developing and
developed countries since the mid 1960's. In this case a genetic
simplification should be seriously noticed if one considers that
mainly two genes Rht1 and Rht2 in bread wheat and one gene sd_1
(DGWG) in rice have been used as the sources of semi-dwarfism for
all breeding programmes of high-yielding varieties. Hu (1987) has
discussed this problem in relation to potential vulnerability of
short-stature rice varieties to brown planthoppers (BPH)
(Nilaparvata lugens). Carefully collected data suggested that architecture of rice plants carrying sd_1 gene is conducive to BPH
propagation, especially at a high level of nitrogen application.
The lack of other available natural sources for semi-dwarfism in
germplasm collections has generated an interest of plant geneticists and breeders in the use of mutation techniques for induction
of new mutant genes, suitable for rice breeding, non-allelic to
sd_1 source(s) of short-stature. There are several semi-dwarf
forms of rice, non-allelic to sd_1 but unsuitable for rice

breeding. Rutger et al. (1986) presented allelic relationships of
induced semi-dwarf mutants and DGWG in California. A new sd mutation, which has economic value for rice breeding, non-allelic to
sd_1 was reported by Reddy et al. (1986) and Awan et al. (1986),
among others. Allelic relations among mutant genes responsible for
semi-dwarfism in rice were presented also by Maluszynski et al.
(1986). Preliminary results of the FAO/IAEA Co-ordinated Research
Programme on "Semi-dwarf mutants for rice improvement in Asia and
the Pacific Region" indicated that by the use of induced mutations
more than 11 non-allelic semi-dwarf lines were obtained. Lately,
Hu (1987) reported on an induced mutant, R-34 from the long grain
variety "California Belle", which is allelic neither to sd_1 or
sd_2 and sd_4. This long grain mutant, as parent, performed an
open style tillering, stiff culms, compact panicle. A good yield
at high level of nitrogen application determined its usefulness in
cross breeding programmes.

With regard to wheat vulnerability - due to the extensive use
of only a few genes as a source of dwarfism, the situation is very
similar to that observed for rice. Konzak (1988) pointed out that
new genetic sources for plant height reduction are needed. This is
not only necessary because of the risk connected with genetic vulnerability but also to complement those presently available because
of associated weaknesses. Two genes Rht1 and Rht2 from Daruma and
additionally Rht8 and Rht9 from Akakomugi are the main dwarfing
genes used in wheat. Mutation techniques have often been applied
to obtain new sources of short, stiff straw but with varying results and promises. The number of genes related to plant height
reduction, their origin and potential value for wheat breeding are
summarized in Table 4. In a review by Dalrymple (1986) the following evaluation was found which seems to be in agreement with
Konzak's (1988) conclusions: "While the prospects of finding additional natural sources of semi-dwarfism are probably slim, induced
mutations are a promising source".

Searching for New, Desirable Gene(s)

An increase in crop plant production very often requires new
special plant characters in order to kindle farmers' interest.
Short duration, heat tolerance, natural defoliation, short
internodes and better plant type are desirable characters for
cotton (Gossypium sp.) cultivation in Pakistan, but are not readily
available in cotton germplasm suitable for local conditions (NIAB,
1988). F_1 seeds from a cross of U.S. cotton variety Deltapine-16
with the local variety AC-134 were irradiated with 30 kR gamma rays
(Khan et al., 1982). From this treatment a line was selected with
a determinate and medium size plant type, heat tolerant, short
duration escaping bollworm attack, and suitability for wheat-cot-
ton-wheat rotation. This mutant was released as variety NIAB-78 in

Table 4. Semi-dwarfing genes in wheat
(from Konzak, 1988; modified)

Gene	Source	Origin*	Initial material	Use in breeding
Rht1	Norin 10	S	Akadaruma?	Widest
Rht2	Norin 10	S	Akadaruma?	Widest
Rht3	Tom Thumb	S	?	Doubtful
Rht4	Burt M937	Mutant	Burt	Uncertain
Rht5	Marfed M1	Mutant	Marfed	Doubtful
Rht6	Burt	S	(Brevor)	Probably wide
Rht7	Bersee Mut	Mutant	Bersee	Doubtful
Rht8	Mara, Sava	S	Akakomugi	Moderate
Rht9	Mara *	S	Akakomugi	Moderate
Rht10	Ai-bian 1	S	Aiganzhao	Uncertain
Rht11	Karlik 1	Mutant	Bezostaja 1	Some
Rht12	Karcag 522M7K	Mutant	Karcag 522	Uncertain
Rht13	Magnif 41M1	Mutant	Magnif 41	Uncertain
Rht14	Castelporziano	Mutant	Cappelli	Some
Rht15	Durox	Mutant	K6800707	Promise
Rht16	Edmore M1	Mutant	Edmore	Promise
Rht17	Chris M1	Mutant	Chris	Uncertain
Rht18	Icaro	Mutant	Anhinga	Promise
Rht19	Vic M1	Mutant	Vic	Promise
Rht20	Burt M860	Mutant	Burt	Promise

* Induced mutant,   S = spontaneous

1983, and became over the next 3 years the most popular cultivar in
Pakistan. Mutant variety introduction has had considerable positive socio-economic impact in all growing areas of Pakistan.
During the seasons 1983/84 to 1985/86 the average yield of cotton
increased from 223 kg/ha to 516 kg/ha and production in the country
was almost doubled from 4.5 million bales in 1983 to 8.50 million
in 1988 (Table 5). From a parallel treatment of 21 other cultivars
from the USA, USSR, Turkey and Iran, a number of various mutants
were selected which significantly increased the cotton gene pool in
Pakistan (Table. 6). Using the H1 mutant in cross breeding a new
heat resistant, high yielding and early maturing variety "Shaheen"
was developed (NIAB, 1988).

The soil-borne Barley Yellow Mosaic Virus (BYMV) causes a
disease which has now very often been detected in Europe and has
already generated serious economic problems in Japan. Infected

Table 5. NIAB-78 growing area (%) in Punjab and the
          increase of cotton production in Pakistan
          (after data presented by NIAB, 1988)

Year	Area under NIAB-78 in Punjab (%)	Production (million Bales)	
		Punjab	Pakistan
1983	20.0	3.25	4.48
1984	24.9	1.69	2.90
1985	30.7	4.45	5.93
1986	44.0	5.65	7.26
1987	65.7	6.52	7.82
1988	70.8	7.20	8.50

barley is easily recognized by the mosaic-yellowing of leaves in
the rosette stage. Some infected plants even reach maturity but
are dwarf and produce only a few seeds. High yield losses are
observed in infected areas, especially since chemical control of
this disease is not possible. The often observed simultaneous
infection of fields by another virus (SBWMV) requires barley
varieties with resistance to both diseases. Todate, no varieties
have been obtained which are resistant to both viruses (Ukai and
Yamashita, 1987).

Table 6. Selected mutants as gene donors for cotton
          breeding in Pakistan
          (after data presented by NIAB, 1988)

Mutant	Desirable character(s)
Mutant B-97	Good plant type, short duration
Mutant-555	High yield potential
Mutant 909	Good plant type, short duration, natural defoliation, heat resistance
Mutant H1	Erect, sympodial, wide adaptability
SANAB (G)	Good plant type, heat resistance, high yield potential
M-59	Single stemmed, short duration, heat resistance
St-3	Uniform boll formation, heat resistance

Resistance to BYMV is determined by two dominant genes Ym_1 and Ym_2 located on chromosomes 4 and 2, respectively (Takahashi et al., 1973). A mutant resistant to both BYMV and SBWMV diseases was found in a collection of 61 early maturing mutants obtained from the winter barley variety Chikurin Ibaraki 1 (Ukai and Yamashita, 1987). The Ea52, a gamma ray induced mutant, later named VRP1, expressed a higher resistance to BYMV than any other known variety. Additionally, this mutant was simultaneously re- sistant to SBWMV. Genetic analysis indicated that one recessive gene ym_3, non-allelic to both Ym_1 or Ym_2, is responsible for resistance to both viruses. As a result of mutagenic treatments of the variety Chikurin Ibaraki 1 with gamma rays, thermal neutrons, ethylene imine or sodium azide, 11 mutants were found. Of these, six were evaluated as having high resistance. The Elisa test con- firmed results of field observations. Mutant VRP1 plants were all free from BYMV particles. SBWMV particles were found only in a few plants of this mutant. Four other mutants were free from both vi- ruses as well (Ukai and Yamashita, 1987).

Chaudhry et al. (1987) observed a high frequency of mutations after treatment of japonica rice fertilized egg cells with NMH (N-methyl-N-nitrosourea). Nine treatment combinations were used in this experiment (3 concentrations with 3 periods of treatment), re- sulting in mutants tolerant to salt stress (Fig. 4). The authors concluded that mutagenic treatment of the fertilized egg cell could be an effective approach to induce salt stress tolerance in salini- ty sensitive rice cultivars.

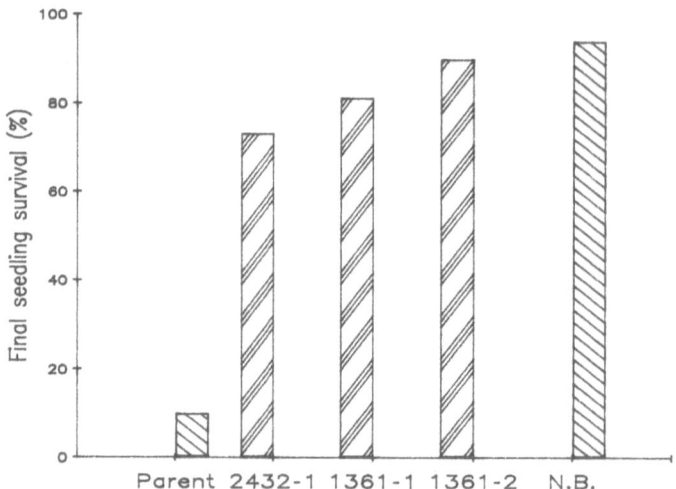

N.B.= Nona Bora, tolerant check

Fig. 4. Response of selected promising M_3 mutants from cv. Taichung 65 to salt stress (0.5% NaCl) (after data presented by Chaudhry et al., 1987)

CROP MUTANTS - BREEDERS' RESOURCES

Rick (1986a) clearly demonstrated that as research intensity increases it is possible to observe an increasing interest in the discovery and genetic analyses of mutants (Fig. 5). The tomato behaves genetically as a strict diploid, which together with auto-matic self-pollination greatly facilitates rapid observation of re-cessive mutants, both spontaneous and induced. The collection of mutants is maintained in the Tomato Genetic Stock Center (TGSC), University of California, Davis. The radiation or chemically in-duced tomato mutants are an important group supplementing the large number of spontaneous mutations.

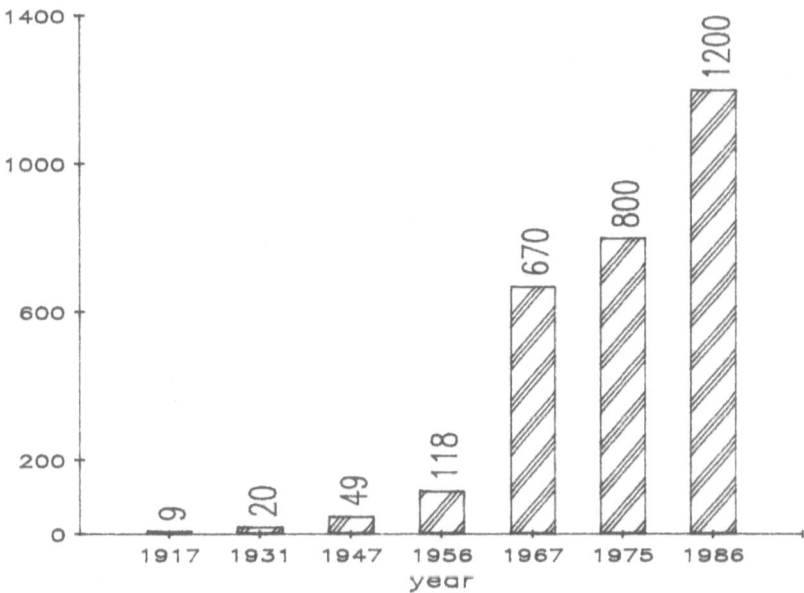

Fig. 5.   Number of known mutant genes in tomato (spontaneous and induced).
(from data presented by Rick, 1986a)

A New Allele in Different Genetic Background

A determinate habit, compact plant type tomato (sp) was discovered in 1914 (Rick, 1986a). This spontaneous mutant completely altered tomato production. Plants carrying this gene are compact with greatly concentrated flowering and fruiting. These characters became essential to develop high yielding varieties, and the lack of other genetic sources for such plant

architecture, made the gene indispensable. Rick (1986b) reported
that by using EMS (ethyl methanesulfonate) it was possible to
induce a new allele at the sp locus in the cultivar VENT Cherry
(LA1221), which is only the second mutation reported at this
locus. It is worth noting that the parent line LA1221 has prolific
flower and fruit production. The new sp version, described as line
LA2705, is exceedingly compact, with determinate growth and flowers
and fruits abundantly.

New Genes for Quantitative Characters

     Forage quality of such crops as maize, sorghum (Sorghum
bicolor L.) or pearl millet (Pennisetum americanum L.) depends to a
great extent on the lignin content and there is a negative relationship between lignin concentration and digestibility. It was
earlier observed that the brown-midrib phenotype (bmr) in maize has
reduced lignin content and increased forage digestibility
(Colenbrander et al., 1973). A phenotypically similar mutant was
obtained in sorghum using DES (diethyl sulfate) (Porter et al.,
1978). A mutant of this same type was also found in an M_2 generation following a 0.05% DES treatment of pearl millet seeds of the
inbred line KS81-1089 (Cherney et al., 1988). Concentrations of
lignin in leaves and stems of mutants were significantly lower than
from normal plants (33 and 25 for leaves or 62 and 31 g/kg of dry
weight for stem, normal plants and mutants, respectively). This
was the reason that in vitro dry matter digestibility was much
higher than normal for the bmr mutant genotype (726 compared to 659
g/kg for normal genotype). The authors concluded that the bmr
trait has excellent potential for improving the forage quality of
pearl millet.

     The quality of edible vegetable oils is dependent mainly on
their fatty acid composition. Low linolenic acid (C18:3) content
in relation to oleic (C18:1) and linoleic (C18:2) is very important
for industry to produce edible oil of longer keeping quality from
linseed. Such genes are present in wild species of the genus
Linum, but unfortunately their transfer to linseed (Linum
usitatissimum L.) was impossible, either by conventional sexual
hybridization or by protoplast fusion (Nichterlein et al., 1988).
As in other oil crops such as rapeseed (Brassica napus L.)
(Röbbelen and Nitsch, 1975) and soybean (Glycine max L.) (Wilcox et
al., 1984), mutation techniques have produced successful results
(Green and Marschall, 1984). Two mutants with altered proportions
of linoleic and linolenic acid were obtained from cv. Glenelg after
EMS treatment. The content of linolenic acid was reduced from 34%
to 22% but the content of linoleic acid was increased from 15% to
27%. Green (1986) used these mutants for genetic analysis of fatty
acid inheritance in flax (Linum usitatissimum L.). It was clearly
demonstrated that the desaturation of linoleic acid to linolenic

acid is controlled by two independently segregating genes Ln1 and
Ln2. Selected mutants M1589 and M1722 are based on independent
mutations in each locus, respectively. The author pointed out that
these mutations express additive gene action (Table. 7). The results of genetic analysis suggest that dominant alleles at either
locus are responsible for an increase of linolenic acid content of
about 10%. As a result of such genetic investigations a recombinant, homozygous for mutant alleles at both loci, was obtained in
the F_2 generation from a cross between the mutants. The so-called "Zero" form has only 1.6% of linolenic acid in comparison to
34.1% in the parent variety "Glenelg".

Table 7.   Linoleic and linolenic acid composition
of all genotypes with different number
of mutated alleles
(after data presented by Green, 1986)

Genotype and origin	Fatty acid composition (%)	
	linoleic	linolenic
AABB Glenelg	14.7	34.1
AaBB Glenelg x M1589	20.7	29.2
AABb Glenelg x M1722	20.3	29.0
AaBb Glenelg x Zero	27.0	22.8
aaBB M1589	27.6	21.6
AAbb M1722	28.9	21.4
aaBb M1589 x Zero	31.7	12.6
Aabb M1722 x Zero	36.8	12.8
aabb Zero	48.2	1.6

Mutants like those described above can become donors of genes
for fatty acid composition in other breeding programmes. In 1988,
in Canada, a summer rape variety "Stellar" was released with low
linolenic but high linoleic acid. This variety has the M11 mutant
obtained by Röbbelen and Nitsch (1975) in its pedigree (Scarth et
al., 1988).

Mutants as a Tool for Genetic Manipulation

In terms of classical plant genetics, mutants can also be used
as unique tools helping in genetic analysis of species and genera.
Induced chromosomal rearrangements leading to aneuploids are commonly used in the "aneuploid methods" for localization of genes in
wheat. Location of genes on one of the chromosomes can be done by

F_1 or F_2 monosomic analysis. The use of telosomics allows for the localization of genes onto specific chromosome arms (Law et al., 1987).

Mutants can help in the introduction of alien genetic variation in wheat, and in this way facilitate the analysis of both donor and recipient species. In hexaploid wheat, Wall et al. (1971) induced with EMS a mutant (10/13), with an increased level of homoeologous pairing. This mutation was later identified as the Ph2 locus on chromosome 3DS and designated ph2b (Sears, 1984). Sears (1977, 1982), using pollen X-irradiation, isolated high-pairing mutations designated ph1b, and an intermediate pairing mutant, ph2a, both on chromosome 5BL. 910 irradiated gametes were tested in this experiment.

Giorgi (1978) reported that, from irradiated durum wheat, cv. Cappelli, a further mutation in the Ph1 locus was isolated. Since 1984 this ph1 mutation is also available in hexaploid triticale (Giorgi and Ceoloni, 1985). These mutants facilitate homoeologous pairing in hybrids of polyploid wheats with related species. Ceoloni (1988, 1989) reviewed a few first successful uses of the ph1 mutation for the transfer into wheat of characters such as resistance to yellow rust, powdery mildew, leaf rust or into 6X triticale, good bread making quality.

Sears (1984), pointed out the possible utility of mutations in gene promoted chromosome pairing. The potential use in wheat breeding of induced mutations in Ph loci, and its advantages in comparison to 5B nullisomics for direct alien introduction was recently discussed by Gale and Miller (1987). Mutants affecting synapsis of homologous chromosomes have been reported in many species. Koduru and Rao (1981) informed about such mutants observed in 126 species. Synaptic mutants have reportedly been induced by chemical or physical mutagens in maize, rice, barley, sorghum, pearl and finger millet (Eleusine coracana Gaertn.), pea (Pisum sativum L.), faba bean, green pepper (Capsicum annuum L.), cabbage (Brassica oleracea L.), watermelon (Citrullus lanatus Mansf.) and jute (Corchorus olitorius L.).

Three synaptic mutants were selected in hexaploid oats (Avena sativa L.) by Rines and Johnson (1988) after mutagenic treatment with sodium azide of seeds from cultivars Stout and Noble. Genetic analysis using monosomic lines indicated that a recessive mutation affects different loci. Mutant Stout 1212 is determined by an allele in locus Syn-5, while mutants Noble 1362 and Noble 1911 are probably mutations in locus Syn-1. These meiotic synapsis-deficient mutants are unique in hexaploid oats. Synaptic deficiencies in oats were previously only observed in nullisomics or derived ditelosomics.

HIGHER PLANT MUTANTS IN CLASSICAL AND MOLECULAR GENETICS

Genetically defined spontaneous or induced mutants are a suitable material for "conventional" genetic analysis. Induced mutation techniques offer the additional possibility to obtain different morphological or biochemical mutants in very similar, or the same, genetic background. Such approach will facilitate to a great extent the future analysis of gene expression in segregating populations. The situation in Arabidopsis thaliana (L.) Heynh. represents an ideal example where the majority of analysed mutants were developed from the Landsberg ecotype, described by Redei (1962).

In crop plants, where mutants have resulted as by-products from specific breeding programmes for particular environments, many different genotypes were involved in creating the genetic marker collections. Among crop plants maize is probably an exception to this point in comparison to other crops.

A very efficient method to induce mutations in maize was developed at the University of Missouri, Columbia (Bird and Neuffer, 1987). Maize is a monoecious species with separate tassel and ear primordia in the seed embryo, thus the treatment of mature pollen grains has been used to simplify the genetic analysis following mutation induction. M_1 plants developed from egg cells fertilized with irradiated pollen will have one mutagenized and one normal genome. Such genetic constitution of M_1 plants facilitates immediate discovery of dominant mutants and provides clear segregations in the M_2 generation. With this technique, EMS treatment gave a very high frequency of mutations. Bird and Neuffer found that a minimum of 78% of pollen grains transferred at least one mutation when 0.06% EMS paraffin oil suspension was applied for 45 minutes. In some experiments the average frequency of mutations per locus was about 1 per 1000.

Pollen treatment methods have allowed the development of a vast collection of different mutants of maize, at the University of Missouri. As far as we know, it is one of the largest known collections of crop plant mutants and about 3,000 induced mutants have been catalogued and preserved. These mutants have greatly helped in the construction of very detailed genetic maps of the maize chromosomes. The diversity of induced mutants has led to the statement that alleles of most of the spontaneously occuring mutants can be obtained by mutagenic treatment (Bird and Neuffer, 1987). The authors conclude: "collection of recessive mutants includes just about every phenotype that has been found in untreated populations of maize". The number of dominant mutants in the collection (over 150 accessions) is more than three times bigger than the number of dominant spontaneous mutants known before in maize.

Genetic Markers

Morphological, biochemical, physiological or conditional-
lethal plant mutants have lately become valuable material for
molecular genetic studies as well as classical analyses.
Arabidopsis thaliana is probably the best example of the use of mutants for detailed genetic analysis. Arabidopsis mutants are induced in many laboratories, and mutagenically treated populations
are now available commercially. The word "production" is used for
mutation technique procedures by some seed companies.

Induced mutants can be classified in various ways since each
morphological character or its heritable change has an appropriate
change in the DNA as well as in the physiological pathway leading
to its morphological expression. Bowman et al. (1988) classified
Arabidopsis mutants into the following groups:
    -   morphological mutants and colour variants
    -   biochemical mutants
    -   developmental mutants
    -   phytohormone mutants
    -   homeotic mutants

Using conventional methods of genetic analysis, Koornneef
(1987) constructed a genetic map which consists of localization and
listing of all types of mutants on five chromosomes. The extent to
which large mutant collections can be utilized in basic studies was
demonstrated during the "Third International Meeting on
Arabidopsis" which was held at Michigan State University, 1987.
The presented papers and posters varied from conventional genetic
analysis, through investigations of metabolic pathways to molecular
genetics.

Sheridan and Neuffer (1980) in maize and Meinke (1986) in
Arabidopsis, have used embryo-lethal mutants for genetic analysis
of plant embryo development. Following previous observations of
variability in embryo development between different lethal mutants
in Arabidopsis (Müller, 1963), they studied in detail unusual patterns of morphogenesis and differentiation as a means to further
plant molecular and developmental biology.

Many induced or spontaneous mutants are used as suitable
markers in molecular genetics. Jefferson (1987, 1989) however,
proposed that the GUS (beta-glucuroni-dase) gene fusion system will
provide a general, non-destructive tool for monitoring and quantifying gene action in transgenic plants, even under field conditions. At present, plant mutants are widely used as selectable
genetic markers. They can be easily produced in mutagenized cells
of in vitro systems or by mutagen treatment of seeds in seed propagated plants. Positive or negative selection systems can be ap-

plied, depending on selecting factor(s), to develop cell lines with
desired genetic marker(s) (Meredith, 1984; Reddy and Kumar, 1988).
Where possible, of course, cell lines obtained and purified in this
way can be regenerated into plants and propagated by seeds. In addition, where mutation techniques are well developed, it is also
possible to screen seedlings of M_2 or following generations for a
particular marker.

Reddy and Kumar (1988) presented a long list of different
types of mutants used as selectable genetic markers (Table. 8).
Depending on the plant species and development of regeneration
methods, some of them are available only as a cell-line. More than

Table 8. Higher plant mutants as selectable genetic markers
          (after Reddy and Kumar, 1988)

Types of mutant(s)	Select-able factors (No.)	Mutated species (No.)	Availability	
			CL	SPL
Antibiotic resistant	9	8	16	3
Purine and pyrimidine analogues resistant	8	8	17	-
Amino acid and amino acid analogue resistant	18	17	52	15
Herbicides resistant	16	13	26	2
Pathotoxins resistant	9	5	9	2
Salt and different metal ions tolerant	6	15	18	1
Chlorate or allyl alcohol resistant	2	9	6	7
Resistant to inhibitory effects of hormones, cold, temperature or UV light	5	6	9	2
Requiring nitrogenous base	4	4	3	2
Amino acid auxotrophs	12	9	15	9
Auxotrophic for vitamins or other growth factor	18	8	9	21
Temperature sensitive and carbohydrate util. negative	4	5	8	-

CL = cell line only, SPL = propagated by seeds

one hundred selective factors, among them, antibiotics, herbicides,
pathotoxins, hormones or salt and temperature, were listed and com-

piled by vitamin and amino-acid auxotrophy. Selectable genetic
markers were found in more than 35 higher plant species.

Mutations as Molecular Markers

The simplicity and ease of availability of electrophoretic
techniques has generated interest in genetic analysis using different plant proteins. In this way many isoenzymatic loci have been
discovered and analyzed. It is obvious that mutagenic agents can
induce considerable polymorphism in structural genes of different
proteins (Dolferus and Jacobs, 1984, 1987; Haughn and Somerville,
1987; Lalonde and Fink, 1987; Last and Fink, 1988). Isoenzyme
polymorphism can also be used to identify genetic differences
between mutants with a similar or even identical morphological
phenotype (Maluszynski et al., 1988).

The rapid development of RFLP (restriction fragment length
polymorphism) technology provides an opportunity for rapid and repeated assessment of variation at many loci. This generates a new
possibility for the use of mutation techniques. Theoretically it
should be possible to purposely induce, in any variety, polymorphisms detectable by a probe already known to be linked to a particular gene (Gale, 1987). The first stage in implementation of
this idea was to assess the potential of various mutation methods
for inducing RFLPs in cereals. It was clearly shown that it is
possible to induce RFLPs in barley using sodium azide, but the rate
of detectable mutations was less than 1/70,000 base pairs. This
estimation was made after analysis of 600 fragments in each of the
27 M_4 homozygous, short strawed mutants from cv. Triumph (Kilian
and Gale, 1989). These results suggest that a different strategy
of mutagenic treatment should be developed to induce a higher level
of RFLPs. In the experiment described above, the chosen dose of
mutagen was optimal to obtain sufficiently large populations of
M_2 plants to select short culm mutants for the purpose of plant
breeding or genetic analysis. A higher dose could induce many more
semi-dwarf mutations but the frequency of multiple mutations will
increase as well. This will have a rather undesirable effect for
future use of such mutants in breeding programmes. The probability
that selected desirable mutants will also carry on other, very
often deleterious mutations, will increase relative to the increase
of mutagen dose. To obtain agronomically useful mutants with high
fertility, a mutagen dose should be used, which is optimal for this
particular purpose. Clearly, a dose which will give the highest
frequency of point mutations is not necessarily the optimal for
plant breeding purposes. Very often an excessive dose can be the
reason for the lack of expected results from mutagenic treatments.
In the case, where RFLPs are expected to be created by mutagenesis,
not only can a higher dose be applied but it is also possible to

use so-called "super-mutagens" (NMH, NEH or EMS) to increase the rate of detectable mutations over 1/70,000 base pairs.

MUTATION TECHNIQUES IN CELL AND TISSUE CULTURE

During the last few years arguments have developed concerning the necessity of using mutagens to induce mutations in in vitro systems. In other words, is somaclonal variation itself sufficient to generate useful mutations? This dilemma does not exist at all if in vitro culturing per se is considered as a mutagenic system, generating somatic variants and mutants with stable changes in the gene base. The use of mutagens can be compared, in such case, with the combined effect of two different mutagenic agents. If the in vitro system generates a sufficient frequency of desirable and genetically stable mutations, there is no need to use additional mutagens. Its addition can increase the frequency of desirable mutant traits in treated cells, however at the same time it could result in a significant increase in the frequency of undesired ones. Thus the necessary first step should be an estimation of the frequency of variation generated by the in vitro system. This problem is relevant to the quantification of variation induced in wheat doubled haploids, and has been discussed by Baenziger et al. (1989) as a breeding strategy. The authors suggested future experiments which should help to establish the proper estimation of variation induced by in vitro culturing.

Induced Mutations

Nevertheless, if chemical or physical mutagens have to be applied in combination with in vitro culture it is necessary that tests of optimal dose rates should be carried out carefully, taking into consideration that various in vitro conditions will affect the susceptibility of particular species and cultivars to particular mutagens.

Experiments performed on maize (Wang et al., 1988; Novak et al., 1988), and on rice (Aldemita and Zapata, 1989; Min et al., 1989) with a large range of doses (X or gamma rays) indicated a very specific reaction of tissue cultures to a particular dose. For example, in maize (Wang et al., 1988) 1 out of 8 applied doses showed a 20% increase in the ability to form embryogenic callus. Similar differences in reaction to gamma-ray irradiation were found in rice.

Haploid cells present an especially attractive object for the use of mutagens. Beversdorf and Kott (1987) developed a mutagenesis/selection system that involves gamma-ray or chemical mutagen treatment of uninucleate, potentially embryogenic micro- spores of Brassica napus L. Doubled haploid mutants, tolerant to the her-

bicide CS (chlorosulfuron), have been obtained in a similar way by
Swanson et al. (1988). In this case, isolated microspores of cv.
Topas, mutagenized with N-ethyl-N-nitroso-urea and gamma-rays have
been used as material for selection.

Mutation Techniques as a Tool in Cell Regeneration

Independent of induced mutations, radiation techniques can
facilitate in vitro plant regeneration in particular species or
varieties. The low response of high-yielding indica varieties to
anther culture makes it impossible to use the doubled haploid techniques for rice improvement. Zapata and Aldemita (1989b) have demonstrated advances using the combined techniques: radiation and
anther culture for doubled haploid production. Basmati 370, an
aromatic indica type cultivar is considered as recalcitrant for
green plant regeneration. With the use of different callus induction media only albino plant regeneration was observed. To avoid
these problems, seeds of Basmati 370 were irradiated with 3 doses
of gamma-rays (15, 20 and 25 kR) and anthers of M_1 plants were
used for in vitro procedure. It was observed that callus production and albino plant regneration decreased with an increased
mutagen dose and, at the same time, one of the doses induced significant green plant regeneration (Fig. 6).

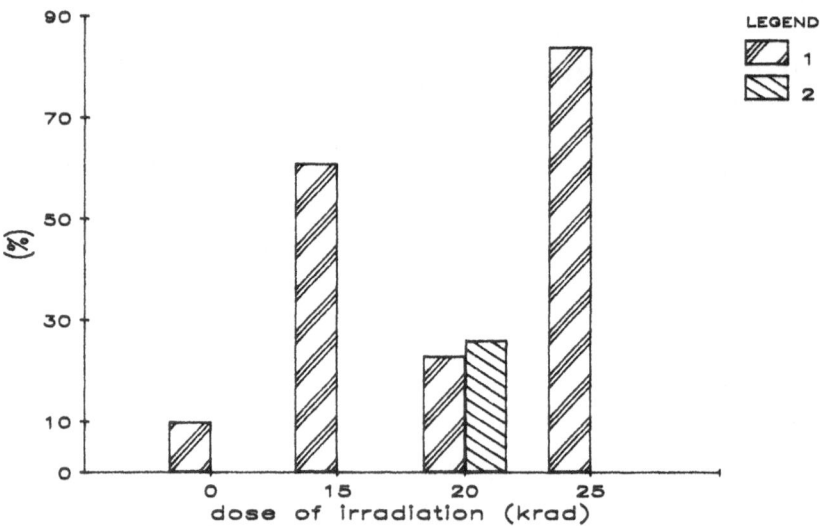

Fig. 6.   Effect of seed irradiation on plant in vitro
          regeneration of rice cv. Basmati 370.
          1 = No. of calli; 2 = calli with green plants
          (from data presented by Zapata and Aldemita, 1989)

Mutation Techniques as a Tool in Cell Fusion

"Gamma fusion" is another method, connected with cell culture, where mutation techniques are used for inactivation or fractionation of the genetic material of a donor cell before fusion with another species as a recipient. Gleba et al. (1988) have obtained asymmetric nuclear hybrids by fusion of mutant cells of Nicotiana plumbaginifolia Viv. and gamma-ray treated protoplast of Atropa belladonna L. Very high doses of irradiation, from 10 to 100 krad, were applied to inactivate the genetic material of the donor species. All regenerated hybrid plants were characterized by a multiple set of N. plumbaginifolia chromosomes plus part of chromosome set of A. belladonna. Phenotypic variability was observed in asymmetric hybrid plants. It is important to note that the degree of elimination of irradiated donor chromosomes was not dependent on the irradiation dose. This experiment indicated that "gamma-asymmetrication" stabilizes the clones and stimulates plant regeneration.

Interspecific hybridization using irradiation techniques was successfully achieved, e.g. between N. repanda Willd. and N. tabaccum L. (Shintaku et al., 1988) or between N. plumbaginifolia and N. tabaccum (Bates et al., 1987). Recently Thomzik and Hain (1988) used X-rays to limit the transfer of nuclear DNA of cv. "Triton" protoplasts to promote the selective transfer of chloroplasts tolerant to the herbicide triazine/triazinone in winter oilseed rapes. In similar experiments, X-rays have been used to inactivate the nucleus of the donor plant in Nicotiana before fusion (Menczel et al., 1982). Sidorov et al. (1987) used gamma irradiated protoplasts of Solanum pinnatisectum Dun. for somatic interspecific hybridization.

CROP PRODUCTION INCREASE BY BACTERIAL MUTAGENESIS

Abelson (1988), commenting on "World Competition in Biotechnology", indicated the introduction of beneficial soil microorganisms as an area of science in which progress is slow. He estimated that "roughly 10^{18} rhizobia improved through mutation by chemicals or radiation are added ... to millions of acres of agricultural soil ... each year".

An interesting approach to biological control of soilborne plant diseases was demonstrated by Toyoda et al. (1988). The authors proposed a new strategy for the control of the fungal disease Fusarium wilt by use of an avirulent strain of bacteria Pseudomonas solanacearum. It is known that fusaric acid is a toxin produced by Fusarium sp. and an important virulence factor in their pathogenesis. Tomato plants were used in this experiment, however this approach may be generally applicable to Fusarium diseases of other important crops.

The stable existence within plants of avirulent bacteria, pro-
tecting to fusaric acid is a basis of this concept. Realization
was possible after detection of a fusaric acid detoxifying mutant
of the bacteria, Pseudomonas solanacearum. An isolated avirulent
strain, U-10A, was mutagenically treated by MNNG (N-methyl-N'-ni-
tro-N-nitrosoguanidine) in concentrations of 50 mg/ml and 100 mg/ml
of fusaric acid to detect resistant mutants. Out of 496 resistant
colonies only three were capable of detoxification. Among them was
the mutant A-16, which was used in subsequent experiments. It was
later demonstrated that resistant mutants incapable of detoxifying
fusaric acid cannot protect tomato cuttings against this toxin.
This detoxification ability of mutant A-16 strain was confirmed by
gas chromatography of culture filtrates. Results clearly indicated
that the "peak" for fusaric acid, added to mineral media, disap-
peared when the mutant A-16 was cultured with toxin.

The protective effect of the mutant A-16 strain was demonstrated by infection of roots of 30 day old tomato plants, variety
"Ponderosa", susceptible to Fusarium oxysporum f. sp. lycopersici
Race 1. One week later inoculated plants were infected by
Fusarium, 3-5 x 10^6 spores/ml. The efficiency of wilt disease
protection by the avirulent Pseudomonas mutant A-16 was determined
30 days after incubation with the pathogen (Fig. 7).

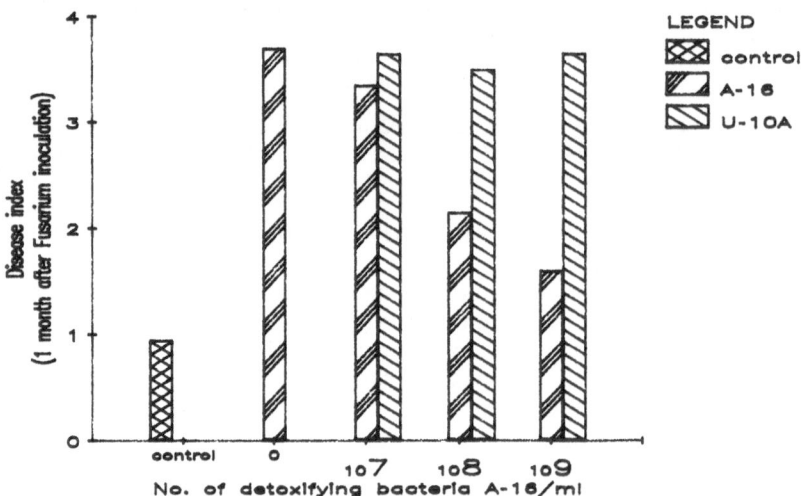

Fig. 7. Protection of tomato plants against fusarium wilt by
        treatment with Pseudomonas solanacearum mutant strain
        A-16.
        (control = non treated, non infected plants;
        U-10A = avirulent strain)
        (after Toyoda et al., 1988, modified)

The protective effect was strongly dependent on the density of
A-16 bacteria. Healthy plants were observed in combination with
infection density 10^9 bacteria per ml. These plants were free of
disease symptoms, as were nontreated and noninfected plants of the
control combination. The next logical step in the implementation
of this strategy to biological control of plant diseases was the
isolation of the fusaric acid-detoxification gene and its cloning
in Escherichia coli. Preliminary results indicated that tomato
plants treated with fusaric acid-detoxifying transformants of
E. coli expressed resistance to fusaric acid, as well. It is still
an open question whether transfer of this gene to the plant genome
is possible.

GENERAL CONSIDERATION

It is very difficult to completely evaluate the impact of the
use of mutation techniques in plant breeding and genetics. Every
year a number of mutant varieties of crops and ornamental species
are released in various countries. The Mutation Breeding Newsletter and related list (Micke et al., 1985) are lagging behind
progress and usually only information concerning the more spectacular results - or those considered to be such by the breeders of
the new mutant variety - are sent to us. Also, it is difficult to
compare the impact of one variety with another in different species
and different countries. We collect the data regarding mutant varieties mainly for the purpose of demonstrating the availability of
these techniques and to make induced mutant germplasm known to
breeders in various crop species.

For example, it is difficult to evaluate the importance of the
four soybean mutants obtained by Sebastian and Chaleff (1987),
which are tolerant to residues of chlorsulfuron. Such tolerance
was previously observed in a tobacco cell line (Chaleff and Ray,
1984) but soybean cell cultures were not suitable for in vitro
selection, therefore seed mutagenesis was used as a simple and
available approach. It remains to be seen: whether this successful
result will lead to new varieties which can compete with those
produced by other plant breeding methods.

Sometimes the economic importance of particular mutagenic
treatments is achieved over long periods of time, sometimes as much
as 30 years. An example of this is the use of X-rays to induce
short culm in a spring barley variety Valticky in the CSSR.
Mutations were induced in 1956, but only in 1965 was the variety
Diamant released. Progenies of Diamant, crossed with sources for
disease resistance, gave rise the variety Trumpf (GDR, 1983) and a
sister line Triumph (grown in the UK). Today, 100% of the spring
barley area in the CSSR (more than 500,000 ha) is sown with vari-

eties having the Diamant genotype in its pedigree (Table 9) and
worldwide there are around 60 of such barley cultivars officially
released.

Table 9.   Spring barley area in Czechoslovakia - 1987
           (Total barley growing area: 589,673 ha)

Variety	Cultivation area (%)	Ancestry (mutant variety or mutant)	Year of release
Bonus	23.0	Diamant	1984
Fatran	0.5	Diamant	1980
Horal	2.0	Diamant	1982
Jarek	1.0	KM1192, Opal *	1987
Jaspis	10.0	ST 6984, Opal *	1986
Karat	2.0	Diamant	1981
Koral	3.0	Hanna *	1978
Kredit	6.0	Nadja *	1984
Krystal	23.0	Koral *, Rapid *	1981
Mars	2.0	Diamant	1983
Opal	1.0	Amethyst *	1980
Orbit	6.0	CE DC/74, Diamant	1986
Perun	6.5	Karat *	1987
Rubin	9.0	Diamant	1982
Zenit	5.0	Karat	1985
TOTAL	100.0		

* Diamant in pedigree
source: Z. Chnoutka (personal communications)

     The results presented here strongly support the second of
Stadler's points: The use of mutations in plant genetics and
breeding has fair and still new possibilities of successful
applications.

ACKNOWLEDGMENTS

     The valuable comments of B. Sigurbjörnsson, A. Micke,
M.D. Gale and P. Gustafson are greatly appreciated. The author
acknowledges Ms. K. Weindl for typing the manuscript.

REFERENCES

Abelson, P.H., 1988, World Competition in Biotechnology, Science, 240:701.

Aldemita, R.R., and Zapata, F.J., 1989, Anther culture of rice: effects of radiation and media components on callus induction and plant regeneration, IAEA (in press).

Ashri, A., 1981, Increased genetic variability for sesame improvement by hybridization and induce mutations, in: Sesame: Status and Improvement, Ashri, A., ed., FAO Plant Production and Protection Paper 29, Rome, pp. 141-145.

Ashri, A., 1988, Sesame breeding: Objectives and Approaches, in: "Oil crops: Sunflower, linseed and sesame," A. Omram, ed., IDRC-MR205e, pp. 152-165.

Awan, M.A., 1986, Semi-dwarf mutants for rice improvement in Asia and the Pacific Region - Progress report, IAEA, Vienna,: 1-5.

Awan, M.A., Cheema, A.A., and Tahir, G.R., 1986, Induced mutations for genetic analysis in rice, in: "Rice Genetics," IRRI, Manila, pp. 697-705.

Bari, B., Mustafa, G., Soomro, A.M., and Baloch, A.W., 1984, Significance of semi-dwarf varieties of rice and their evolution through induced mutation, in: "Semi-dwarf cereal mutants and their use in cross-breeding II," IAEA-TECDOC-307, Vienna, pp. 219-224.

Bates, G.W., Hasenkampf, C.A., Contolini, C.L., and Piastuch, W.C., 1987, Asymmetric hybridization in Nicotiana by fusion of irradiated protoplasts, Theor. Appl. Genet., 74:718-726.

Baenziger, P.S., Peterson, C.J., Morris, M.R., Mattern, P.J., 1989, Quantifying gametoclonal variation in wheat doubled haploids, in: "Current Options for Cereal Improvement," M. Maluszynski, ed., Kluwer Academic Publishers, Dordrecht, pp. 1-9.

Beversdorf, W.D., and Kott, L.S., 1987, An in vitro mutagenesis/ selection system for Brassica napus, Iowa State Journal of Research, 61(4):435-443.

Bird, R. McK., and Neuffer, M.G., 1987, Induced mutations in maize, Plant Breeding Reviews, 5:139-180.

Bowman, J.L., Yanofsky, F., and Meyerowitz, E.M., 1988, Arabidopsis thaliana: A review, Oxford Surveys of Plant Molec. & Cell Bio., 5:57-87.

Ceoloni, C., Del Signore, G., Bitti, O., 1989, Use of high pairing wheat mutants for the transfer of useful traits from alien species into cultivated wheats, in: "Current options for cereal improvement," M. Maluszynski, ed., Kluwer Academic Publishers, Dordrecht, pp. 19-30.

Ceoloni, C., 1988, Transfer of alien genes into cultivated wheat and triticale genotypes by the use of homoeologous pairing mutants, in: "Semi-dwarf cereal mutants and their use in cross-breeding III," IAEA-TECDOC 455, Vienna, pp. 165-176.

Chaleff, R.S., and Ray, T.B., 1984, Herbicide-resistant mutants from tobacco cell cultures, Science, 223:1145-1151.

Chaudhry, M.A., Yoshida, S., and Vergara, B.S., 1987, Induced
    variability for salt tolerance in rice (Oryza sativa L.) after
    N-methyl-N-nitrosaurea treatment of fertilized egg cells,
    Environ. and Exp. Bot., 27(1):29-35.
Cheema, A.A., and Awan, M.A., 1985, Linkage and inheritance
    studies of heading date and plant height in induced mutants of
    rice, Egypt. J. Genet. Cytol., 15:307-312.
Cherney, J.H., Axtell, J.D., Hassen, M.M., and Anliker, K.S.,
    1988, Forage quality characterization of a chemically induced
    brown-midrib mutant in pearl millet, Crop. Sci., 28:783-787.
Colenbrander,V.F., Lechtenberg, V.L., and Bauman, L.F., 1973,
    Digestibility and feeding value of brown midrib corn stover
    silage, J. Anim. Sci., 37:294-295.
Dalrymple, D.G., 1986, Development and spread of high-yielding
    wheat varieties in developing countries, Agency for Inter-
    national Development, Washington, D.C., pp.1-99.
Dolferus, R., and Jacobs, M., 1984, Polymorphism of alcohol
    dehydrogenase in Arabidopsis thaliana (L.)Heynh.: genetical
    and biochemical characterization, Biochem. Genet., 22:817-838.
Dolferus, R., and Jacobs, M., 1987, Characterization of the Arabi-
    dopsis ADH gene and analysis of EMS induced ADH-null mutants,
    Third Inter. Meeting on Arabidopsis, MSU, Abstract No. 108.
Gale, M.D., 1987, Project Proposal, IAEA, Vienna.
Gale, M.D., and Miller, T.E., 1987, The introduction of alien
    genetic variation in wheat, in: "Wheat Breeding," F.G.H.
    Lupton, ed., Chapman and Hall, London, pp. 173-210.
Giorgi, B., 1978, A homoeologous-pairing mutant isolated in
    Triticum durum cv. Cappelli, Mut. Breed. Newsl., 11:4-5.
Giorgi, B., and Ceoloni, C., 1985, A ph1 hexaploid triticale:
    production, cytogenetic behaviour and use for intergeneric
    gene transfer, in: "Proc. Eucarpia Meeting: Genetics and
    breeding of triticale," pp. 105-117.
Gleba, Y.Y., Hinnisdaels, S., Sidorov, V.A., Kaleta, V.A.,
    Parokonny, A.S., Boryshuk, N.V., Cherep, N.N., Negrutiu, I.,
    and Jacobs, M., 1988, Intergeneric asymmetric hybrids between
    Nicotiana plumbaginifolia and Atropa belladonna obtained by
    "gamma-fusion", Theor. Appl. Genet., 76:760-766.
Green, A.G., 1986, Genetic control of polyunsaturated fatty
    acid biosynthesis in flax (Linum usitatissimum) seed oil,
    Theor. Appl. Genet., 72:654-661.
Green, A.G., and Marshall, D.R., 1984, Isolation of induced
    mutants in linseed (Linum usitatissimum) having reduced
    linolenic acid content, Euphytica, 33:321-328.
Haughn, G.W., Somerville, C.R., 1987, An Arabidopsis acetolactate
    synthase gene in tobacco confers resistance to sulfonylurea
    herbicides, Third Inter. Meeting on Arabidopsis, MSU, Abstract
    No. 42.
Hoyt, E., 1988, Conserving the wild relatives of crops, IBPGR,
    IUCN, WWF, Rome, pp. 45.

Hu, C.H., 1987, Modernization and diversification of rice
    varieties in California and a new semi-dwarf mutant of
    possible economic use, in: "Crop exploration and utilization
    of genetic resources," Sung-Ching Hsieh, ed., Taiwan
    Provincial Taichung District, Agricultural Improvement
    Station, pp. 77-90.
Jacobsen, P., 1966, Demarcation of mutant-carrying regions in
    barley plants after ethylmethane-sulfonate seed treatment,
    Radiation Bot., 6:313-328.
Jefferson, R.A., 1987, Assaying chimeric genes in plants: The
    GUS gene fusion system, Plant Molec. Bio. Reporter,
    5(4):387-405.
Jefferson, R.A., 1989, The GUS gene fusion system as a versatile
    new tool for agricultural molecular biology, IAEA, (in press).
Khan, S.I.M., Chaudhry, B.M., Aslam, M., and Bandesha,A.A., 1982,
    Mutation breeding of cotton, Mutation Breeding Newsletter,
    20:11-12.
Kilian, A. and Gale, M.D., 1989, RFLP mapping in cereals and the
    production of induced RFLPs for use in breeding, IAEA, (in
    preparation).
Kirby, E.J.M. and Appleyard, M., 1987, Development and structure
    of the wheat plant, in: "Wheat Breeding," F.G.H. Lupton, ed.,
    Chapman and Hall, London, pp. 287-311.
Koduru, P.R.K., and Rao, M.K., 1981, Cytogenetics of synaptic
    mutants in higher plants, Theor. Appl. Genet., 59:197-214.
Konzak, C.F., 1984, Role of induced mutations, in: Crop Breeding,
    a Contemporary Basis, P.B. Vose and S.G. Blixt, ed., Pergamon
    Press, Oxford, pp. 216-292.
Konzak, C.F., 1988, Genetic analysis, genetic improvement and
    evaluation of induced semi-dwarf mutants in wheat, in:
    "Semi-dwarf cereal mutants and their use in cross-breeding
    III," IAEA-TECDOC-455, Vienna, pp. 77-94.
Koornneef, M., 1987, Linkage map of Arabidopsis thaliana (2n-10),
    in: "Genetic maps 1987: A compilation of linkage and restric-
    tion maps of genetically studied organisms," S.J. O'Brien,
    ed., CSH, pp. 742-745.
Lalonde, B.A., and Fink, G.R., 1987, Analysis of histidine bio-
    synthesis in Arabidopsis thaliana, Third Inter. Meeting on
    Arabidopsis, MSU, Abstract No. 109.
Last, R.L., and Fink, G.R., 1988, Tryptophan-requiring mutants of
    the plant Arabidopsis thaliana, Science, 240:305-310.
Law, C.N., Snape, J.W., and Worland, A.J., 1987, Aneuploidy in
    wheat and its uses in genetic analysis, in: "Wheat Breeding,"
    F.G.H. Lupton, ed., Chapman and Hall, London, pp. 109-128.
Maluszynski, M., 1982, The high mutagenic effectiveness of MNUA in
    inducing a diversity of dwarf and semi-dwarf forms of spring
    barley, Acta Societatis Botanicorum Poloniae, 51:429:440.
Maluszynski, M., Micke, A., and Donini, B., 1986, Genes for semi-
    dwarfism in rice induced by mutagenesis, in: "Rice Genetics,"
    IRRI, Manila, pp. 729-737.

Maluszynski, M., Sigurbjörnsson, B., Micke, A., 1988, Gene mani-
    pulation by mutation techniques, IAEA-TECDOC-455, Vienna,
    pp. 19-30.
Maluszynski, M., Szarejko, I., Madajewski, R., Fuglewicz, A.,
    Kucharska, M., 1988, Semi-dwarf mutants and heterosis in
    barley. I. The use of barley sd-mutants for hybrid breeding,
    IAEA-TECDOC-455, Vienna pp. 193-206.
Meinke, D.W., 1986, Embryo-lethal mutants and the study of plant
    embryo development, Oxford Surveys of Plant Molecular & Cell
    Biology, 3:122-165.
Menczel, L., Galiba, G., Nagy, F., Maliga, P., 1982, Effect of
    radiation dosage on efficiency of chloroplast transfer by
    protoplast fusion in Nicotiana, Genetics, 100:487-495.
Meredith, C.P., 1984, Selection better crops from cultured cells,
    in: "Gene manipulation in plant improvement," J.P. Gustafson,
    ed., Plenum Press, New York, pp. 503-528.
Micke, A., Maluszynski, M., Donini, B., 1985, Plant cultivars
    derived from mutation induction or the use of induced mutants
    in cross breeding, Mutation Breeding Review, 3, IAEA, Vienna.
Micke, A., Donini, B., and Maluszynski, M., 1987, Induced mutations
    for crop improvement - a review, Trop. Agric. (Trinidad),
    64:259-278.
Min, S., Wang, C., Wang, G., Xiong, Z. and Qi, X., 1989, Rice
    improvement (involving altered flower structure more suitable
    to cross-pollination) using in vitro techniques in combination
    with mutagenesis, in: "Current Options for Cereal
    Improvement," M. Maluszynski, ed., Kluwer Academic Publishers,
    Dordrecht, pp. 147-152.
Mullenax, R.H., and Osborne, T.S., 1966, Normal and gamma-rayed
    resting plumule of barley, Radiation Bot., 7:273-282.
Müller, A.J., 1963, Embryonentest zum Nachweis rezessiver Letal-
    faktoren bei Arabidopsis thaliana, Biol. Zentralbl.,
    82:133-163.
NIAB, 1988, Successful application of nuclear techniques for the
    improvement of cotton crop and role of NIAB-78 in cotton
    production, Nuclear Institute for Agriculture and Biology,
    Faisalabad, pp. 1-6.
Nichterlein, K., Marquard, R., and Friedt, W., 1988, Breeding for
    modified fatty acid composition by induced mutations in
    linseed (Linum usitatissimum L.), Plant Breeding, 101:
    190-199.
Novak, F.J., Daskalov, S., Brunner, H., Nesticky, M., Afza, R.,
    Dolezelova, M., Lucretti, S., Herichova, A., and Hermelin, T.,
    1988, Somatic embryogenesis in maize and comparison of genetic
    variability induced by gamma radiation and tissue culture
    techniques, Plant Breeding, 101:66-79.
Neuffer, M.G. and Chang, M.T., 1989, Induced mutations in
    biological and agronomic research, in: Proc. XIIth Eucarpia
    Congress, (1989), Göttingen, (in press).

Plucknett, D.L., Smith, N.J.H., Williams, J.T., and Anishetty, N.M., 1987, "Gene banks and the world's food," Princeton University Press, Princeton N.J.

Porter, K.S., Axtell, J.D., Lechtenberg, V.L., and Colenbrander, V.F., 1978, Phenotype, fiber composition and in-vitro dry matter disappearance of chemically induced brown midrib (bmr) mutants of sorghum, Crop. Sci., 18:205-208.

Reddy, S.S., and Kumar, S., 1988, Selectable genetic markers in higher plants, Indian J. Exp. Biol., 26:567-582.

Reddy, T.P., Vaidyanath, K., and Reddy, G.M., 1986, Evaluation and genetic analysis of semi-dwarf mutants in indica rice, Progress report, IAEA, Vienna, pp. 1-9.

Redei, G.P., 1962, Supervital mutants of Arabidopsis, Genetics, 47:443-460.

Rick, C.M., 1986a, Tomato mutants: freaks, anomalies and breeders's resources, Hort. Science, 21(4):918-919.

Rick, C.M., 1986b, New mutant at the sp locus, TGC Report, 36:33.

Rines, H.W., and Johnson, S.S., 1988, Synaptic mutants in hexaploid oats (Avena sativa L.), Genome, 30:1-7.

Röbbelen, G., and Nitsch, A., 1975, Genetical and physiological investigations on mutants for polyenoic fatty acids in rape seed, Brassica napus L. 1. Selection and description of new mutants. Z. Pflanzenzücht., 75:93-105.

Rutger, J.N., Azzini, L.E., and Brookhouzen, P.J., 1986, Inheritance of semi-dwarf and other useful mutant genes in rice, in: "Rice Genetics," IRRI, Manila, pp. 261-271.

Scarth, R., McVetty, P.B.E., Rimmer, S.R., and Stefansson, B.R., 1988, Stellar low linolenic-high linoleic acid summer rape, Can. J. Plant Sci., 68:509-511.

Sears, E.R., 1977, An induced mutant with homoeologous pairing in common wheat, Can. J. Genet. Cytol., 19:585-593.

Sears, E.R., 1982, A wheat mutation conditioning an intermediate level of homoeologous chromosome pairing, Can. J. Genet. Cytol., 24:715-719.

Sears, E.R., 1984, Mutations in wheat that raise the level of meiotic chromosome pairing, in: "Gene manipulation in plant improvement," J.P. Gustafson, ed., Plenum Press, New York, pp. 295-300.

Sebastian, S.A., and Chaleff, R.S., 1987, Soybean mutants with increased tolerance for sulfonylurea herbicides, Crop Sci., 27:948-952.

Sheridan, W.F., and Neuffer, M.G., 1980, Defective kernel mutants of maize. II. Morphological and embryo culture studies, Genetics, 95:945-960.

Shintaku, Y., Yamamoto, K., and Nakajima, T., 1988, Interspecific hybridization between Nicotiana repanda Willd. and N. tabaccum L. through the pollen irradiation technique and the egg cell irradiation technique, Theor. Appl. Genet., 76:293-298.

Sidorov, V.A., Zubko, M.K., Kuchko, A.A., Komarnitsky, I.K.,
    and Gleba, Y.Y., 1987, Somatic hybridization in potato: use of
    gamma-irradiated protoplasts of Solanum pinnatisectum in
    genetic reconstruction, Theor. Appl. Genet., 74:364-368.
Sigurbjörnsson, B., 1983, Induced mutations, in: "Crop Breeding,"
    D.R. Wood, ed., American Society of Agronomy and Crop Science
    Society of America, Madison, Wisconsin, pp. 153-176.
Singh, M.P., and Sinha, P.K., 1986, Induced mutagenesis in native
    rices, in: "Rice Genetics," IRRI, Manila, pp. 719-727.
Stadler, L.J., 1930, Some genetic effects of X-rays in plants,
    J. of Heredity, 30:3-19.
Swaminathan, M.S., 1986, Integration of the tools of mendelian
    and molecular genetics in crop improvement, Indian J. Genet.,
    46(Suppl.):12-29.
Swanson, E.B., Coumans, M.P., Brown, G.L., Patel, J.D., and
    Beversdorf, W.D., 1988, The characterization of herbicide
    tolerant plants in Brassica napus L. after in vitro selection
    of microspores and protoplasts, Plant Cell Reports, 7:83-87.
Takahashi, R., Hayashi, J., Inoue, T., Moriya, I., and Hirao, C.,
    1973, Studies on resistance to yellow mosaic disease in
    barley, I. Tests for varietal reactions and genetic analysis
    of resistance to the diseases, Ber. Ohara Inst. Landw. Biol.,
    Okayama Univ., 16:1-17
Thomzik, J.E., and Hain, R., 1988, Transfer and segregation of
    triazine tolerant chloroplasts in Brassica napus L., Theor.
    Appl. Genet., 76:165-171.
Toyoda, H., Hashimoto, H., Utsumi, R., Kobayashi, H., and
    Ouchi, S., 1988, Detoxification of fusaric acid by a fusaric
    acid-resistant mutant of Pseudomonas solanacearum and its
    application to biological control of fusarium wilt of tomato,
    Phytopathology, 78(10):1307-1311.
Ukai, Y., and Yamashita, A., 1987, Induced mutants highly resis-
    tant to barley yellow mosaic virus, in: "Barley Genetics V,"
    S. Yasuda and T. Konishi, eds., Okayama, pp. 279-286.
Wall, A.M., Riley, R., and Chapman, V., 1971, Wheat mutants
    permitting homoeologous meiotic chromosome pairing, Genet.
    Res., 18:311-328.
Wang, A.S., Cheng, D.S.K., Milcic, J.B., and Yang, T.C., 1988,
    Effect of X-ray irradiation on maize inbred line B73 tissue
    cultures and regenerated plants, Crop. Sci., 28:358-362.
Wilcox, J.R., Cavins, J.F., Nielsen, N.C., 1984, Genetic alter-
    ation of soybean oil composition by a chemical mutagen, J. Am.
    Oil. Chem. Soc., 61:97-100.
Zapata, F.J., and Aldemita, R.R., 1989, Induction of salt toler-
    ance in high-yielding rice varieties through mutagenesis and
    anther culture in: "Current Options for Cereal Improvement,"
    M. Maluszynski, ed., Kluwer Academic Publishers, Dordrecht,
    pp. 193-202.

IN VITRO CULTURE OF RICE:

TRANSFORMATION AND REGENERATION OF PROTOPLASTS

Thomas K. Hodges, Jianying Peng, Lisa Lee, and David S. Koetje

Department of Botany and Plant Pathology
Purdue University
West Lafayette, Indiana 47907

INTRODUCTION

Rice (*Oryza sativa* L.) is the primary food source for about 40% of the people of the world (Yamada and Loh, 1984). Based upon population growth projections, rice yields must increase by over 5 million tons per year just to maintain current levels of rice consumption per person (IRRI, 1985). Because of the importance of rice production, extensive research efforts and progress have been made during the last 30 years to improve yields. Most of this research has been carried out in China, Japan, India, and at the International Rice Research Institute (IRRI) in the Philippines on the two major subspecies of *O. sativa*, the japonica and indica varieties. Japonica lines are grown primarily in temperate regions of the world, especially northern China and Japan. Researchers in these countries have improved the yield of these lines considerably during the past two decades. The indica varieties, which are more adapted to the humid tropics and are thus an important food source to a larger proportion of the human population, have been improved markedly at IRRI, primarily through conventional plant breeding. It is apparent that even higher performance rice cultivars will be needed in the future. Some improvements will continue through conventional breeding programs, but these programs will be most effective when coupled to the current advances in *in vitro* cell culture of rice and genetic engineering using the tools of cellular and molecular biology.

Genetic engineering of rice (as well as all other important cereals) has not been possible until very recently because of fundamental problems associated with the introduction of genes into cells and in the regeneration of mature and fertile plants following genetic transformation. In the japonica varieties, the techniques for regeneration of plants from protoplasts were developed from 1985 to 1987 (Fujimura et al., 1985; Abdullah et al., 1986; Coulibaly and Demarly, 1986; Toriyama et al., 1986; Yamada et al., 1986; Kyozuka et al., 1987), and for the indica varieties these procedures were developed within the last year (Lee et al.,

Gene Manipulation in Plant Improvement II
Edited by J. P. Gustafson
Plenum Press, New York, 1990

1989). Genetic transformation of protoplasts by direct uptake of plasmid DNA into monocot cells (Potrykus et al., 1985), including rice (Baba et al., 1986; Uchimiya et al., 1986), has also been developed. Thus, the primary obstacles to genetic engineering of rice have now been eliminated. In this presentation, our recent work regarding the procedures for regenerating indica rice from protoplasts will be presented in the historical context of rice improvement through breeding and *in vitro* cell culture. In addition, the current status of transformation of rice protoplasts will also be discussed.

GREEN REVOLUTION TO GENE REVOLUTION

The first major improvement in rice yields came in 1960 with the release of a dwarf strain of *O. sativa*, Taichung Native 1 (TN1) (Yamada and Loh, 1984). Although TN1 was widely planted in Taiwan and India, it did not gain wide recognition as the prototype for modern rice. This distinction belongs to IR8. In 1962 breeders at IRRI crossed Peta, a tall Indonesian rice producing high yields, with a dwarf Taiwanese variety called Dee-Geo-Woo-Gen (DGWG) to produce what became known as IR8, a semi-dwarf strain of indica rice that possessed high-yield potential. It was the first in a long series of improved indica lines released by IRRI. The initial production of these improved dwarf rice lines, along with the high-yielding dwarf wheat (*Triticum aestivum* L. em Thell.) lines developed at the International Maize and Wheat Improvement Center (CIMMYT) in Mexico, represented the golden age of what became known as the Green Revolution in plant variety development. Now it is anticipated that the introduction of specific genes for tolerance or resistance to salts, herbicides, insects, viruses, improved yields, etc. through cellular and molecular biology techniques will provide the next major development of new rice (and other plant) varieties. Given the present rate of progress, this era could appropriately be called the Gene Revolution.

IN VITRO CULTURE OF RICE

Primary Callus

In vitro culture of somatic tissues of rice (seeds, embryos, roots) resulting in organ differentiation was first described in 1968 (Nishi et al., 1968; Tamura, 1968; Kawata and Ishihara, 1968). Nishi et al. (1968) accomplished the first recovery of whole rice plants from callus. This gave rice the distinction of being the first cereal to be regenerated *in vitro*. In their experiments, callus was induced from seeds of the japonica variety Kyoto Asahi upon incubation in a Linsmaier and Skoog (1965) medium (LS) containing 4 mg/l 2,4-dichlorophenoxyacetic acid (2,4-D). Also in 1968, Niizeki and Oono (1968) reported rice to be the first cereal to produce haploid plants via anther culture. Since then, several other explants including immature endosperm, roots, shoots, nodes, inflorescences, leaves, and leaf sheaths were used to induce callus and to regenerate plants, generally by an organogenic differentiation (Henke et al., 1978; Conger, 1981). Regeneration via somatic embryogenesis has been reported for callus obtained from either immature regions of rice leaves (Bhattacharya and Sen, 1980; Wernicke et al., 1981), seed (Siriwardana and Nabors, 1983; Heyser et al., 1983), or immature embryos

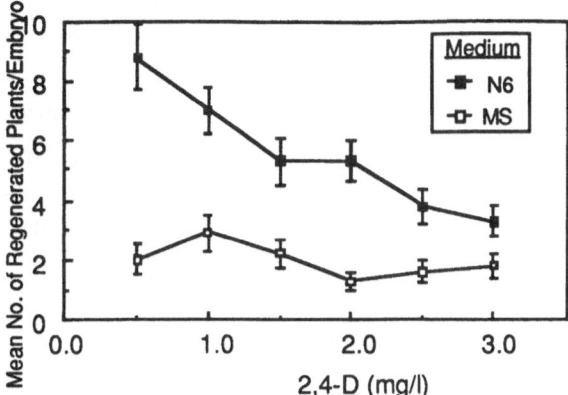

Fig. 1. Differential effect of medium and 2,4-D concentration on plant regeneration of IR54 rice. Calli were grown on MS-type medium or N6-type medium for three weeks and then transferred to identical media lacking phytohormones for plant regeneration. The number of plants per embryo-derived callus was counted six weeks after the transfer. Bars represent standard errors of the means (after Koetje et al., 1989).

(Heyser et al., 1983; Lai and Liu, 1982, 1986; Koetje et al., 1989). Initially, the nutrient medium used for the *in vitro* culture of rice was either the LS or MS (Murashige and Skoog, 1962) media. However, a significant advance occurred in rice cell culture with the development of the N6 medium by Chu et al. (1975). This medium contains a lower concentration of ammonium than most other media. Even though the N6 medium was developed initially for haploid cell culture (see also Chen et al., 1985), it has been found to be effective for regenerating plants from somatic tissues of several cereals including rice (Wernicke et al., 1981; Peng and Hodges, 1989). In addition to the nutrient medium, a hormone or growth regulator is required for callus induction; 2,4-D is generally used. There is, however, an interaction between the type of nutrient medium used and 2,4-D. For example, the N6 medium, in combination with low concentrations of 2,4-D for inducing callus formation from immature embryos, was found to be much superior to the MS medium for regenerating plants from the indica variety IR54 (Fig. 1; Koetje et al., 1989). This result illustrates just one of the many complex interactions that influence *in vitro* culture of plant cells. The nutrient medium and 2,4-D interaction is further complicated by an interesting effect of 2,4-D on the induction of divisions in somatic cells of monocots. Wernicke and Brettell (1980) and Wernicke et al. (1981, 1986) have shown that as monocot cells become older (in going from the base to the tip of a grass leaf), progressively higher concentrations of 2,4-D are required to induce callus formation. When the monocot cells are sufficiently old, i.e. at an advanced developmental stage, the 2,4-D is ineffective in callus induction. Thus, it appears that as monocot cells gradually become older and more differentiated, it is still possible to reverse the differentiation process with

Table 1. Effect of 2,4-D concentration in callus-induction medium
on plant regeneration of various indica genotypes.

	Number of Plants Regenerated*	
Genotype	0.5 mg/L 2,4-D	2.0 mg/L 2,4-D
IR8	3.2 ± 0.5	4.0 ± 0.6
IR36	0.2 ± 0.1	0.0 ± 0.0
IR38	4.4 ± 0.7	0.9 ± 0.3
IR45	2.4 ± 0.8	3.4 ± 0.6
IR52	4.3 ± 0.6	4.2 ± 0.5
IR54	7.8 ± 0.7	4.3 ± 0.4
IR64	0.3 ± 0.2	0.2 ± 0.1

* Data represent mean number of plants per embryo-derived callus ± standard
error after three weeks on N6 medium containing 2,4-D and an additional six
weeks of culture on hormone-free medium (from Koetje et al., 1989).

progressively higher levels of 2,4-D until some advanced state of differentiation has
occurred.   Once this advanced developmental stage has been attained, the
dedifferentiation process cannot be triggered by 2,4-D. By comparison, fully
differentiated dicot cells, e.g. tobacco mesophyll cells, can be induced by 2,4-D to
divide and form callus (Flick and Evans, 1984). The apparent difference in the
ability of 2,4-D to reverse differentiation in monocots as compared to dicots
undoubtedly represents one of the main reasons that *in vitro* culture and
regeneration of monocots has been more difficult to achieve than has been the case
for dicots. Thus, efficient regeneration in rice requires a proper balance between the
nutrient medium, the 2,4-D concentration, and the age or developmental stage of the
cells in the explant. Understanding the basis for these interesting interactions
would certainly help to clarify *in vitro* culture of rice and other monocots.

In most cereals, the genotype influences regeneration markedly (Guha-
Mukherjee, 1973; Green and Phillips, 1975; Sears and Deckard, 1982; Hodges et
al., 1985, 1986; Lai and Liu, 1986; Abe and Fatsuhara, 1986a); this is also true for
rice.   Abe and Futsuhara (1984) found that 11 of 28 japonica varieties and only 2
of 16 indica varieties had good regeneration capacity. Miah et al. (1985) also
observed that the genotype was important in callus formation from anthers; and in a
diallel study, they found that japonica varieties had higher general combining
abilities than the indica lines. Recently, however, several indica lines have been
regenerated from callus using explants of either mature seeds (Siriwardana and
Nabors, 1983; Heyser et al., 1983) or immature embryos (Heyser et al., 1983; Lai
and Liu, 1982, 1986; Peng and Hodges, 1989; Koetje et al., 1989). Regeneration
efficiencies of various improved indica varieties are shown in Table 1. IR54 was
found to be a good regenerator while IR36 and IR64 were quite poor. These three
varieties along with IR8 were crossed reciprocally in order to assess the genetic
control of regeneration in rice (Peng and Hodges, 1989). Regeneration from
primary callus derived from immature embryos of the parents and hybrids (Tables 2
and 3) indicated that regeneration is controlled by nuclear genes (exhibiting both

Table 2. *In vitro* culture response of four indica varieties of rice. Immature embryos (10-12 days post anthesis) were excised and plated on N6 medium supplemented with 2 mg/L 2,4-D to induce callus formation. Three-week-old calli were transferred to N6 regeneration medium lacking 2,4-D but supplemented with 0.5 mg/L of 6-benzyladenine.

| Genotype | (%) Callus production* | No. plants regenerated / callus | | |
		Range	Mean ±SE	Regeneration frequency (%)
IR54	100 (96/96)	1 - 21	6.0 ± 0.47	85.7
IR36	96.3 (73/76)	0	0.0 ± 0.00	0.0
IR8	90.0 (72/80)	1 - 11	2.0 ± 0.35	46.3
IR64	100 (88/88)	1 - 8	1.2 ± 0.31	46.9

* Numbers in parentheses are number of calli produced / number of immature embryos plated with no contamination. Regeneration frequency is the percentage of calli that produced plants (from Peng and Hodges, 1989).

Table 3. Plant regeneration of a diallel rice population. The four parental varieties used in Table 2 were crossed to produce all combinations of F1 hybrids. Immature embryos of both the F1 hybrids and the parents were excised and cultured as described in Table 2.

| Female parent | Male parent | | | |
	IR54	IR36	IR8	IR64
		(plants / embryo)†		
IR54	6.0 a	4.5 b	1.3 de	1.1 defg
IR36	5.6 a	0.0 g	3.3 c	0.0 g
IR8	1.1 defg	0.1 fg	2.0 d	3.1 c
IR64	0.8 efg	0.0 g	2.1 d	1.2 def

† Means followed by different letter(s) are significantly different at 5% level by Duncan's multiple range test (from Peng and Hodges, 1989).

dominant and additive effects) as well as cytoplasmic genes (Peng and Hodges, 1989). Recognizing the existence of regeneration genes and understanding the environmental conditions (e.g.nutrient media, growth regulators, etc.) governing their expression in heterogeneous explant tissues such as immature embryos, seeds, roots, and leaves remains a difficult problem. However, using cellular and molecular biology techniques, these problems could now be investigated much more rigorously than has been possible before.

Cell Suspension Cultures

Liquid cultures of plant cells are of interest because the cells are more amenable to biochemical and morphological studies, including somatic embryogenesis or the selection of mutant cell types, than are callus cultures grown in semi-solid medium. Suspension cultures are also unique in that the cell population frequently becomes more homogeneous with time in culture especially if care is taken to transfer only the most rapidly growing cells. The cells in such cultures are generally small and have a dense cytoplasm. It is significant that nearly all successes at regenerating monocot plants from protoplasts have started with cells grown in liquid suspension cultures.

Although rice cells have been reported to grow in liquid media for some time (Maeda, 1965, 1969, 1973; Ohira et al., 1973), regeneration of plants from cells grown in liquid cultures was not reported until 1984 (Ye, 1984), and there are still only a few such reports (Abe and Futsuhara, 1986b; Zimny and Lorz, 1986). Table 4 summarizes the studies on rice cells cultured in liquid media, the regeneration of plants from cells grown in liquid cultures, as well as cell growth and plant regeneration from protoplast cultures. In this table we have identified the genotype, the explant, the nutrient media at different stages of *in vitro* culture, and the growth regulators used. The composition of the various media are presented in Table 5 to illustrate the substantial differences that exist between them and to simplify the identification of the critically important components. As mentioned above, suspension cultured cells have served as the source of protoplasts in all of the successful protoplast-regeneration studies to date, but even in these reports, there was little information presented concerning regeneration from the cell suspension cultures themselves.

Protoplast Cultures

Japonica varieties. During the last decade, researchers throughout the world have been striving to develop conditions that would allow for regeneration of mature and fertile plants from protoplasts of both japonica and indica varieties of rice (as well as other cereals). Successful regeneration of rice plants from protoplasts of japonica genotypes was reported at nearly the same time from six different laboratories (Fujimura et al., 1985; Abdullah et al., 1986; Coulibaly and Demarly, 1986; Toriyama et al., 1986; Yamada et al., 1986; Kyozuka et al., 1987). Some comparisons of these studies are presented in Table 4. One of the most surprising results from these reports is that a wide variety of conditions were suitable for growing plants from protoplasts. For example, different genotypes were used in each laboratory. Explants used included seeds, leaf bases, immature embryos, and anthers. Callus-induction media were either LS, MS, or N6 (see

Table 4. Characteristics of liquid cell cultures and protoplast cultures of rice.

genotype[a]	explant	callus culture medium[b]	liquid culture medium[b]	liquid regeneration medium[b]	protoplast culture medium[b]	protoplast regeneration medium[b]	reference
GROWTH OF LIQUID CELL CULTURES							
Tan-ginbozu(J)seed Te-Tep (I) Konansen (I)		MS(2D)	MS(2D)	----	----	----	Maeda, 1965; 1969; 1973
Norin (J)	seed	B5(2D)	R2(2D)	----	----	----	Ohira et al., 1973
REGENERATION FROM LIQUID CELL CULTURES							
Reiho (J) Fujisaka 5 (J) Taichung 65 (J)	seed	MS(2D)	N6 or B5 (2D)	MS	----	----	Ye, 1984
Chyokoto (I)	root	MS(2D)	MS(1-2D)	MS(0.5K)	----	----	Abe and Futsuhara, 1986b
Taipei (J)	root tips	MS (var. auxins)	MS(2D)	MS(1K)	----	----	Zimny and Lorz, 1986
GROWTH OF PROTOPLAST CULTURES							
Unknown	leaf sheath	SH(0.5D, 0.5I, 0.1K)	----	----	SH(1.4D)	SH(0.5D, 0.1K, 0.25N) roots only	Deka and Sen, 1976
Unknown	pollen	MS-type(2D)	N6type(1D)	----	MS(1D)	----	Cai et al., 1978
A-58 (J) + 23 others	hypocotyl	MS(2D)	----	----	MS(4D, 2B) colonies	----	Niizeke and Kita, 1981
Norin 8 (J)	anther	N6(2D)→ AA/MS(2D)	AA(2D), MS(2D)	----	AA(2D)	cell colonies	Wakasa et al., 1984
Chinsurah Boro cytopl. + Taichung 65(J)	anther	B5(2D)→ AA(1D,	AA(1D, 0.2K, 0.1G) 0.2K, 0.1G)	----	B5/AA (several combinations)	roots	Toriyama and Hinata, 1985
C5924 (J)	root (Oc line)	MS(1D, 0.1K)	MS(1D)	----	MS(0.5D, 0.3K)	transformed cells	Baba et al., 1986
Taipei 309 (J)	seed	LS(2.5D)	LS(2.5D)→ AA(2D,0.2K, 0.1G)	----	Kao*(0.5D, 1N, 0.5Z)	albino embryoids	Thompson et al., 1986
C5924 (J)	root (Oc line)	MS(1D, 0.1K)	MS(1D)	----	MS(1D)	transformed cells	Uchimiya et al., 1986

(continued)

Table 4 (continued)

genotype[a]	explant	callus culture medium[b]	liquid culture medium[b]	liquid regeneration medium[b]	protoplast culture medium[b]	protoplast regeneration medium[b]	reference

REGENERATION FROM PROTOPLAST CULTURES

genotype[a]	explant	callus culture medium[b]	liquid culture medium[b]	liquid regeneration medium[b]	protoplast culture medium[b]	protoplast regeneration medium[b]	reference
Nihonbare (J) Sasanishiki (J)	seed embryo	not given	R2/B5(1D)	----	R2/B5(2D)→ N6(2D,0.5B)	N6	Fujimura et al., 1985
Taipei 309 (J) Fujisaka 5 (J)	leaf or seed	LS(2.5D)	LS(2.5D)→ AA(2D)	MS, N6 no plants	Kao*(0.5D)	MS, N6	Abdullah et al., 1986
			[primary callus could also be placed directly into AA(2D)]				
Moroberekan (J)	seed	MS(2D)	Kao(2D) (modif.)	----	Kao(2D)→ MS(2D, 0.1B)	MS	Coulibaly and Demarly, 1986
Yamahoushi (J)	anther panicle	N6(2D), AA ----	AA(1D, 0.2K, 0.1G)	----	B5/AA→ N6(2D,1K)	N6 (1K, 0.2I)→ MS, N6	Toriyama et al., 1986
A-58 ms (J) Fujiminori (J)	seed	LS(4D)	LS(4D)	----	RY2(4D)	N6→ LS(0.9B)	Yamada et al., 1986
Nipponbare (J) Iwaimochi (J) Norin 14 (J) Fujisaka 5 (J) Koshihikari (J)	seed or imm. emb.	MS(2D)	R2(1D)	----	R2(2D) (nurse)→ N6(2D)	N6(with 2-10B, K, Z)	Kyozuka et al., 1987; Ogura et al., '87 Shimamoto et al., '89
IR54 (I)	imm. emb. seed	LS, MS, N6(2D)	N6(4D)	MS	Kao*(0.5D) (feeder)→ LS(0.5D)	N6, MS (2B)	Lee et al., 1989

[a] Letters in parentheses identify genotype as japonica (J) or indica (I), respectively.

[b] For a listing of media ingredients and references refer to Table 5. Hormone concentration in mg/L and identification indicated in parentheses: B, benzyladenine; D, 2,4-dichlorophenoxyacetic acid; G, gibberellic acid; I, indole-3-acetic acid; K, kinetin; N, naphthaleneacetic acid; Z, zeatin.

Table 5). Suspension culture growth media were LS, MS, N6, AA, B5, Kao, or R2. The initial protoplast culture media were either Kao, Kao*, AA, B5, R2, or RY2. The final regeneration media was either LS, MS, or N6, and included different types and concentrations of growth regulators. In different laboratories various additives were used, including different vitamin mixtures and undefined media components such as coconut water or calf serum. In one laboratory (Kyozuka et al., 1987), a nurse culture of suspension cells was required for protoplast growth. Thus, regeneration of several japonica varieties from protoplasts has been achieved under several different *in vitro* culture conditions. These successes represented a dramatic breakthrough as they relate to the potential for genetically engineering the rice plant. However, the indica varieties still remained recalcitrant until this past year when we (Lee et al., 1989) were able to develop suitable conditions for regeneration of this important rice group.

Indica varieties. Regeneration of mature and fertile plants from protoplasts of

Table 5. Composition of various media used for *in vitro* cell and protoplast culture.

Compound	Concentration in Medium (mg/L unless otherwise stated)									
	AA	B5	Kao	Kao*	LS	MS	N6	R2	RY2	SH
Macronutrients:										
$Na_2H_2PO_4 \cdot 2H_2O$	----	170	----	----	----	----	----	310	----	----
KH_2PO_4	170	----	170	170	170	170	400	----	170	----
$NH_4H_2PO_4 \cdot H_2O$	----	----	----	----	----	----	----	----	----	300
KNO_3	----	2500	1900	1900	1900	1900	2830	4000	1900	2500
$(NH_4)_2SO_4$	----	134	----	----	----	----	463	335	67	----
NH_4NO_3	----	----	600	600	1650	1650	----	----	----	----
$MgSO_4 \cdot 7H_2O$	370	250	300	300	370	370	185	250	370	400
$CaCl_2 \cdot 2H_2O$	440	150	600	600	440	440	166	150	440	200
$FeSO_4 \cdot 7H_2O$	27.8	----	----	----	27.8	27.8	27.8	----	----	15.0
Na • EDTA	37.3	----	----	----	37.3	37.3	37.3	----	----	20.0
Fe • EDTA	----	----	----	----	----	----	----	16	19	----
Sequestrene 330 Fe	----	28	28	28	----	----	----	----	----	----
KCl	2940	----	300	300	----	----	----	----	----	----
Micronutrients:										
$MnSO_4 \cdot H_2O$	16.9	10	10	10	16.9	16.9	3.3	1.7	16.9	10.0
$ZnSO_4 \cdot 7H_2O$	8.6	2.0	2.0	2.0	8.6	8.6	1.5	2.2	8.6	1.0
H_3BO_3	6.2	3.0	3.0	3.0	6.2	6.2	1.6	2.9	6.2	5.0
$CuSO_4 \cdot 5H_2O$	0.025	0.025	0.025	0.025	0.025	0.025	----	0.11	0.025	0.2
$Na_2MoO_4 \cdot 2H_2O$	0.25	0.25	0.25	0.25	0.25	0.25	----	0.13	0.25	0.1
$CoCl_2 \cdot 6H_2O$	0.025	0.025	0.025	0.025	0.025	0.025	----	----	0.025	0.1
KI	0.83	0.75	0.75	0.75	0.83	0.83	0.83	----	0.83	1.0
Vitamins:										
Nicotinic Acid	0.5	1.0	----	----	----	0.5	0.5	----	----	5.0
Thiamine • HCl	0.4	10.0	1.0	10.0	0.4	0.1	1.0	1.0	10.0	5.0
Pyridoxine • HCl	0.1	1.0	1.0	1.0	----	0.5	0.5	----	1.0	0.5
Nicotinamide	----	----	1.0	1.0	----	----	----	----	1.0	----
D-Ca Pantothenate	----	----	1.0	0.5	----	----	----	----	0.5	----
Folic Acid	----	----	0.4	0.2	----	----	----	----	0.2	----
p-Aminobenzoate	----	----	0.02	0.01	----	----	----	----	0.01	----
Biotin	----	----	0.01	0.005	----	----	----	----	0.005	----
Choline Chloride	----	----	1.0	0.5	----	----	----	----	0.5	----
Riboflavin	----	----	0.2	0.1	----	----	----	----	0.1	----
Ascorbic Acid	----	----	2.0	1.0	----	----	----	----	1.0	----
Vitamin A	----	----	0.01	0.005	----	----	----	----	0.005	----
Vitamin D_3	----	----	0.01	0.005	----	----	----	----	0.005	----
Vitamin B_{12}	----	----	0.02	0.01	----	----	----	----	0.005	----
Organic Acids:										
Na-pyruvate	----	----	20	5.0	----	----	----	----	5.0	----
Citric Acid	----	----	40	10.0	----	----	----	----	10.0	----
Malic Acid	----	----	40	10.0	----	----	----	----	10.0	----
Fumaric Acid	----	----	40	10.0	----	----	----	----	10.0	----
Sugars and Analogs:										
Sucrose (variable)	20 g	20 g	250	125	30 g	30 g	50 g	20 g	----	30 g
Glucose	----	----	68.4 g	68.4 g	----	----	----	----	90 g	----
Fructose	----	----	250	125	----	----	----	----	----	----
Ribose	----	----	250	125	----	----	----	----	----	----
Xylose	----	----	250	125	----	----	----	----	----	----
Mannose	----	----	250	125	----	----	----	----	----	----
Rhamnose	----	----	250	125	----	----	----	----	----	----
Cellobiose	----	----	250	125	----	----	----	----	----	----
Sorbitol	----	----	250	125	----	----	----	----	----	----
Mannitol	----	----	250	125	----	----	----	----	----	----
Myo-inositol	20	100	100	100	100	100	100	----	100	1000
Nucleic Acid Bases:										
Adenine	----	----	0.10	----	----	----	----	----	----	----
Guanine	----	----	0.03	----	----	----	----	----	----	----
Thymine	----	----	0.03	----	----	----	----	----	----	----
Uracil	----	----	0.03	----	----	----	----	----	----	----
Hypoxanthine	----	----	0.03	----	----	----	----	----	----	----
Xanthine	----	----	0.03	----	----	----	----	----	----	----

(continued)

Table 5 (continued)

Compound	Concentration in Medium (mg/L unless otherwise stated)									
	AA	B5	Kao	Kao*	LS	MS	N6	R2	RY2	SH
Amino Acids:										
Glutamine	877	----	5.6	----	----	----	----	----	----	----
Alanine	----	----	0.6	----	----	----	----	----	----	----
Glutamate	----	----	0.6	----	----	----	----	----	----	----
Cysteine	----	----	0.2	----	----	----	----	----	----	----
Aspartic Acid	266	----	0.1	----	----	----	----	----	----	----
Arginine	174	----	0.1	----	----	----	----	----	----	----
Glycine	7.5	----	0.1	----	----	2.0	2.0	----	----	----
All other L-amino acids	----	----	0.1	----	----	----	----	----	----	----
Phytohormones:										
2,4-D (variable)	1.0	2.0	0.5	0.2	2.0	2.0	2.0	2.0	4.4	0.5
CPA	----	----	----	----	----	----	----	----	----	2.0
Zeatin	----	----	0.5	0.5	----	----	----	----	----	----
NAA	----	----	1.0	1.0	----	----	----	----	----	----
GA$_3$	0.1	----	---	----	----	----	----	----	----	----
Kinetin	0.2	----	----	----	----	----	----	----	----	0.1
Miscellaneous:										
Casamino Acids	----	----	250	125	----	----	----	----	----	----
Coconut Water	----	----	20 ml	10 ml	----	----	----	----	----	----
Calf Serum	----	----	----	----	----	----	----	----	8 ml	----
pH	5.8	5.5	5.6	5.7	5.6	5.7	5.8	6.0	5.6	5.9

AA - Müller and Grafe, 1978   MS - Murashige and Skoog, 1962
B5 - Gamborg et al., 1968   N6 - Chu et al., 1975
Kao - Kao and Michayluk, 1975   R2 - Ohira et al., 1973
Kao* - Kao, 1977   RY2 - Yamada et al., 1986
LS - Linsmaier and Skoog, 1965   SH - Shenk and Hildebrandt, 1972

suspension cultured cells has been obtained for the indica genotype IR54 (Lee et al., 1989). Because of the interest in being able to genetically engineer the indica rice varieties, a detailed description of the regeneration protocol used for IR54 is as follows:

1. Immature embryos (about 0.75-1 mm long) are cultured in Petri plates containing LS medium with 3% sucrose, 2.5 mg/l 2,4-D, and 250 mM tryptophan in 0.4% (w/v) agarose for three to four weeks.

2. Suspension cultures are initiated by carefully teasing away embryogenic regions from three- to four-week-old primary callus. Approximately 1 gFW of callus is placed into 40 ml of medium containing N6 basal salts and vitamins, 20 mM L-proline, 250 mM tryptophan, 2% sucrose, pH 5.8 before autoclaving and 2 to 4 mg/l 2,4-D. Suspensions are shaken on a gyrotary shaker at 120 rpm in the dark at 25 C. During the establishment of the cell suspensions, the medium is changed without removing any callus every two to three days for two to four weeks. Subsequently, the cell suspensions are subcultured twice each week using 5 ml packed-cell-volume (pcv) per 40 ml of the same medium for the next three months, and thereafter the subculture protocol is the same but is done only once per week.

3. Protoplasts are isolated from 1 gFW of cells by incubating them in 10 ml of a solution containing 1% (w/v) cellulase "Onosuka" RS (Yakult Honsha Co.,

Japan) and 0.1 % (w/v) Pectolyase Y-23 (Seishin Pharmaceutical Co., Japan) in CPW salts (Frearson et al. 1973) with 0.4 M mannitol, 5 mM MES, pH 6.0. After four hours of shaking at 40 rpm in the dark at 25 C, the mixture is filtered through a 30 μm nylon mesh, washed with two volumes of the CPW medium, and centrifuged at 80 x g for 15 min. Protoplasts are partially purified by layering over a 0.6 M sucrose solution and centrifuging at 40 x g for 10 min. Finally, protoplasts are suspended in Kao* medium with 50 mg/l arginine, 0.5 mg/l 2,4-D, and 7 % glucose prior to culturing.

4. Protoplasts are cultured in the presence of a nurse culture consisting of cells from a suspension culture (our most useful nurse culture has been cells of the japonica variety Calrose 76, but suspension cells of IR52 have also been effective). The nurse cells are prepared at a density of 1.5 ml pcv per 20 ml of Kao* medium. Five ml of a mixture containing the nurse cells in 0.8% low melting point agarose and Kao* medium are placed into a Petri dish (60 x 15 mm). A 0.8 μm Millipore filter is placed on top of the solidified agarose containing the nurse cells, and to this are added 0.2 ml of protoplasts at a density of 5 x 10^5 protoplasts per ml. Protoplasts are cultured in a dark chamber at 25 C for three weeks, at which time they were transferred to larger Petri plates (100 x 15 mm) containing LS basal medium with 2% (w/v) sucrose, 0.5 mg/L 2,4-D, and 0.4% (w/v) agarose (no nurse cells). Three weeks later callus colonies one to two mm in diameter with embryo-like structures are transferred to regeneration medium.

5. Regeneration medium consists of either MS or N6 basal salts medium supplemented with 3% w/v sucrose, 0.4 % agarose, 0.03 μM NAA, and 9 μM benzylaminopurine (or kinetin). Cultures are kept in a dark chamber for one week and then transferred into light (75 μmole $m^{-2}s^{-1}$) with a 16 hour photoperiod and at 25 C. Plantlets with shoots at least 5 cm long are transferred to jars containing MS basal salts and 3 % sucrose to promote root development. After the establishment of a vigorous root system, the plants are transferred to soil and moved to the greenhouse or growth chamber for further development.

Using the above procedures, it has been possible to regenerate mature plants from protoplasts obtained from suspension cultures of IR54 and IR52 derived from callus of immature embryos, a suspension culture of IR54 derived from callus of seeds, and even from the three-week-old primary callus of IR54 (Lee et al., 1989). Critical factors contributing to regeneration were the use of selected cell cultures that were still highly embryogenic and regenerable (from the original callus and from cells in liquid suspensions), optimized conditions to obtain high yields of viable protoplasts (in excess of 10^6 protoplasts per gFW cells), the use of nurse cells (Table 6), the selection of small embryogenic calli at an early stage of protoplast-callus formation which are then transferred to regeneration medium, and finally, the use of a cytokinin in the regeneration medium (Table 7). All of the plants shown in Table 7 have reached maturity, although they were about 20% shorter than seed-grown plants. Otherwise, they appeared to be normal phenotypically. Of the 76 protoplast-regenerated plants of IR54, 62 flowered and set seed with 30 to 80% of the caryopses being fertile. Thus, it is now possible to regenerate indica varieties as well as the japonica varieties of rice from protoplasts into mature, fertile plants.

Table 6. Effect of feeder (nurse) cells on the plating efficiency of
rice protoplasts obtained from IR54[a] suspension cells
when the culture was 7, 10, or 12 months old. Isolated
protoplasts were plated onto a 0.8 μm Millipore filter at a
density of 5 x 10^5 per Petri plate. The number of visible
colonies under a 4x dissection microscope at two weeks
was counted, and plating efficiency was calculated as the
percentage of colonies formed divided by the number of
protoplasts per plate times 100.

Nurse cells	Culture age (mo.)	Protoplast plating efficiency (%)
none	7	0
Calrose 76	7	2.5
Calrose 76	10	2.6
Calrose 76	12	1.4
IR54-4	7	<0.1
IR52	7	<0.001
IR54-5	7	<0.01

[a] The IR54 cell line for obtaining protoplasts was IR54-1 and the IR54-4 and
IR54-5 are other selected lines (Lee et al. 1989), and all of them were maintained
in N6 medium. The Calrose 76 suspension line was maintained in AA medium.

Table 7. Effect of cytokinins on plant regeneration from protoplasts
obtained from IR54[a] suspension cells.

Regeneration medium[a]	Regeneration frequency (%)	Plants regenerated	Plants per regenerating callus
N6	0 (0)	0	0
N6-B	5 (9)	21	4.2
N6-Z	2 (4)	5	2.5
N6-K	1 (2)	3	3.0
MS	0 (0)	0	0
MS-B	17 (34)	33	1.9
MS-Z	5 (8)	6	1.5
MS-K	6 (12)	8	1.3

[a] The IR54 line is IR54-1 from Lee et al. (1989). All treatments contained 55
calli initially except for MS-B which contained 50. Regeneration media were
either N6 or MS with benzylaminopurine (B), zeatin (Z), or kinetin (K).

TRANSFORMATION

Protoplast Transformation Yielding Transgenic Callus

Transformation of monocot protoplasts can be achieved via direct uptake of DNA (Potrykus et al., 1985), and this is generally mediated by some treatment such as polyethylene glycol or electroporation that increases the permeability of cell membranes to DNA . In rice, transient expression of the enzyme, chloramphenicol acetyl transferase (CAT), was expressed in leaf protoplasts 48 hours after electroporating plasmid DNA containing the CAT coding region flanked by the 35S promoter from cauliflower mosaic virus (35SCaMV) and a 3' noncoding sequence of the small subunit gene of ribulose 1, 5-bisphosphate carboxylase (Ou-Lee et al., 1986). Stable transformation of japonica rice protoplasts that were capable of growing into callus, but not into plants, was first reported by Baba et al. (1986) and Uchimiya et al. (1986) (see Table 4). In both studies the protoplasts were derived from a suspension culture using four-year-old callus (line Oc) obtained initially from roots of variety C5924. The transformation procedures used by Baba et al. (1986) employed spheroplasts of virulent strains A208 and A227 of *Agrobacterium tumefaciens*, and they reported that callus colonies contained opines as well as nopaline synthase activity. Uchimiya et al. (1986) transformed line Oc protoplasts with circular plasmid DNA consisting of the structural gene *neo* encoding aminoglycoside phosphotransferase II [APH(3')II] (Beck et al., 1982), which confers resistance to kanamycin, and the nopaline synthase promoter from the Ti plasmid of *A. tumefaciens*. Polyethylene glycol (PEG) was used to facilitate the direct uptake of DNA into these protoplasts. Protoplasts grew into colonies for one month prior to transfer to medium containing 100 µg/ml kanamycin sulfate. The transformation efficiency was about 2.5×10^{-5}, even though the *neo* gene and kanamycin may not be the most useful selectable system for monocot cells (Hauptmann et al., 1988). After two months, 31 of 34 colonies resistant to kanamycin exhibited APH(3')II activity. Southern blot hybridization (Southern, 1975) indicated that the callus line exhibiting the highest enzyme activity also had the highest number of copies of the gene. Similar results have been reported (Yang et al., 1988) for Taipei 309 protoplasts transformed with the same *neo* gene but with the 35SCaMV promoter and a *nos* polyadenylation region (Fromm et al., 1986). Either electroporation or PEG were used to facilitate transformation of the protoplasts. In this study, the transformation efficiencies were also about 2×10^{5}.

We have cotransformed IR54 protoplasts with two separate plasmids; one (pKAN) contained the *neo* gene with a 35SCaMV promoter and a T-DNA poly A tail region from the Ti plasmid of *Agrobacterium tumefaciens* (Bytebier et al., 1987). The second plasmid (pPUR) contained the *uid*A gene (Jefferson et al., 1987) that encodes ß-glucuronidase (GUS), the 35SCaMV promoter, and a *nos* polyadenylation region (Lyznik et al., 1989). Cotransformation was performed by incubating the protoplasts with circular forms of both plasmids in the presence of 20% PEG. Putative transformants surviving seven weeks of growth (Fig. 2) on kanamycin (100 µg/ml) were analyzed for the *neo* gene by Southern hybridization and for GUS activity resulting from expression of the *uid*A gene. The *neo* gene was detected in nine of 11 kanamycin-resistant colonies. Three of these 9 lines (lanes 1, 8, and 9 of Fig. 3) also had high levels of GUS activity (Fig. 3) and the *neo* gene from pKAN was confirmed to be present in these lines by Southern

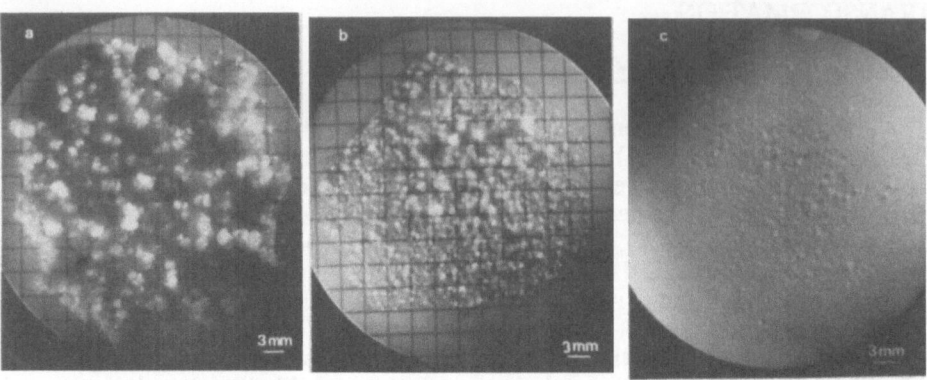

Fig. 2. IR54 callus growth on medium with or without kanamycin seven weeks
after protoplast isolation and PEG-mediated transformation. A).
Protoplasts were treated with PEG in the absence of plasmid DNA and
grown in medium without kanamycin. B). Protoplasts were treated with
PEG along with 50 μg pPUR and 100 μg pKAN plasmids and grown in
medium with kanamycin. C). Control (no plasmid DNA) protoplasts
grown in medium with kanamycin. Protoplasts were treated with 20%
PEG in F solution (Krens et al., 1982) with or without plasmid DNA
additions as stated above, then added to Millipore filters placed over
Calrose 76 feeder cells in LS medium with 1.0 mg/l 2,4-D. Filters with
protoplasts were transferred to fresh medium without or with kanamycin
(100 mg/l) after three weeks. The medium was renewed after two weeks.

analysis (Fig. 3). In five separate experiments, GUS activity was detected in 13 of
52 kanamycin-resistant lines analyzed for a cotransformation efficiency of 25%.
Since the proficiency for selecting lines transformed with the *neo* gene was about
80%, the cotransformation efficiency would be somewhat higher than 25%. These
efficiencies are only slightly lower than the cotransformation efficiencies reported
for tobacco protoplasts (Schocher et al., 1986). Thus, cotransformation makes it
feasible to use the *neo* gene for selection of transformants to increase the probability
of identifying callus colonies transformed with other genes that may not be as easily
detected, e.g. genes controlled temporally or by tissue-specific promoters.

Protoplast Transformation Yielding Transgenic Rice Plants

The first successes in obtaining transgenic rice plants were reported by
Toriyama et al. (1988) and Zhang et al. (1988). In both studies circular forms of
plasmids carrying the *neo* gene under control of the 35SCaMV promoter were
introduced into protoplasts with the aid of electroporation, and transformed calli
were selected in the presence of G418 (Toriyama et al., 1988) or kanamycin (Zhang
et al., 1988). Toriyama et al. (1988) obtained five transgenic plants; one was a
diploid, three were triploids, and the ploidy of the fifth one was not identified.
Leaves from three of the plants exhibited APHII activity, and DNA from four of the
plants hybridized with the gene fragment by Southern analysis. Zhang et al. (1988)

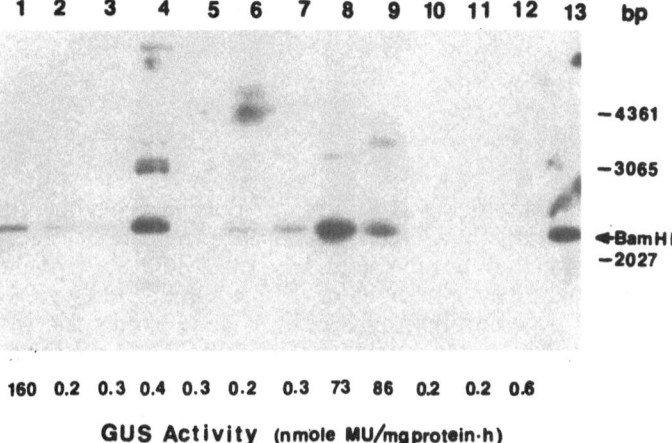

Fig. 3. Southern blot hybridization analysis of DNA for *neo* gene and GUS activities from IR54 callus lines resistant to kanamycin. DNA was isolated according to Mettler (1987). Ten µg DNA from each sample, digested with *Hind*III and *Eco*RI, were hybridized to a ^{32}P-labeled 2.2 Kb *Bam*HI fragment of pKAN. Lanes 1-11 were loaded with DNA from 11 different kanamycin resistant calli, lane 12 was loaded with DNA isolated from control callus, and Lane 13 was loaded with 2.2 Kb *Bam*HI fragment of pKAN equivalent to five copies per haploid genome (1.2 x 10^9).

regenerated six plants, all of which contained the *neo* gene, but only two expressed enzyme activity. It was not clear in either paper whether the transgenic plants flowered and set seed.

More recently, Shimamoto et al. (1989) reported the regeneration of large numbers of transgenic rice plants. They electroporated protoplasts of the japonica line, Nipponbare, in the presence of a plasmid containing the bacterial *hph* gene that encodes for resistance to the antibiotic hygromycin. The *hph* gene was flanked by the 35SCaMV promoter and a polyadenylation site from CaMV. Transformed calli were selected in the presence of 20 µg/ml of hygromycin. Approximately 100 plants were regenerated from hygromycin-resistant calli, and Southern blot analysis of DNA from leaves demonstrated the presence of the *hph* gene. Regenerated plants were fertile; selfed progeny were shown to be resistant to hygromycin and to contain the *hph* gene. They also obtained cotransformation with the GUS gene in four out of six hygromycin-resistant plants. These results clearly illustrate that the basic problems related to genetically engineering the rice plant via protoplast transformation and regeneration have been resolved.

As the utility of protoplasts for use in producing transgenic rice plants was being realized, it has also become evident that other techniques such as DNA delivery to somatic cells via the particle bombardment technique (Klein et al., 1988) and the transfer of DNA down excised pollen tubes (Zhou et al., 1983; Luo and

Wu, 1988) will also be useful methods for producing transgenic rice plants (see other articles in this volume).

SUMMARY

In the 20 years since rice was first reported to be regenerated from *in vitro* callus cultures, conditions have been developed to regenerate plants from many different genotypes including representatives from both the japonica and indica groups. Various studies have demonstrated that somatic embryo formation and plant regeneration are controlled by factors of gene expression which are very sensitive to the nutrients in the culture medium, the 2,4-D concentration used to enhance cell divisions, and the age or developmental state of cells in the explant. Conditions have been established which allow growth of rice cells in liquid medium and regeneration of plants from these cells, although there has been a surprising paucity of such reports. Efforts to regenerate mature and fertile plants from single protoplasts, which first met with success in 1985 for the japonica varieties, have now extended this capacity to the indica lines. Techniques have also been developed for transforming protoplasts via direct uptake of DNA that is facilitated by either electroporation or PEG, and there are now three reports of the production of transgenic plants of japonica varieties. Thus, the primary barriers have been overcome for the genetic transformation of protoplasts (albeit uncontrolled and unpredictable) and the regeneration of mature, fertile plants (although at a low frequency) of both the japonica and the indica varieties of rice. There are still many problems regarding genetic engineering of the rice plant such as the efficiency of plant regeneration, the directing of foreign genes to specific sites in the genome, understanding the regulation of expression of the inserted genes, and the identification of genes that will confer desired traits to the rice plant; nevertheless, the major barriers have been overcome and progress should now be rapid.

ACKNOWLEDGEMENTS

This is journal paper 11,921 of the Purdue University Agricultural Experiment Station. Research was supported by The Rockefeller Foundation.

ADDENDUM

After this manuscript was submitted for publication, two additional papers relevent to this topic were published. W. Zhang and R. Wu (1988, Theor. Appl. Genet., 76: 835-840) reported regeneration of transgenic japonica rice from protoplasts. GUS activity was detected under control of maize alcohol dehydrogenase regulatory sequences. J. Kyozuka, E. Otoo, and K. Shimamoto (1988, Theor. Appl. Genet., 76: 887-890) reported regeneration of plants from protoplasts of the indica genotype, Chyokoto.

REFERENCES

Abdullah, R., Cocking, E. C., and Thompson, J. A., 1986, Efficient plant regeneration from rice protoplasts through somatic embryogenesis, Bio/Technology, 4: 1087-1090.

Abe, T. and Futsuhara, Y., 1984, Varietal difference of plant regeneration from root callus tissues in rice, Jpn. J. Breed., 34: 147-155.

Abe, T. and Futsuhara, Y., 1986a, Genotypic variability for callus formation and plant regeneration in rice (Oryza sativa L.), Theor. Appl. Gen. 72: 3-10.

Abe, T. and Futsuhara, Y., 1986b, Plant regeneration from suspension cultures of rice (Oryza sativa), Jpn. J. Breed., 36: 1-6.

Baba, A., Hasezawa, S., and Syono, K., 1986, Cultivation of rice protoplasts and their transformation mediated by Agrobacterium spheroplasts, Plant Cell Physiol., 27: 463-471.

Beck, E., Ludwig, G., Auerswald, E. A., Reiss, B., and Schaller, H., 1982, Nucleotide sequence and exact localization of the neomycin phosphotransferase gene from transposon Tn5, Gene, 19: 327-336.

Bhattacharya, P. and Sen, S.K., 1980, Potentiality of leaf sheath cells for regeneration of rice (Oryza sativa L.) plants, Theor. Appl. Genet., 58: 87-90.

Bytebier, B., Deboeck, F., De Greve, H., Van Montagu, M., and Hernalsteens, J-P., 1987, T-DNA organization in tumor cultures and transgenic plants of the monocotyledon Asparagus officinalis, Proc. Natl. Acad. Sci. USA, 84: 5345-5349.

Cai, Q., Zhou, Y., and Wu, S., 1978, A further study on the isolation and culture of rice (Oryza sativa L.) protoplasts, Acta Bot. Sinica, 20: 97-102.

Chen, T. H., Lam, L., and Chen, S. C., 1985, Somatic embryogenesis and plant regeneration from cultured young inflorescences of Oryza sativa, Plant Cell Tissue Culture, 4: 51-54.

Chu, C. C., Wang, C. C., Sun, C. S., Hsu, C., Yin, K. C., Chu, C. Y., and Bi, F. Y., 1975, Establishment of an efficient medium for anther culture of rice through comparative experiments on the nitrogen sources, Sci. Sin., 18: 659-668.

Conger, B. V., 1981, Agronomic Crops, in: "Cloning Agricultural Plants via In Vitro Techniques", Conger, B. V., ed., CRC Press, Boca Raton, Fla., pp. 165-215.

Coulibaly, M. Y. and Demarly, Y., 1986, Regeneration of plantlets from protoplasts of rice, Oryza sativa L., Z. Planzenzüchtg, 96: 79-81.

Deka, P. C. and Sen, S.K., 1976, Differentiation of calli originated from isolated protoplasts of rice (Oryza sativa L.) through plating techniques, Mol. Gen. Genet., 145: 239-243.

Flick, C. E., and Evans, D. A., 1984, Tobacco in:, "Handbook of Plant Cell Culture (vol. 2)", Sharp, W. R., Evans, D. A., Ammirato, P. V., and Yamada, Y., eds., Macmillan Publishing Co., New York, NY, pp. 606-630.

Frearson, E. M., Power, J. B., and Cocking, E. C., 1973, The isolation, culture and regeneration of Petunia leaf protoplasts, Dev. Biol., 33: 130-137.

Fromm, M.E., Taylor, L. P., and Walbot, V., 1986, Stable transformation of maize after gene transfer by electroporation, Nature, 319: 791-793.

Fujimura, T., Sakurai, M., Akagi, H., Negishi, T., and Hirose, A., 1985, Regeneration of rice plants from protoplasts, Plant Tiss. Cult. Lett., 2: 74-75.

Gamborg, O. L., Miller, R. A. and Ojima, K., 1968, Nutrient requirements of suspension cultures of soybean root cells, Exp. Cell Res., 50: 151-158.

Green, C. E., and Phillips, R. L., 1975, Plant regeneration from tissue cultures of maize, Crop Sci., 15: 417-421.

Guha-Mukherjee, S., 1973, Genotypic differences in the *in vitro* formation of embryoids from rice pollen, J. Exp. Bot., 21: 139-144.

Hauptmann, R. M., Vasil, V., Ozias-Akins, P., Tabaeizadeh, Z., Rogers, S.G., Fraley, R. T., Horsch, R. B., and Vasil, I. K., 1988, Evaluation of selectable markers for obtaining stable transformants in the gramineae, Plant Physiol., 86: 602-606.

Henke, R. R., Mansur, M. A., and Constantin, M. J., 1978, Organogenesis and plantlet formation from organ- and seedling-derived calli of rice (*Oryza sativa*), Physiol. Plant., 44: 11-14.

Heyser, J. W., Dykes, T. A., DeMott, K. J., and Nabors, M. W., 1983, High frequency, long-term regeneration of rice from callus culture, Plant Sci. Lett., 29: 175-182.

Hodges, T. K., Kamo, K. K., Becwar, M. R., and Schroll, S., 1985, Regeneration of maize, in: "Biotechnology in Plant Science: Relevance to Agriculture in the Nineteen Eighties", Zaitlin, M., Day, P., and Hollaender, A., eds., Academic Press, Orlando, FL, pp. 15-33.

Hodges, T. K., Kamo, K. K., Imbrie, C. W., and Becwar, M. R., 1986, Genotype specificity of somatic embryogenesis and regeneration in maize, BioTechnology, 4: 219-223.

IRRI, 1985, "International Rice Research: 25 Years of Partnership", International Rice Research Institute, Los Banos, Laguna, Philippines.

Jefferson, R. A., Kavanagh, T. A., and Bevan, M. W., 1987, GUS fusions: ß-glucuronidase as a sensitive and versatile gene fusion marker in higher plants, EMBO J., 6: 3901-3907.

Kao, K. N., 1977, Chromosomal behavior in somatic hybrids of soybean-*Nicotiana glauca*, Mol. Gen. Genet., 150: 225-230.

Kao, K. N. and Michayluk, M. R., 1975, Nutritional requirements for growth of *Vicia hajastana* cells and protoplasts at a very low population density in liquid media, Planta, 126: 105-110.

Kawata, S. and Ishihara, A., 1968, The regeneration of rice plant, *Oryza sativa* L., in the callus derived from the seminal root, Proc. Jpn. Acad., 44: 549.

Klein, T. M., Harper, E. C., Svab, Z., Sanford, J. C., Fromm, M. E., and Maliga, P., 1988, Stable genetic transformation of intact *Nicotiana* cells by the particle bombardment process, Proc. Natl. Acad. Sci. USA, 85: 8502-8505.

Koetje, D. S., Grimes, H. D., Wang, Y. C., and Hodges, T. K., 1989, Regeneration of indica rice (*Oryza sativa* L.) from primary callus derived from immature embryos, J. Plant Physiol., in review.

Krens, F. A., Molendijk, L., Wullems, G. J., and Schilperoort, R. A., 1982, *In vitro* transformation of plant protoplasts with Ti-plasmid DNA, Nature, 296: 72-74.

Kyozuka, J., Hayashi, Y., and Shimamoto, K., 1987, High frequency plant regeneration from rice protoplasts by novel nurse culture methods, Mol. Gen. Genet., 206: 408-413.

Lai, K. L., and Liu, L. F., 1982, Induction and plant regeneration of callus from immature embryo of rice plants (*Oryza sativa* L.), Jpn. J. Crop Sci., 51: 70-74.

Lai, K. L., and Liu, L. F., 1986, Further studies on the variability of plant regeneration from young embryo callus cultures of rice plants (*Oryza sativa* L.), Jpn. J. Crop Sci., 55: 41-46.

Lee, L., Schroll, R. E., Grimes, H. D., and Hodges, T. K., 1989, Plant regeneration from indica rice (*Oryza sativa* L.) protoplasts, Planta, in review.

Linsmaier, E. M. and Skoog, F., 1965, Organic growth factor requirements of tobacco tissue cultures, Physiol. Plant., 18: 100-127.

Luo, Z.X. and Wu, R., 1988, A simple method for the transformation of rice via the pollen-tube pathway, Plant Molec. Biol. Rept., 6: 165-174.

Lyznik, L. A., Ryan, R., Ritchie, S., and Hodges, T. K., 1989, Stable PEG-mediated cotransformation of maize protoplasts with *uid*A and *neo* genes, in review.

Maeda, E., 1965, Callus formation and isolation of single cells from rice seedlings, Proc. Crop Sci. Jpn., 34: 139-147.

Maeda, E., 1969, Multiplication of rice cells freely suspended *in vitro*, Proc. Crop Sci. Jpn., 38: 535-546.

Maeda, E., 1973, Proliferation and properties of rice cells subcultured in a liquid medium, Proc. Crop Sci. Jpn., 42: 110-115.

Mettler, I. J., 1987, A simple and rapid method for minipreparation of DNA from tissue cultured plant cells, Plant Molec. Biol. Rept., 5: 346-349.

Miah, M. A. A., Earle, E. D., and Khush, G. S., 1985, Inheritance of callus formation ability in anther cultures of rice, *Oryza sativa* L., Theor. Appl. Genet., 70: 113-116.

Müller, A. J. and Grafe, R., 1978, Isolation and characterization of cell lines of *Nicotiana tabacum* lacking nitrate reductase, Mol. Gen. Genet., 161: 67-76.

Murashige, T. and Skoog, F., 1962, A revised medium for rapid growth and bioassays with tobacco tissue cultures, Physiol. Plant., 15: 473-497.

Niizeki, H. and Oono, K.l, 1968, Induction of haploid rice plant from anther culture, Proc. Jpn. Acad., 44: 554-557.

Niizeki, H. and Kita, F., 1981, Cell division of rice and soybean and their fused protoplasts, Jpn. J. Breed., 31: 161-167.

Nishi, T., Yamada, Y., and Takahashi, E., 1968, Organ redifferentiation and plant regeneration in rice callus, Nature, 219: 508-509.

Ogura, H., Kyozuka, J. Hayashi, Y., Koba, T., and Shimamoto, K., 1987, Field performance and cytology of protoplast-derived rice (*Oryza sativa*): high yield and low degree of variation of four japonica cultivars, Theor. Appl. Genet., 74: 670-676.

Ohira, K., Ojima, K., and Fujiwara, A., 1973, Studies on the nutrition of rice cell culture I. A simple, defined medium for rapid growth in suspension culture, Plant & Cell Physiol., 14: 1113-1121.

Ou-Lee, T.M., Turgeon, R., and Wu, R., 1986, Expression of a foreign gene linked to either a plant-virus or a Drosophila promoter, after electroporation of protoplast of rice, wheat, and sorghum, Proc. Natl. Acad. Sci. USA, 83: 6815-6819.

Peng, J. and Hodges, T. K., 1989, Genetic analysis of plant regeneration in rice (*Oryza sativa* L.), *In Vitro*: Cellular and Devel. Biol., in press.

Potrykus, I., Saul, M.W., Petruska, J., Paszkowski, J., and Shillito, R. D., 1985, Direct gene transfer to cells of a graminaceous monocot, Molec. Gen. Genet., 199: 183-188.

Schocher, R J., Shillito, R. D., Saul, M. W., Paszkowski, J., and Potrykus, I., 1986, Co-transformation of unlinked foreign genes into plants by direct gene transfer, Bio/Technology, 4: 1093-1096.

Sears, R. G. and Deckard, E. L., 1982, Tissue culture variability in wheat: Callus induction and plant regeneration, Crop Sci., 22: 546-550.

Shenk, R. V. and Hildebrandt, A. C., 1972, Medium and techniques for induction and growth of monocotyledonous and dicotyledonous plant cell cultures, Can. J. Bot., 50: 199-204.

Shimamoto, K., Terada, R., Izawa, T., and Fujimoto, H., 1989, Transgenic rice plants: expression and transmission of foreign genes introduced by electroporation, in press.

Siriwardana, S, and Nabors, M. W., 1983, Tryptophan enhancement of somatic embryogenesis in rice, Plant Physiol., 73: 142-146.

Southern, E., 1975, Detection of specific sequences among DNA fragments separated by gel electrophoresis, J. Mol. Biol., 98: 503-517.

Tamura, S., 1968, Shoot formation in calli originated from rice embryo, Proc. Jpn. Acad., 44: 544-548.

Thompson, J. A., Abdullah, R., and Cocking, E. C., 1986, Protoplast culture of rice (*Oryza sativa* L.) using media solidified with agarose, Plant Sci., 47: 123-133.

Toriyama, K. and Hinata, K., 1985, Cell suspension and protoplast culture in rice, Plant Sci., 41: 179-183.

Toriyama, K., Hinata, K., and Sasaki, T., 1986, Haploid and diploid plant regeneration from protoplasts of anther callus in rice, Theor. Appl. Genet., 73: 16-19.

Toriyama, K., Arimoto, Y., Uchimiya, H., and Hinata, K., 1988, Transgenic rice plants after direct gene transfer into protoplasts, Bio/Technology, 6: 1072-1074.

Uchimiya, H., Fushimi, T., Hashimoto, H., Harada, H., Syono, K., and Sugawara, Y., 1986, Expression of a foreign gene in callus derived from DNA treated protoplasts of rice (*Oryza sativa* L.), Mol. Gen. Genet., 16: 204-207.

Wakasa, K., Kobayashi, M., and Kamada, H., 1984, Colony formation from protoplasts of nitrate reductase-deficient rice cell lines, J. Plant Physiol., 117: 223-231.

Wernicke, W. and Brettell, R., 1980, Somatic embryogenesis from *Sorghum bicolor* leaves, Nature, 287: 138-139.

Wernicke, W., Brettell, R.I.S., Wakizuka, T., and Potrykus, I., 1981, Adventitious embryoid and root formation from rice leaves, Z. Pflanzenphysiol., 103: 361-365.

Wernicke, W., Gorst, J., and Milkovits, L., 1986, The ambiguous role of 2,4-dichlorophenoxyacetic acid in wheat tissue culture, <u>Physiol. Plant.</u>, 68: 597-602.

Yamada, Y. and Loh, W. H., 1984, Rice, <u>in</u>: "Handbook of Plant Cell Culture", eds. Ammirato, P. V., Evans, D. A., Sharp, W. R. and Yamada Y., MacMillian Co., vol. 3, pp. 151-170.

Yamada, Y., Yang, Z. Q., and Tang, D. T., 1986, Plant regeneration from protoplast derived callus of rice (<i>Oryza sativa</i>), <u>Plant Cell Rept.</u>, 5: 85-88.

Yang, H., Zhang, H.M., Davey, M. R., Mulligan, B. J., and Cocking, E. C., 1988, Production of kanamycin resistant rice tissues following DNA uptake into protoplasts, <u>Plant Cell Rept.</u>, 7: 421-425.

Ye, H., 1984, Studies on cell suspension culture and plant regeneration in rice, <u>Acta Botanica Senica</u>, 26: 52-59.

Zhang, H.M., Yang, H., Rech, E. L., Golds, T. J., Davis, A. S., Mulligan, B. J., Cocking, E. C., and Davey, M. R., 1988, Transgenic rice plants produced by electroporation-mediated plasmid uptake into protoplasts, <u>Plant Cell Rept.</u>, 7: 379-384.

Zhou, G.Y., Weng, J., Zeng, Y., Huang, J., Qian, S., and Liu, G., 1983, Introduction of exogenous DNA into cotton embryos, <u>Meth. Enz.</u>, 101: 432-455.

Zimny, J., and Lorz, H., 1986, Plant regeneration and initiation of cell suspensions from root-tip derived callus of <i>Oryza sativa</i> (rice), <u>Plant Cell Rept.</u>, 5: 89-92.

IN VITRO MANIPULATION OF BARLEY

AND OTHER CEREALS

Horst Lörz, Reinhold Brettschneider, Sabine
Hartke, Ravinder Gill[1], Erhard Kranz, Peter
Langridge[2], Andrzej Stolarz and Paul Lazzeri

Max-Planck-Institut für Züchtuchtungsforschung,
D-5000 Köln 30, Fed. Rep. of Germany

[1] Bio-Organic Division, Plant Biotechnology Sect.
Bhabha Atomic Research Centre, Bombay, India

[2] Agricultural Biochemistry Department, Waite
Institute, Glen Osmond, South Australia

INTRODUCTION

Already, it can be seen and more so in the
future, it is expected, that plant cell biology and
molecular biology will have a major impact on
agriculture by supplementing the present activities of
plant breeders in expanding and diversifying the gene
pool of crop species and in speeding up the breeding
process. Different strategies are used which apply in
vitro methods to generate diversity within existing
populations, to identify rare, but desired individual
plants, and to broaden the genetic pool of breeding
material. Of major interest are the production of
homozygous lines by anther- and micropsore-culture,
the in vitro selection of cultured cells to create
stress or disease resistant plants, protoplast fusion
and somatic hybridization to overcome the natural
barriers of incompatibility or to establish new
combinations of organellar and nuclear genomes in
somatic hybrid or cybrid plants, and finally the
transfer of isolated genes to achieve a directed,
highly defined genetic modification of a specific crop
plant. Progress with cereals and grasses in the past
has been rather slow when compared to species such as
Nicotiana tabacum, Solanum tuberosum or Brassica
napus. The field has been reviewed recently in nume-

rous articles (Göbel and Lörz, 1988; Lörz et al., 1988; Ozias-Akins and Vasil, 1988), thus mostly recent experiments from our laboratory and new developments will be discussed.

PRODUCTION AND MANIPULATION OF HAPLOIDS

Anther culture as a means to produce haploid plants and derived thereof homozygous lines can today be considered a routine procedure. It has been integrated in the breeding schemes of Hordeum vulgare, Triticum aestivum and Oryza sativa (Dunwell, 1986). Especially in barley the efficiency has been improved significantly by using strictly defined culture conditions from the donor plants, applying a defined regime of cold treatment and using improved culture media with different carbohydrates in replacement for sucrose which originally has been used exclusively. For some barley genotypes or cultivars such as "Igri" or "Dissa" the efficiency has been optimized up to a yield of 10 homozygous green plants from 1 anther plated (M. Jäger-Gussen, pers. comm.).

For the purpose of genetic manipulation haploids are of interest as a source for the isolation of haploid protoplasts, and if isolated micropsores are cultured successfully, as in barley and wheat (Köhler and Wenzel, 1987; Datta and Wenzel, 1987), for direct manipulation of the unicellular microspore. The isolated microspores are suitable target cells for microinjection as has been demonstrated with rapeseed (Neuhaus et al., 1987). This should be also possible with graminaceous micropsores and is now being studied intensively, however the low efficiency of culture response are the main obstacles to obtain stably transformed plants with this method.

The natural pathway of micropsore development leads to the formation of mature pollen grains. Recently it has been shown with Nicotiana tabacum microspores, that the complete development from the uninuclear microspore to the mature pollen grain can be obtained under in vitro culture conditons (Benito-Moreno et al., 1988). This culture system is being studied also in wheat and Zea mays. In combination with microinjection this system should allow one to deliver DNA into an isolated haploid cell which subsequently will be used for pollination and ferti-lisation giving rise to genetically modified seeds. The bottlenecks at present are the reproducibility of the system, and the methods for pollination with <u>in</u> <u>vitro</u> matured pollen have to be developed.   Figure 1 shows a schematic summary of the approaches being studied at present for <u>in</u> <u>vitro</u> pollination and delivery at DNA into pollen grains of maize.

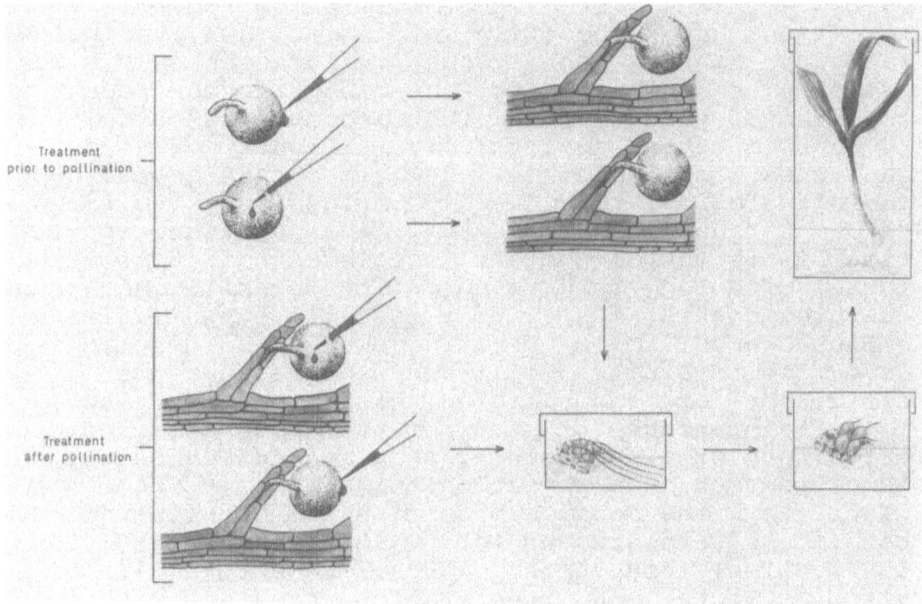

Fig. 1. In vitro manipulation, pollination and ferti-
        lization of maize. Treatment of pollen grains
        can be performed prior to pollination or after
        pollination. <u>In</u> <u>vitro</u> pollinated ovules
        develop in high frequency to seeds and
        thereafter germinate normally.

      In maize good progress has been made in the pro-
duction of gynogenetic haploid plants from cultured,
unpollinated ovules (Lashermes and Beckert, 1988).
Like the microspore-derived embryoids, non-fertilized
egg cells and embryoids developing thereof resemble
attractive target cells for direct DNA delivery by
microinjection or microprojectiles. The severe

restrictions are due to the genotype and a rather low
efficiency of plant regeneration.

PROTOPLAST, CELL AND TISSUE CULTURE

    The procedure of in vitro plant regeneration from
multicellular explants and regeneration via somatic
embryogenesis or shoot-root organogenesis provides a
suitable means for rapid multiplication, in vitro
selection, and induction of somaclonal variation.
Immature embryos and young inflorescences are most
commonly used as explants, but other tissues such as
the base of young leaves, the mesocotyl, and root tips
have been used successfully. Plants of numerous
genotypes and practically all important cereal crops
including maize, rice, wheat, barley, rye, Triticosecale, Sorghum and Pennisetum. For practical applications it is most important to establish in vitro cultures with long term regeneration capabilities. Such
cultures consist mostly of a specific type of friable
callus which forms somatic embryos. These in turn
give rise to single cell derived regenerants. The
efficiency of in vitro culture initiation, the long
term maintenance of totipotent cultures, and the
efficiency of plant regeneration are influenced by the
genotype used, the physiological stage of the donor
plant, by the media composition and physical culture
conditions (Stolarz and Lörz, 1986; Lührs and Lörz,
1987; Zimny and Lörz, 1989). Recent results of in
vitro culture and regeneration studies with rice
genotypes from the breeding program of the
International Rice Research Institute (Los Banos,
Philippines) indicate that there are different,
non-linked genes which influence tissue culture
response, namely I) callus formation, II) somatic
embryogenesis, III) shoot formation and IV) plant
regeneration (Hartke and Lörz, in prep.).

    Limitations with respect to application are seen
mainly in the inability to induce somatic
embryogenesis and to regenerate plants from many or
all important genotypes with sufficient efficiency
applying the same protocol. The culture experiments
are still mainly trial and error. Although culture
conditions are well defined, little information is
available on the molecular events that cause and
accompany the developmental process of somatic
embryogenesis, in vitro organogenesis and differentia-

tion. Understanding these developmental processes on a molecular level could help to explain and to overcome frequently encountered difficulties, such as genotype dependence of embryogenic response. In the *Gramineae*, embryogenic suspensions are, in contrast to callus, extremely difficult to obtain.

Totipotent suspension cultures with long term regenerating capacity have been obtained from a few genotypes only (for reviews see Vasil, 1987; Lazzeri and Lörz, 1988). Suspensions are of importance in providing suitable material for *in vitro* selection, but also as source of material for the isolation of protoplasts. So far, all culture and regeneration experiments with mesophyll protoplasts isolated directly from leaves of graminaceous species have failed. Either no divisions at all, or only one or two cell divisions could be induced. Instead of isolating protoplasts directly from the plant, *in vitro* cultures can also be used as a source for the preparation of protoplasts. It is assumed that protoplasts express the same competence with respect to regeneration as the cells from which they have been isolated. Therefore the establishment of embryogenic and regenerating suspensions is a critical step in a program for in vitro manipulation of cereals.

Experimental conditions for the isolation of cereal protoplasts from suspension cells, culture and regeneration to callus have been established for several species and different genotypes of wheat, maize, rice, barley and triticale. Formation of shoots and 'plantlets' have also been observed in a few cases, but regeneration of fertile plants has so far been achieved reproducibly from rice protoplasts only (Fujimura et al., 1985; Abdullah et al., 1986; Yamada et al., 1986; Kyozuka et al., 1987). Plant regeneration from protoplasts has been described in preliminary reports also for maize (Rhodes et al., 1988). Albino plantlets (Lührs and Lörz, 1988) and more recently also green plantlets (Lazzeri and Lörz, unpublished) were obtained from barley protoplasts. The results obtained with rice, maize and barley demonstrated the suitability of embryogenic cell cultures for the isolation of totipotent protoplasts. However, further improvements towards higher efficiency, better reproducibility, extension to other genotypes, and most important, to all cereal species are necessary.

PROTOPLAST TRANSFORMATION

The incubation of freshly isolated protoplasts with naked DNA is a direct way of gene transfer, and has been shown first with tobacco protoplasts and subsequently also with protoplasts of different graminaceous species. DNA-uptake is stimulated either by PEG-treatment, induced by electroporation, or by treatments utilising both stimuli in slightly modified techniques. Transformation of Gramineae was achieved first with protoplasts of Triticum monoccocum and Lolium multiflorum, and meanwhile also with protoplasts of Zea mays, Oryza sativa, Pennisetum americanum, Panicum maximum and Hordeum vulgare (Table 1). In all cases plasmids have been used which contained a selectable marker gene (coding, e.g. for kanamycin or hygromycin resistance) and a constitutive promoter (35S gene of CaMV or nopalin synthase gene).

In respect to direct gene transfer, cereal protoplasts are considered not to be different from protoplasts of other species. However, the limitation concerning cereal crops still remains the difficulty of regenerating plants from protoplasts. Whereas in maize and rice protoplast-derived and transformed plants have been recently obtained, in the other species mentioned DNA transfer to protoplasts has resulted in transformed callus only. Further progress in plant regeneration from protoplasts will contribute in the future to increase the number of transgenic cereal plants.

MICROINJECTION AS A TOOL FOR DNA TRANSFER

Microinjection systems have been used for transformation in animal culture systems, and recently attention has been given to their use in plants. Technical aspects of microinjection have been covered in detail in recent reviews (Schweiger et al., 1987; Mathias, 1987), so we will consider mainly potential application in grass species.

Table 1. Transformation of protoplasts from Poaceae

species	tissue	method	DNA	test material	regeneration
Lolium	non-morpho suspension cultures	electro	NPT II	callus	no plants
Triticum monococcum	suspension cultures	PEG	NPT II	callus	no plants
maize	embryogenic suspension cultures	electro	NPT II	callus plants	all plants sterile
barley	embryogenic suspension cultures	PEG	NPT II	callus	no plants so far
rice	embryogenic suspension	electro	NPT II	plants	5 plants

1) Potrykus et al. (1985)
2) Lörz et al. (1985)
3) Rhodes et al. (1988), Fromm et al. (1986)
4) Lazzeri and Lörz (1988)
5) Toriyama et al. (1988)

Transformation via microinjection requires that target cells be immobilised, injected, and that they subsequently divide, and ideally regenerate plants. For injection, protoplasts are preferred to intact cells as fine micropipettes are liable to damage in penetrating cell walls. To date, transformation via microinjection of DNA has been reported in tobacco and Medicago sativa (Crossway et al., 1986; Reich et al., 1986).

Grass protoplasts can be immobilised and injected using standard techniques, but very few have resulted high frequencies of transformed cells. Furthermore, they do not thrive under the low-density conditions used for post-injection culture. The latter problem may be overcome using nurse-culture techniques, but the present rates of division and regeneration make the process impracticable. A more attractive alternative is the injection of cells in the small aggregates found in morphogenic suspensions. Microinjection into cells is generally less efficient than injection into protoplasts, because difficulty in locating the nucleus means DNA is usually delivered into the cytoplasma (Crossway et al., 1987), but in grasses this would be offset by the increased chance of injecting competent cells. A problem with the injection of multicellular structures is that not all cells would be accessible, resulting in a chimeric tissue and probably necessitating the use of selectable markers. This removes one of the major attractions of microinjection; that individual cells can be injected and monitored, without the need for mass selections.

Pollen grains and microspore-derived embryos have been mentioned in the previous chapter as "candidates" for microinjection, but all these structures as well as ovules and small zygotic embryos pose problems of nuclear location and low efficiency of post-injection culture.

BIOLISTIC TRANSFORMATION

The biolistic transformation is a recently developed method which employs high velocity microprojectiles to deliver DNA into cells and tissues. This method has great potential for transformation of intact cells and tissues, especially for species which are not transformable by other methods

such as protoplast transformation or Agrobacterium-
mediated gene transfer. Like capillary microinjection, the particle gun method allows for the delivery
of DNA into cells where a cell wall is present.
Therefore the number of possible target cells is
unlimited and theoretically all type of cells can be
transformed. Several examples of transformation
applying the particle gun or a process of electric
discharge to acclerate a DNA-coated particles into the
cells are listed in Table 2.

Suspension cells of different cereal species and
scutellum tissue of maize have been treated and
expression of the introduced genes have been
demonstrated by the activity either of neomycin
phosphotransferase (NPT II) or glucoronidase (GUS)
(Table 2; Brettschneider and Lörz, unpublished). The
major efforts now have to be directed toward
transformation of specific target cells, which are
competent for plant regeneration and provide a high
probability that the foreign genes are stably
integrated and consequently are stably inherited.
Such cells can be found in the zygotic embryos, in
young stages of floral tillers, in pollen grains, in
haploid embryoids, in somatic embryos, or any cells
which give rise to a fertile plant originating from a
single, successfully transformed cell.

POLLEN-MEDIATED GENE TRANSFER

Pollen-mediated transformation has been discussed
for many years as an alternative gene transfer
procedure not restricted by any host range or tissue
culture limitations. The method of incubating pollen
with total genomic DNA or recombinant DNA plasmids,
followed by pollination and seed production, has been
applied to Petunia, Nicotiana, maize (De Wet et al.,
1985; Ohta, 1986), wheat (Hess, 1987; Picard et al.,
1988) and rice (Luo and Wu, 1988). While in the
earlier reports evidence for transformation was based
mostly on phenotypic changes and formal genetic
analyses the recent publications included preliminary
molecular evidence.

In an approach as described by Picard (1988) and
outlined in Fig. 2, we have treated bread wheat, durum
wheat, barley and rice with Agrobacterium tumefaciens
carrying NPT, HYG and GUS as selectable and screenable
markers. In durum wheat 209 seeds derived from

Table 2. BIOLISTIC TRANSFORMATION EXPERIMENTS

species	target tissue	gene	effect	reference
Chlamydomonas	cells	chloroplast ATP-synthetase	restoring photosynthetic activity	Boynton et al. (1988)
yeast	cells	subunit of cytochrome oxidase	restoring mitochondrial function	Johnston et al. (1988)
tobacco	leaves suspensions	GUS and NPT II	expression of reporter gene	Klein et al. (1988)
soybean	zygotic embryos	NPT II	expression of reporter gene	McCabe et al. (1988)
rice	cells	GUS and CAT	expression of reporter gene	Wang et al. (1988)
T. monococcum	cells	GUS and CAT	expression of reporter gene	Wang et al. (1988)
maize	suspension cells	NPT II and GUS	expression of reporter gene	Klein et al. (1988)
maize	zygotic embryos	GUS	expression of reporter gene	Klein et al. (1988)

treated plants were tested on kanamycin and 1055 seeds
on hygromycin containing mediums, 4 and 55 plantlets,
respectively, showed some antibiotic resistance.
However, in none of the plants could any gene activity
nor a positive Southern hybridisation be found. This
indicates that the selection scheme applied was not
very reliable and numerous escapes were found.
Experiments with wheat, barley and rice are in
progress and preliminary results have shown positive
Southern hybridisation in several of the analysed
wheat plants (Langridge et al., in prep.). Even so,
further and more detailed molecular proof and evidence
for reproducibility and general applicability of the
system needs to be provided.

DIRECT DNA DELIVERY TO EMBRYOS AND PLANTS

Another transformation method which does not
involve any tissue culture technique was developed for
Secale cereale (De la Pena et al., 1987). Plasmid DNA
was injected into young floral tillers of rye at a
specific stage during floral development. Although
only a small number of transformants were obtained and
confirmed by biochemical and molecular evidence this
approach provided clear-cut evidence that
transformation of cereals is possible without using
specific vector systems, without using in vitro
cultured cells, and it is independent of isolated
single cells. Present limitations are seen mainly in
the low efficiency, and it has to be seen whether this
procedure can also be applied to other species, such
as wheat, barley, maize or rice.

A highly attractive and simple method of
introducing genes directly into cereal cells was
applied by incubating mechanically isolated wheat
embryos in DNA solution (Töpfer et al., 1989). Mature
embryos isolated from dry seeds take up DNA rapidly by
imbibition, and transient expression of chimeric genes
was demonstrated by assaying NPT- or CAT-activity in
embryos of wheat, barley, rice, triticale, Avena
sativa and maize. Embryos treated this way can
efficiently and easily be regenerated to plants.
Whether the foreign genes are stably integrated and
transmitted to the progeny is presently under
investigation. Instead of regenerating plants
directly from the treated embryos, we have used the
treated embryos of rice and durum wheat to establish
embryogenic, scutellum derived callus cultures. In a

subsequent step these cultures were kept on a
selective medium to select for transformed cells and
eventually transformed regenerants. So far, however,
this approach has not led to stably transformed in
vitro cultures or plants.

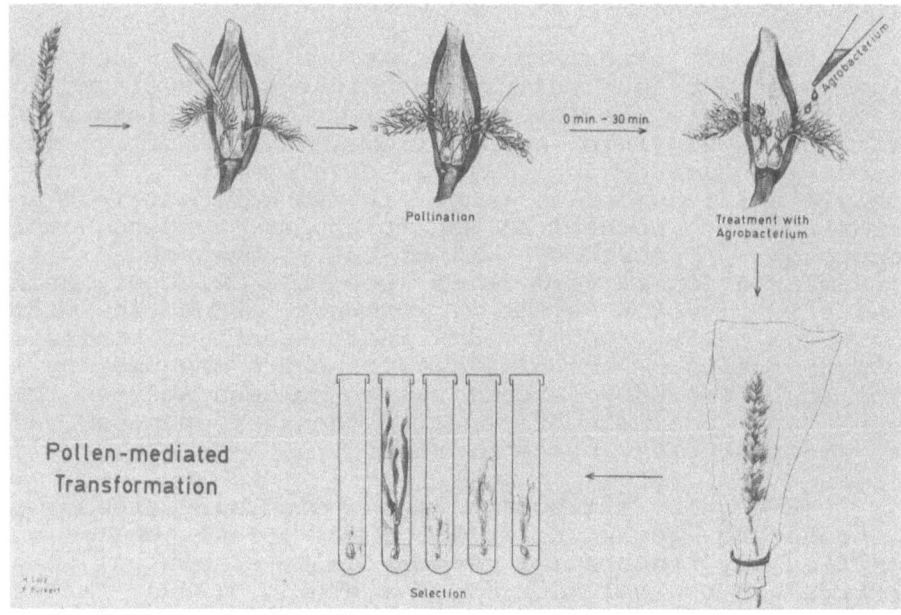

Fig. 2. Pollen-mediated transformation of wheat.
        Florets are emasculated, hand pollinated and
        about 30 min later droplets containing either
        plasmid DNA or agrobacteria carrying the gene
        of interest are added onto the stigma. Ears
        are bagged after the treatment, and seeds
        derived thereof are screened on selection
        medium containing e.g. kanamycin or hygro-
        mycin.

CONCLUSIONS

Different methods are presently under investigation to develop an efficient and widely applicable transformation technique for cereals. One major direction is the development of protocols for plant regeneration from isolated protoplasts. Protoplasts can be transformed efficiently and in addition protoplasts allow one to study all aspects of somatic cell genetics. Transfer of DNA to plants and cultured plant cells has been achieved now with several methods including microinjection, biolistic processes, pollen-mediated transformation, etc. However, the transformation efficiency as well as reproducibility when applied to other graminaceous species are still limited. It can be foreseen, however, that the experimental difficulties will be overcome soon and the transformation technique will not be any longer the limiting step for biotechnological improvement of cereal crops.

ACKNOWLEDGEMENTS

The authors gratefully acknowledge the financial support by the Bundesministerium für Forschung und Technologie, Bonn (grant BCT-365), the European Community, Biological Action Program (grant BAP-0013-D), and the Rockefeller Foundation, Rice Biotechnology Program (grant RF-86058 52). We thank Mrs. Furkert for art work and Mr. Bock for photographic work.

REFERENCES

Abdullah, R., Cocking, E.C., and Thompson, J.A., 1986, Efficient plant regeneration from rice protoplasts through somatic embryogenesis, Bio/Technology, 4: 1087-1090.

Benito-Moreno, R.M., Macke, F., Alwen, A., and Heberle-Bors, E., 1988, In-situ seed production after pollination with in-vitro-matured, isolated pollen, Planta, 176: 145-148.

Boynton, J.E., Gillham, N.W., Harris, E.H., Hosler, J.P., Johnson, A.M., Jones, A.R., Randolph-Anderson, B.L., Robertson, D., Klein, T.M., Shark, K., and Sanford J.C., 1988, Chloroplast transformation of <u>Chlamydomonas</u> using high velocity microprojectiles, <u>Science</u>, 240: 1534-1538.

Crossway, A., Hauptli, H., Houck, C.M., Irvine, J.M., Oakes, J.V., and Perani, L.A., 1986, Micromanipulation techniques in plant biotechnology, <u>Bio/Technology</u>, 4: 320-334.

Datta, S.K., and Wenzel, G., 1987, Isolated microspore derived plant formation via embryogenesis in <u>Triticum</u> aestivum L., <u>Plant Science</u>, 48: 49-54.

De la Pena, A., Lörz, H., and Schell, J., 1987, Transgenic rye plants obtained by injecting DNA into young floral tillers, <u>Nature</u>, 325: 274-276.

De Wet, J.W.J., Bergquist, R.R., Harlan, J.R., Brink, D.E., Cohen, C.E., Newell, C.A., and de Wet, A.E., 1985, Exogenous gene transfer in maize (<u>Zea</u> <u>mays</u>) using DNA-treated pollen, in: "Experimental manipulation of ovule tissues", Chapman, G.P., Mantell, S.H., Daniels, R.W. (eds.), Longman, London, pp. 197-209.

Dunwell, J.M. 1986, Barley. in: "Handbook of Cell Culture", Evans, D.A., Sharp, W.R., and Ammirato, P.V. eds., McMillan, New York, pp. 339-369.

Fromm, M., Taylor, L.P., and Walbot, V., 1986, Stable transformation of maize after gene transfer by electroporation, <u>Nature</u>, 319: 791-793.

Fujimura, T., Sakurai, M., Negishi, T., and Hirose, A., 1985, Regeneration of rice plants from protoplasts, <u>Plant Tissue Culture Letters</u>, 2: 74-75.

Göbel, E. and Lörz, H., 1988, Genetic manipulation of cereals, <u>Oxford Surveys of Plant Molecular and Cell Biology</u>, 5: 1-22.

Graves, A.C.F., and Goldman, S. L., 1986, The transformation of <u>Zea</u> <u>mays</u> seedlings with <u>Agrobacterium tumefaciens</u>. Detection of T-DNA specific enzyme acti- vities, <u>Plant Mol. Biol.</u>, 7: 43-50.

Hauptmann, R.M., Vasil, V., Ozias-Akins, P., Tabaeizadeh, Z., Rogers, S.G., Fraley, R.T., Horsch, R.B., and Vasil, I.K., 1988, Evaluation of selectable markers for obtaining stable transformants in the Gramineae, <u>Plant Physiol.</u>, 86: 602-606.

Hess, D., 1987, Pollen based techniques in genetic
    manipulation,   in: "Pollen-Cytology   and
    Development" Giles, K.L. and Prakash, J., eds.,
    Academic Press, Orlando, pp. 367-395.

Johnston, S.A., Butow, R., Shark, K., and Sanford,
    J.C., 1988, Transformation of yeast mitochondria by
    bombardement of cells with microprojectiles,
    Science, 240: 1538-1541.

Klein, T.M., Fromm, M.E., Gradziel, T., and Sanford,
    J.C., 1988, Gene transfer into Zea mays cells by
    high- velocity microprojectiles is monitored with
    B-glucuronidase marker, Bio/Technology, 6: 559-563.

Klein, T.M., Harper, E.C., Svab, Z., Sanford, J.C.,
    Fromm, M.E., and Maliga, P., 1988, Stable genetic
    transformation of intact Nicotiana cells by the
    particle bombardement process. ˙Proc. Natl. Acad.
    Sci. USA, 85: 8502-8505.

Köhler, F., and Wenzel, G., 1985, Regeneration of
    isolated barley microspores in conditioned media
    and trials to characterise the responsible fac-
    tor, J. Plant Physiol., 121: 181-191.

Kott, L.S., and Kasha, K.J., 1984, Initiation and
    morphological development of somatic embryoids from
    barley cell cultures, Can. J. Bot., 62: 1245-1249.

Kyozuka, J., Hayashi, Y., and Shimamoto, K., 1987,
    High frequency plant regeneration from rice proto-
    plasts by novel nurse culture methods, Mol. Gen.
    Genet., 206: 408-413.

Lashermes, P., Gaillard, A., and Beckert, M., 1988,
    Gynogenetic haploid plants analysis for agronomic
    and enzymatic markers in maize (Zea mays L.),
    Theor. Appl. Genet., 76: 570-572.

Lazzeri, P.A., and Lörz, H., 1988, In vitro genetic
    manipulation of cereals and grasses, Advances in
    Cell Culture, 6: 291-325.

Lörz, H., Baker, B., and Schell, J., 1986, Gene
    transfer to cereal cells mediated by protoplast
    transformation, Mol. Gen. Genet., 199: 178-182.

Lörz, H., Göbel, E., and Brown, P.T.H., 1988, Advances
    in culture and progress towards genetic transforma-
    tion of cereals, Plant Breeding, 100: 1-25.

Lührs, R., and Lörz, H., 1988, Initiation of morpho-
    genic cell suspension and protoplast cultures of
    barley, Planta, 175: 71-81.

Luo, Z.-X., and Wu, R., 1988, A simple method for the
    transformation of rice via the pollen-tube pathway,
    Plant Mol. Biol. Rep., 6: 165-174.

McCabe, D.E., Swain, W.F., Martinell, B.J., and
    Christou, P., 1988, Stable transformation of soy-
    bean (Glycine max) by particle acceleration, Bio/
    Technology, 6: 923-926.

Neuhaus, G., Spangenberg, G., Mittelsten-Scheid, O.,
    Schweiger, H.G., 1987, Transgenic rapeseed plants
    obtained by the microinjection of DNA into micro-
    spore-derived embryoids, Theor. Appl. Genet., 75:
    30-36.

Ohta, Y., 1986, High-efficiency genetic transformation
    of maize by a mixture of pollen and exogenous DNA,
    Proc. Natl. Acad. Sci. USA, 83: 715-719.

Ozias-Akins, P., and Vasil, I.K., 1988, In vitro
    regeneration and genetic manipulation of grasses,
    Physiol. Plant., 73: 565-569.

Picard, E., Jacquemin, J.M., Granier, F., Bobin, M.,
    and Forgeois, P., 1988, Genetic transformation of
    wheat (Triticum aestivum L.) by plasmid DNA uptake
    during pollen tube germination; 7th Int. Wheat
    Genetics Symposium, Cambridge, U.K., (1): 779-781.

Potrykus, I., Saul, M.W., Petruska, J., Paszkowski,
    J., and Shillito, R., 1985, Direct gene transfer to
    cells of a graminaceous monocot, Mol. Gen. Genet.,
    199: 181-188.

Reich, T.J., Iyer, V.N., and Miki, B.L., 1986, Effi-
    cient transformation of alfalfa protoplasts by the
    intranuclear microinjection of Ti plasmids, Bio/
    Technology, 4: 1001-1004.

Rhodes, C.A., Lowe, K.I.S., and Ruby, K-L., 1988,
    Plant regeneration from protoplasts isolated form
    embryogenic maize cell cultures, Bio/Technology,
    6: 56-60.

Rhodes, C.A., Pierce, D.A., Mettler, I.J., Mascaren-
    has, D., and Detmar, J.J., 1988, Genetically trans-
    formed maize plants from protoplasts, Science, 240:
    204-207.

Schweiger, H.-G., Dirk, J., Koop, H.-U., Kranz, E., Neuhaus, G., Spangenberg, G., and Wolff, D., 1987, Individual selection, culture and manipulation of higher plant cells, Theor. Appl. Genet., 73: 769-783.

Stolarz, A., and Lörz, H., 1986, Somatic embryogenesis, in vitro multiplication and plant regeneration from immature embryos of hexaploid Triticale (X Triticosecale Wittmack), Z. Pflanzenzüchtung, 96: 353-362.

Töpfer, R., Gronenborn, B., Schell, J., and Steinbiss, H.-H., 1989, Uptake and transient expression of chimeric genes in seed-derived embryos, The Plant Cell, 1: 133-139.

Toriyama, K., Arimoto, Y., Uchimaya, H., and Hinata, K., 1988, Transgenic rice plants after direct gene transfer into protoplasts, Bio/Technology, 6: 1072-1074.

Vasil, I.K., 1987, Developing cell and tissue culture systems for the improvement of cereal and grass crops, J. Plant Physiol. 128: 193-218.

Wang, Y.C., Klein, T.M., Fromm, M., Cao, J., Sanford, J.C., and Wu, R., 1988, Transformation of rice, wheat and soybean by the particle bombardement method, Plant Molecular Biology, 11: 433-439.

Yamada, Y., Yang, Z.Q., and Tang, D.T., 1986, Plant regeneration from protoplast derived callus of rice, (Oryza sativa), Plant Cell Reports, 5: 85-88.

Zimny, J., and Lörz, H., 1989, High frequency of somatic embryogenesis and plant regeneration of rye (Secale cereale L.), Plant Breeding, 102: 89-100.

TRANSFORMATION AND REGENERATION OF NON-

SOLANACEOUS CROP PLANTS

Maud A. W. Hinchee, Christine A. Newell, Dannette V. Connor-Ward, Toni A. Armstrong, W. Randy Deaton, Shirley S. Sato, and Renee J. Rozman

Monsanto Agricultural Company
An Operating Unit of Monsanto Company
700 Chesterfield Village Parkway
Chesterfield, Missouri 63198

INTRODUCTION

Currently, the production of transgenic tobacco (*Nicotiana tabacum* L.) and petunia *(Petunia hybrida)* plants is considered routine using *Agrobacterium*-mediated transformation methods (Gasser and Fraley, 1989). The basic protocol involves the inoculation of leaf pieces with disarmed *Agrobacterium tumefaciens* strains which confer kanamycin resistance. After a period of *Agrobacterium* and explant co-culture of several days, the growth of the bacterial population is inhibited by bacteriostatic antibiotics and the leaf tissue is induced to regenerate. The induction and development of shoots on leaf explants occurs in the presence of kanamycin, and the majority of the callus and shoots produced contain the gene for kanamycin resistance and express that phenotype. Large numbers of transgenic plants can be produced using this method.

There are several reasons as to why tobacco and petunia are easily genetically engineered using *Agrobacterium* transformation methods. First, these species regenerate quite readily. Many cell and tissue types are competent for regeneration. Regeneration can be via shoot organogenesis directly from single or multiple epidermal and subepidermal cells, or from a intermediate callus initiated from these same cells (Hinchee, personal observations). The regeneration response is therefore very plastic since shoot organogenesis can occur from different cell types, cell numbers, and at different times following explanting of the tissue onto culture medium. Tobacco and petunia are also good hosts for the plant pathogen *Agrobacterium. Agrobacterium* induces tumors readily on tobacco and petunia, and on explant tissues *in vitro* (Horsch, et al., 1985).

Until recently, it was difficult to visualize the transformation process histologically and determine where and when it occurred. If an explant tissue was

Gene Manipulation in Plant Improvement II
Edited by J. P. Gustafson
Plenum Press, New York, 1990

inoculated with *Agrobacterium* but the result was no production of transformed callus or shoots, it was difficult to determine what had blocked the recovery of transgenic tissue. The interaction between regeneration and the transformation process can now be visualized using a histochemical marker system. The ß-glucuronidase (GUS) gene has been inserted into *Agrobacterium* vectors, and transformed cells which express this gene can be identified histochemically since the GUS enzyme will convert a colorless substrate (5-bromo-4-chloro-3 indolyl glucuronide) into an insoluble blue indigo precipitate in the cell cytoplasm (Jefferson, et al., 1987). In histological examinations which we have conducted, it appeared that GUS positive cells can be found anywhere in *Agrobacterium*-inoculated leaf pieces. Any cell type appeared susceptible to transformation and the cells which underwent rapid shoot organogenesis were also frequently GUS-positive and apparently transformed. Thus the "targeting" of *Agrobacterium* transformation was to the regenerable cell population. Kanamycin selection made the petunia transformation procedure more efficient. When the transformed cells also expressed an introduced neomycin phosphotransferase gene, they were capable of growth in the presence of kanamycin. Kanamycin selection pressure applied during the first cell divisions of shoot organogenesis halted the development of shoots from non-transformed cells. Kanamycin was a relative benign selective agent since it inhibited growth without immediately killing the non-transformed tissue. Transformed cells were able to undergo regeneration normally despite the growth inhibition placed on the non-transformed cell population.

The petunia example can be used to identify key requirements in the development of a transformation/regeneration protocol. Identification of the processes which make tranformation of petunia easy also potentially identifies the steps which are problematic in more recalcitrant species. The following is a list of processes which are critical to consider for developing a successful transformation/regeneration protocol. First, the regeneration process itself has to have the potential to be linked with *Agrobacterium* tranformation. The number of cells competent to regenerate and the rate of regeneration are important factors in target identification. In addition, the mechanism by which regeneration occurs influences the transformation efficiency. For example, if only one cell in an explant is transformed, then the chances of recovering a transgenic plant from this cell differs depending on its mode of regeneration. If the regeneration process requires multicellular initiation, then there is a relatively small chance that all cells involved in shoot initiation are transformed and produce a non-chimeric transgenic shoot. The opposite situation exists when a single cell is competent to form a callus, and that callus can be induced later to form shoots. In this latter case, there is an excellent chance of obtaining a transgenic plant from a single transformed cell.

The targeting of transformation to regeneration competent cells requires that the *Agrobacterium* has access to these cells. In some cases, these cells are not close to the wound site through which *Agrobacterium* typically enters the plant tissues. Cell accessibility is an important aspect in transformation efficiency. Knowing which cells regenerate, and when they do so, is important in determining the best method for preparing the explant for transformation. The bacterial inoculation can be performed so that wounding and timing of *Agrobacterium* introduction corresponds with the location and phase of initiation that will be most efficient for targeting. The best strategy is to ensure that the bacteria have the greatest chance of reaching their cellular targets prior to, or during, the first cell divisions in regeneration.

Also critical is the effectiveness of *Agrobacterium* in transforming the target cell population. *Agrobacterium* is more pathogenic on some species than others. Plants such as large seeded legumes and cereals have an apparently low transformation efficiency (Grimsley, et al., 1987; Hobbs, et al., 1989; Byrne et al., 1987), while it appears that species such as tobacco and petunia are quite amenable to tranformation by the plant pathogen *Agrobacterium*.   Differences in *Agrobacterium* pathogenicity translates to differences in *in vitro* transformation efficiency. Transformation efficiency is influenced by the *Agrobacterium*/host plant interaction, and this interaction differs not only between *Agrobacterium* strains and different plant species, but also between *Agrobacterium* and different cultivars of a single species (Hinchee, et al., 1988). If the transformation efficiency is low, then the chance of transforming a regeneration competent cell found at low frequency in the inoculated tissue is also quite low.

The successful selection for transformed regenerating cells and tissues is dependent upon the above factors in addition to the characteristics of the particular selective agent being used. If efficiency of tranformation and of targeting is low, then the chance of developing an efficient selection system relies on the effectiveness of the selectable marker gene in conferring resistance. Selection efficiency is also influenced by the interaction of the transformed cell population with surrounding cells in the selected tissue. If the selective agent is very detrimental to the health of non-transformed cells within a population of non-transformed cells, then it sometimes becomes difficult to recover a small number of transformed cells. In these cases, the concentration of the selective agent needs to be carefully modulated to ensure that the transformed, regeneration competent cells are physiologically supported during the regeneration process. If the selective agent is relatively benign and provides primarily a growth advantage to the transformed cells over the non-transformed cells, then infrequent and small transformed cell populations are much more capable of subsequent growth and development.

In order to develop a transformation/regeneration system for species which have appeared to be recalcitrant to *Agrobacterium*-mediated transformation, the described processes should be taken into account. A thorough knowledge of the mechanism by which regeneration occurs coupled with an analysis of the *Agrobacterium* transformation efficiency and targeting is critical for the necessary linking of the transformation and regeneration components. The GUS histochemical marker system has facilitated investigations into the processes involved in transformation. GUS positive cells can be used as an early histological marker which tells the location and number of cells which are transformed. It also can be used to determine if transformation is targeted to the appropriate cells, and demonstrate if transformed cells are responding positively to selection pressure.

TRANSFORMATION OF RECALITRANT SPECIES

Soybean

The concept outlined above for dissecting a transformation/ regeneration protocol into several critical processes, allowed us to develop an *Agrobacterium-* mediated transformation system for soybean.

   Soybean (*Glycine max (L.) Merr.*) has been a difficult species to genetically engineer using the *Agrobacterium* transformation method.   Soybean is a relatively poor host for *Agrobacterium*. Wang and co-workers (1984) reported that only 2.5% of nearly 1000 soybean cultivars were susceptible to tumor induction by *Agrobacterium*. In addition, the transformation of soybean cells does not always result in the formation of a tumor (Faccioti, et al., 1985). The protocol described below is the only current method for the production of genetically engineered soybean plants using *Agrobacterium*- mediated transformation. The successful development of this protocol was due to the critical evaluation of three parameters in the transformation process. These were: 1) the use of cultivars susceptible to *Agrobacterium* transformation, 2) the development of a regeneration response from soybean cotyledons, and 3) the enrichment for transformed tissue by kanamycin selection. Transformation efficiency, targeting and selection were roadblocks in the tranformation process and optimiziation in these areas was undertaken. The use of the GUS gene, which acts as a histochemical marker for transfomed cells, aided in overcoming these roadblocks.

   Several regeneration methods for soybean were investigated in order to determine their suitability in producing transgenic plants. Somatic embryogenesis was found to be a reliable regeneration system, but slow and inefficient in shoot production (Lazzeri, et al., 1985). The cotyledonary node explant which is highly prolific in shoot organogenesis (Wright, et al., 1986) was not very efficient when adapted to an *Agrobacterium* transformation protocol (Pierson, et al., 1986). Soybean cotyledons were found to be an explant source which were highly efficient in the production of regenerated plants and also to be amenable to *Agrobacterium* transformation protocols. Approximately 100% of all explants were capable of regeneration when placed on an MS or B5 medium containing 5 μM BA, and approximately 70% of the shoots which elongated were rooted and grown to maturity in the green house. This regeneration protocol produced shoots from several soybean genotypes although some genotypic variation in the frequency of the regeneration response was observed.

   The cotyledon regeneration system proved to be a vehicle which allowed for the targeting of transformation to the cells responsible for regeneration. Other regeneration systems for soybean [cotyledonary node (Cheng, et al., 1980) , primary leaf (Wright et al., 1987), and immature embryo (Ranch, et al., 1985)] have not yet been reported as yielding transgenic soybean plants from *Agrobacterium*-mediated transformation. In these systems, *Agrobacterium* transformation may be difficult to target regenerable cells. Although transformation of protoplasts eliminates the targeting issue associated with organ explants, plant regeneration from soybean protoplasts has only been documented once (Wei and Xu, 1988).   Using an *Agrobacterium tumefaciens* strain pTiT37-SE :: pMON9749 which confers GUS activity to transformed cells, it was possible to identify which cells were transformed in a variety of soybean explants. Out of the explant systems mentioned above, only the soybean cotyledon demonstrated a relatively high number of transformed sectors 3 weeks after inoculation.   Each individual GUS positive sector was the result of several rounds of division from an intitial GUS positive cell or cells. In addition, some of the GUS-positive sectors were observed in the regeneration competent region which was located adjacent to an incipient shoot primordium. In other explants such as immature embryos, leaves and cotyledonary nodes, none of the transformed GUS positive sectors were localized in regions known to be competent for regeneration based on histological analysis. The utility of the cotyledon explant in targeting transformation to cells competent to regenerate was verified when several GUS positive shoot primordia were also observed at the same 3 week time point (Hinchee, et al., 1988).

Soybean is not easily transformed with *Agrobacterium* (Byrne, et al., 1987, Kudirka, et al., 1986). In order to maximize the transformation of cells in cotyledon explants, a screen for susceptibility to *Agrobacterium* infection was undertaken. One hundred soybean cultivars, chosen for their genetic diversity, were screened for their *in vitro* response to *A. tumefaciens* infection. This screen involved taking hypocotyl explants from the different soybean cultivars and selecting for kanamycin resistant callus after transformation with an *A. tumefaciens* strain pTiT37-SE :: PMON273 conferring resistance to kanamycin. This assay provided information as to which soybean cultivars were capable of producing the greatest amount of transformed callus after *Agrobacterium* inoculation. Out of 100 cultivars examined, only three were determined to be susceptible in repeated tests. The cultivars 'Delmar', 'Maple Presto' and 'Peking' produced nearly 4 times the number of transgenic callus as did the majority of the cultivars after *Agrobacterium* transformation and selection with kanamycin (Hinchee, et al., 1988). The cultivar 'Peking' had been previously identified as a genotype which was susceptible to *Agrobacterium* based on tumor formation (Byrne, et al., 1987), and this cultivar's *in vitro* response to *Agrobacterium* correlated with it's *in vivo* response.

Although several susceptible cultivars of soybean were identified, not all responded equally in terms of regeneration. The 'Delmar' and 'Maple Presto' explants did not produce as many shoots after *Agrobacterium* infection as did the cultivar 'Peking', so 'Peking' was chosen for linking *Agrobacterium* transformation with the regeneration response.

First, transformation of cotyledon explants was demonstrated and optimized. It was difficult to use the method of counting kanamycin resistant calli per inoculated explant as a way of measuring transformation efficiency in soybean cotyledons. This was because the large cotyledons explants were not sensitive to kanamcyin, and produced approximately the same amount of callus regardless of antibiotic concentration. In order to determine the efficacy of *Agrobacterium* transformation of cotyledon explants, the GUS gene was used as a histochemical marker for presence and growth of transformed cells. Soybean cotyledon explants were inoculated *Agrobacterium* containing pMON9749, and GUS activity was localized 3 weeks after *Agrobacterium* infection. In free-hand sections of the cotyledons, multiple GUS-positive callus sectors were identified in callus associated with the excision wound site. The number of these GUS positive sectors increased when the cotyledons were cultured on medium containing 200-300 mg/l kanamcyin instead of on medium without kanamycin (Hinchee, et al., 1988). This demonstrated that the number of actively growing, transformed cells could be enriched by kanamycin selection, even though the total callus mass remained the same. This enrichment for the growth of transformed cells increased the competitiveness of the transformed cells relative to the nontransformed cells, and allowed the growth of more transformed cells than would have been obtained without selection.   This kanamycin enrichment aided in the production of transgenic shoots by increasing the chance that a transformed cell could grow into a tissue capable of multicellular initiation of shoot primordia.

The successful targeting of transformation to regeneration, coupled with kanamycin enrichment, proved beneficial for obtaining transgenic soybean plantlets. Transgenic soybean plantlets were identified in 6 separate experiments in which immediate kanamycin selection was utilized following *Agrobacterium* co-culture. Of 128 soybean plantlets produced from kanamcyin-selected soybean cotyledons, 6% were identified as being transgenic. This was not the case when inoculated cotyledons were cultured on medium without kanamycin. One hundred shoots produced independently from 202 non-selected cotyledon explants did not yield a single transgenic plantlet.

Using the transformation/regeneration protocol described above, Roundup® tolerant soybean plants have been produced. Roundup® is an herbicide developed at Monsanto which is highly effective and non-selective. Soybean cotyledon explants which were inoculated with *A. tumefaciens* pTiT37-SE::pMON894. The vector pMON894 confers Roundup® tolerance because it contains a chimeric gene consisting of a coding sequence for a mutant form of petunia EPSP synthase cDNA driven by the CaMV 35S promoter with a duplication of the enhancer region (Hinchee et. al., 1988). Leaves of pMON894 transgenic soybean plants produced the altered EPSP synthase and showed enzyme activity which was resistant to Roundup®. When sprayed in the greenhouse with Roundup®, the plants demonstrated complete vegetative protection at field dose rates.

The basic principles outlined in the introduction, and exemplified in the development of the soybean transformation protocol, can be applied to other recalcitrant species. Not all species require as rigorous an approach to the development of a transformation system as did soybean. The rigor depends on what roadblocks to successful transformation are presented. The approach taken for developing transformation systems for potato (*Solanum tuberosum* L.), cotton (*Gossypium hirsutum*) and flax (*Linum usitatissimum* L.) using *Agrobacterium* will be described to illustrate that certain aspects need to be optimized in some crops and not in others.

Potato

Not all solanaceous plants have the same plasticity in culture as do tobacco and petunia. Potato, especially key commercial cultivars such as 'Russet Burbank', has proved less amenable to *Agrobacterium*-mediated tranformation. Regeneration from 'Russet Burbank' can be obtained relatively easily from several explants such as stems, leaves and tubers (Silva, et al., 1985; De Block, 1988; Visser, et al., 1989). The regeneration response, however, relies on the formation of shoots directly initiated from the explant or from a determinate (not capable of sustained growth) intermediate callus. Therefore the transformation has to be directly targeted (such as in soybean) to the cells which are competent to regenerate. In our hands, 'Russet Burbank' tubers and leaves do produce much transformed callus after *Agrobacterium* inoculation. This appeared to be the result of low transformation efficiency and poor tissue health after *Agrobacterium* introduction. As a result, the leaf and tuber explant sources did not produce transgenic plants after *Agrobacterium* inoculation. A successful transformation/regeneration protocol for 'Russet Burbank' was developed by identifying a regenerable explant with less sensitivity to *Agrobacterium*, and by taking steps to improve the transformation efficiency of the *Agrobacterium* infection process. The stem explant proved to be more readily transformed by *Agrobacterium* and less sensitive to inoculation with this plant pathogen. Approximately 50% of the explants produced callus and shoots on kanamycin containing medium after inoculation. However, the transformation targeting was not very precise. Not all calli were competent to form shoots, and not all shoots were transgenic. This can be attributed to inaccessibility of the regeneration competent cells to *Agrobacterium*. Shoots can be formed directly from the stem, and the cells involved in this process were rarely targeted by *Agrobacterium* transformation. Kanamycin selection improved the efficiency of transgenic plant production, but the antibiotic at the concentrations tested did not completely inhibit shoot organogenesis from non-transformed cells. Although transgenic 'Russet Burbank' plants can be routinely produced, targeting and and selection efficiency are two aspects that still require optimization in the 'Russet Burbank' transformation protocol.

The 'Russet Burbank' transformation/regeneration protocol has been used to produce potato plants which are resistant to potato viruses PVX and PVY. This was achieved by transferring chimeric genes containing the CaMV 35S regulatory sequence in front of coding sequences for the respective coat proteins for these two viruses. The potato plants expressing these coat proteins are completely resistant to PVY and PVX infection and systemic spread, and field tests are currently being conducted to examine virus protection in the field and resulting tuber yield.

Cotton

Transformation of cotton with *Agrobacterium* at first appeared problematic since seedling explants inoculated with *Agrobacterium* showed significant browning and poor health subsequent to inoculation. The *Agrobacterium* strain first utilized was derived from GV3111. When we utilized the A208-derived disarmed strain pTiT37-SE, a higher transformation efficiency and improved tissue health was observed. When this improved *Agrobacterium* inoculation procedure was applied to explants which callused readily, a single transgenic callus was produced on approximately 30 - 50 % of the explants placed on medium containing 25-100 mg/l kanamycin. Plant regeneration can be obtained from certain cotton cultivars via somatic embryogenesis from a maintainable callus (Trolinder and Goodin, 1988). When transformation was combined with this regeneration protocol for the cotton cultivar 'Coker 310', transgenic callus and somatic embryos were obtained. Regeneration from a callus compensated for an initally low transformation efficiency. This was because any cell at the cut edge of a cotton hypocotyl segment could be transformed and subsequently regenerated. This resulted in a high probability for producing transgenic plants using this system. However, low efficiency production of transgenic callus after *Agrobacterium* infection was the primary roadblock in the protocol. It was found that the method by which the *Agrobacterium* was introduced and removed from the explant tissue after inoculation significantly altered the transformation efficiency . There is a delicate balance between the size of the tissue, the amount of bacteria introduced to the tissue, and the thoroughness by which the bacteria are removed from the explants. In transformation/regeneration protocols in which the explants exhibit extreme sensitivity to *Agrobacterium* infection, precise optimization of the inoculation procedures are needed in order to obtain uniformly good results.

Using the above transformation/regeneration protocol, transgenic cotton plants have been produced which contain genes of agricultural value. Roundup® tolerant cotton plants have been produced; and plants which demonstrate vegetative protection against field dose rates of Roundup® are currently being tested in the field. Additionally, cotton plants have been produced which contain a chimeric gene encoding for the *Bacillus thuringiensis* endotoxin. These plants have demonstrated vegetative resistance to the lepidopteran class of insects and will be field tested in the near future.

Flax

Flax is a good natural host for *Agrobacterium*. None of the detrimental responses to bacterial inoculation seen in cotton and potato were observed in inoculated flax explants. A highly efficient shoot organogenesis protocol for flax also existed (McHughen and Schwartz, 1984). Therefore it seemed simple to link a highly prolific regeneration system with *Agrobacterium*-mediated gene transfer in order to obtain transgenic flax plants. However, despite utilization of several *Agrobacterium* strains and different inoculation procedures, no transgenic shoots were obtained. This was because the regeneration system involved direct shoot organogenesis along the surface of a seedling hypocotyl segment. The regeneration competent cells were not being targeted by *Agrobacterium* in this explant system. The regeneration competent cells were epidermal cells and the initiation of the regeneration response occurred within the first day after explant preparation and inoculation. In order to address this targeting issue, the method by which regeneration occurred was altered. A different explant, the cotyledon, provided a method by which the shoot formation could be linked to transformation. The cotyledon provided an intermediate callusing step, which allowed more cells to form shoots in addition to delaying the shoot initiation process. This increased the chance that a kanamycin selected callus had the potential to regenerate. The transformation/regeneration protocol using cotyledons allowed for the recovery of transgenic flax plants. This same targeting issue has been addressed in another way. Epidermal peels were used as the explant tissue in order to provide access to the regeneration competent cells of the epidermis to *Agrobacterium* and transgenic plants were produced (Jordan and McHughen, 1988). By using either of these two transformation protocols, transgenic plants were produced which contained genes for Roundup® tolerance. The plants have demonstrated vegetative tolerance to field dose rates of Roundup and have been field tested in Canada.

SUMMARY

Not all species will be efficiently transformed using *Agrobacterium*-mediated methods. However, there is a good chance that careful identification of processes which limit the use of *Agrobacterium* will allow these limitations to be overcome and result in improved transformation efficiency. Other methods for transfomation such as free DNA transfer via the particle gun (Klein et al. 1988) or by electroporation into protoplasts (Rhodes et al. 1988) are being developed for the transformation of species or genotypes highly recalcitrant to *Agrobacterium* tranformation, but as yet these methods have not been developed to the extent that they are routine or highly efficient. *Agrobacterium*-mediated plant transformation is used routinely in several species because it has proven reliable and efficient. In the future, *Agrobacterium*-mediated transformation will provide the most common avenue for the introduction of value-added agronomic traits into crop species.

REFERENCES

Byrne, M. C., McDonnell, R. E., Wright, M. S., and Carnes, M. G., 1987, Strain and cultivar specificity in the *Agrobacterium*-soybean interaction, Plant Cell, Tissue and Organ Cult., 8: 3-15.

De Block, M., 1988, Genotype-independent leaf disc transformation of potato (*Solanum tuberosum*) using *Agrobacterium tumefaciens*, Theor. Appl. Genet., 76: 767-774.

Grimsley, N., Hohn, T. Davies, J. W., and Hohn, B., 1987, *Agrobacterium*-mediated delivery of infectious maize streak virus into maize plants, Nature, 325: 177-179.

Hinchee, M. A. W., Connor-Ward, D. V., Newell, C. A., McDonnell, R. E., Sato, S. E., Gasser, C. S., Fischhoff, D. A., Re, D. B., Fraley, R. T., and Horsch, R. B., 1988, Production of transgenic soybean plants using *Agrobacterium*-mediated DNA transfer, Bio/Technology, 6: 915-922.

Hobbs, S. L. A., Jackson, J. A., and Mahon, J. D., 1989, Specificity of strain and genotype in the susceptibility of pea to *Agrobacterium tumefaciens*, Plant Cell Rep., 8: 274-277.

Horsch, R. B., Fry, J. E., Hoffmann, N. L., Eichholtz, D., Rogers, S. G., and Fraley, R. T., 1985, A simple and general method for transferring genes into plants, Science, 227: 1229-1231.

Jefferson, R. A., 1987, Assaying chimeric genes in plants: the GUS gene fusion system. Plant Mol. Biol. Rep., 5: 387-405.

Jordan, M. C., and McHughen, A. 1988, Glyphosate tolerant flax plant from *Agrobacterium* mediated gene transfer, Plant. Cell Rep., 7: 285-287.

Klein, T. M., Fromm, M., Weissinger, A., Tomes, D., Schaaf, S., Sletten, M., and Sanford, J. C., 1988, Transfer of foreign genes into intact maize cells with high-velocity microprojectiles, Proc. Natl. Acad. Sci. USA, 85: 4305-4309.

Kudirka, D. T., Colburn, S. M., Hinchee, M. A., and Wright, M. S., 1986, Interactions of *Agrobacterium tumefaciens* with soybean (*Glycine max* (L.) Merr.) leaf explants in tissue culture, Can. J. Genet. Cytol., 28: 808-817.

Lazzeri, P. A., Hildebrand, D. F., and Collins, G. B., 1985, A procedure for plant regeneration from immature cotyledon tissue of soybean, Plant Mol. Biol. Rep., 5: 160-167.

McHughen, A., and Swartz, M., 1984, A tissue-culture derived salt-tolerant line of flax (*Linum usitatissimum*), J. Plant Physiol.,117: 109-117.

Pierson, P. E., Schaefer, T. J., Wright, M. S., White, E. L., Hinchee, M. A., McDonnell, R. E., and Carnes, M. G., 1986, In Vitro transformation of soybean, (*Glycine max* L.) Merrill, using *Agrobacterium tumnefaciens*, in: "VI International Congress of Plant Tissue and Cell Culture", D. A. Somers, B. G. Gengenbach, D.D. Biersboer, W. P. Hackett and C.E. Green, eds., Intern. Assoc. for Plant Tissue Culture, Minneapolis, pp.128.

Ranch, J. P., Oglesby, L, and Zielinski, A. C., 1985, Plant regeneration from embryo-derived tissue cultures of soybeans, In Vitro Cell. & Dev. Biol., 21: 653-658.

Rhodes, C. A., Pierce, D. A., Mettler, I. J., Mascarenhas, D., and Detmer, J. J., 1988, Genetically transformed maize plants from protoplasts, Science, 240: 204-207.

Silva, G. H., 1985, *In vitro* storage of potato tuber explants and subsequent plant regeneration, HortSci., 20: 139-140.

Trolinder, N. L, and Goodin, J. R., 1988, Somatic embryogenesis in cotton (*Gossypium*) I. Effects of source of explant and hormone regime, Plant Cell, Tissue and Organ Cult., 12: 31-42.

Visser, R. G. F., Jacobsen, E., Hesseling-Meinders, A., Schans, M. J., Witholt, B., and Feenstra, W. J. 1989, Transformation of homozygous diploid potato with an *Agrobacterium tumefaciens* binary vector system by adventitious shoot regeneration on leaf and stem segments, Plant Mol. Biol., 112: 329-337.

Wang, L., Yin, G., Luo, J., Lei, B., Wang, J., Yao, Z., Li, X., Shao, Q., Jiang, X. and Zhou, Z., 1984, Tumor induction and gene transfer in annual species of *Glycine* by *Agrobacterium tumefaciens*, in "Proceedings Second U.S. - China Soybean Symposium", D. Wong, D. Bothel, R. Nelson, W. Nelson, and W. Wolf, eds., Office of International Cooperation and Development, USDA, Washington, D.C., pgs. 195-198.

Wei, Z., and Xu, Z., 1988, Plant regeneration from protoplasts of soybean (*Glycine max* L.), Plant Cell Reports, 7: 348-351.

HAPLOIDS IN CEREAL IMPROVEMENT: ANTHER AND MICROSPORE CULTURE

K.J. Kasha, A. Ziauddin and U.-H. Cho

Crop Science Dept., University of Guelph, Guelph, Ontario,
Canada,   N1G 2W1

I.   INTRODUCTION

For years, cereals and monocots in general have been poor
responders in in vitro haploid production systems when compared to
many dicot crops. The response frequency was low with very strong
genotype effects and often large numbers of albino progeny.
However, we are now seeing evidence of a breakthrough that can
provide embryo frequencies equivalent to those obtained in some
Brassicas and other dicot crops. These results are with specific
genotypes of barley (Hordeum vulgare L.) (Hunter, 1988; Hunter et
al., 1989; Olsen, 1987), where green plant production has increased
up to 100 fold. As a result, further improvements in wheat
(Triticum aestivum L.) and other cereals will likely follow.   In
addition, plants have been obtained from isolated microspore
culture in barley, wheat, rice (Oryza sativa L.) and corn (Zea mays
L.). These cultures open up new possibilities for transformation,
mutation and selection in cereals.

Relative to other haploid production systems, it has been
found that some of the diploid plants in the "hap" gene system in
barley are actually spontaneously doubled haploids (Hagberg and
Hagberg, 1987), making the system more useful for breeding.   In
relation to wide hybridization followed by chromosome elimination,
haploids or polyhaploids have been produced for seven additional
Hordeum species after hybridization with Hordeum bulbosum (2x or
4x) (Subrahmanyam and von Bothmer, 1987). Perhaps more intriguing
is that polyhaploids of wheat have been produced by chromosome
elimination when pollinated with such divergent species as sorghum
(Sorghum bicolor L.) and corn (Laurie and Bennett, 1988;
O'Donoughue et al., 1988; Comeau et al., 1988). Chromosomally

Gene Manipulation in Plant Improvement II
Edited by J. P. Gustafson
Plenum Press, New York, 1990

unstable hybrids from such wide crosses are also being examined as
a means of transferring genetic material (Jörgensen et al., 1986;
Linde-Laursen and von Bothmer, 1986).

Numerous reviews are available and may be referred to for
details on methods of haploid production in cereals (Kasha and
Seguin-Swartz, 1983; Jensen, 1986; Liang and McHughen, 1987), more
specifically, on in vitro culture methods (Dunwell, 1985; Hu and
Yang, 1986; Hu, 1986; Wenzel and Foroughi-Wehr, 1984; Heberle-Bors,
1985; Prakash and Giles, 1987), on pollen morphogenesis and
ultrastructure (Sunderland and Huang, 1987; Sangwan and Sangwan-
Norreel, 1987; Huang, 1986) and on applications (Choo et al., 1985;
Baenziger et al., 1984). Therefore, in this paper we will
summarize recent advances in anther and isolated microspore culture
and discuss three areas of concern; namely pollen embryogenesis,
albinism and genotype effects.

II. ANTHER CULTURE

When examining the development of anther culture systems in
the different cereals, there is a striking similarity of factors
that have been shown to be important to response. However, equally
striking is the variability in response among genotypes within a
species and between different species under similar conditions.
These differences would appear to be due to both inherited and
environmentally influenced developmental differences. The latter
no doubt have a genetic base reflected in levels of hormones,
enzymes etc. in the maternal anther tissue and microspores. Thus,
the first limiting factor in anther culture success is how well the
anther donor plants are grown. Good fertility, growth environment
and pest control are critical as noted in many reports.

In this section we shall examine the factors involved in the
improvement of anther cultures in barley, wheat and rice, and by
drawing parallels to their evolvement, pinpoint factors that either
have or could be examined in other monocots. The recent
improvements in barley with changes to nitrogen source, sugars and
starch provide an obvious new area for investigation.

A. Barley

The recent improvement in anther culture success (Hunter,
1988; Hunter et al., 1989; Olsen, 1987) has been achieved by the
right combination of media components, environment and genotype.
An order of magnitude change in frequencies means that success is
measured in green plants per anther as opposed to the previous rate
of per 100 anthers (Table 1). Albino plant frequencies are still
high and success with a broad range of genotypes remains to be
demonstrated as the winter barley cultivar Igri has been used in

Table 1. Maximum frequencies of green plants reported via anther culture in barley and wheat based on a few selected references.

Reference	Medium	Yield per 100 anthers	
		Calli (embryos)	Green plants
BARLEY:			
Foroughi-Wehr & Freidt (1982)	mod. MS	7.3	1.3
Kao & Horn (1982)	Ficoll	309	10
Szarejko & Kasha (1989)	BAC 3	414	14.4
Sorvari and Schieder (1987)	EDAM	1730	43.3
Olsen (1987)	Mod. MS	--	464
Hunter et al. (1988)	FHG	--	600
WHEAT:			
Jones & Petolino (1987)	N6	8.6	4.2
Picard et al. (1987)	Miller	28	3.8
Szakacs et al. (1988)	Potato	19.6	5.9
Liang et al. (1987)	85D3	--	71.5
Kasha & Oro (unpub.)	BAC 1	71	46.7
Ouyang et al. (1987)	Potato II	1983	70.6

recently published reports. Close comparisons are not valid because of differences in plant growth environment, media and genotypes. The first evidence of a breakthrough came from the 1987 European patent application No. 87200773.7 by Shell International for C.P. Hunter. The patent focused attention on the form and concentration of sugar in culture media. As we understand it, the

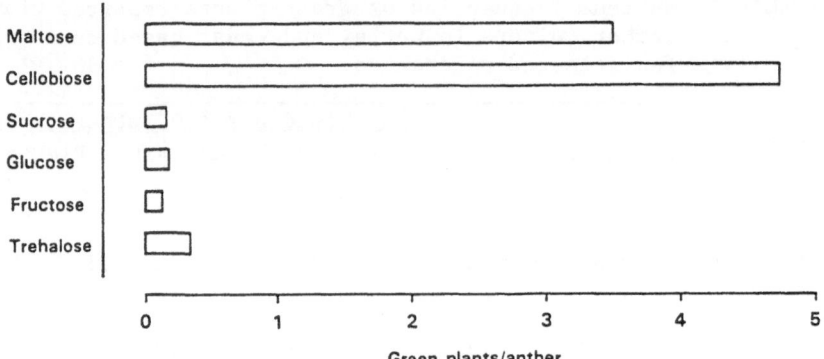

Fig. 1. Comparison of 6 sugars as carbon source.
Sugar concentration 150 mM; 8 replicates; 12 anthers/10 ml
medium. (from Hunter et al., 1989)

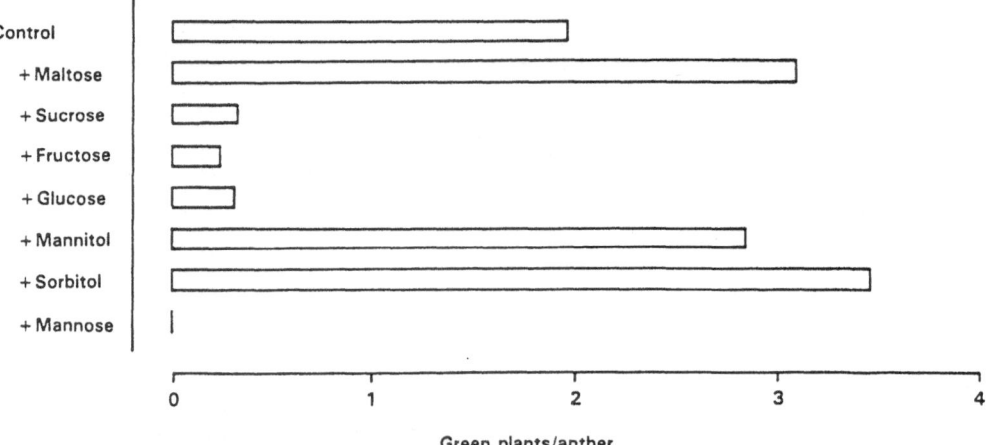

Fig. 2. Effects of additional sugars on plant yield on 100 mM
maltose. Each treatment contained 100 mM maltose plus the
named sugar at 50 mM; 8 replicates; 12 anthers/10 ml
medium. (from Hunter et al., 1989).

Table 2. Basic media used in cereal anther culture, hormones omitted.

| Constituents | Medium (mg/l) | | | | | |
	MS	FHG	N6	BAC3 [*]	Kao. [**]	Pot.II [***]
KNO_3	1900	1900	2830	2600	2200	1000
NH_4NO_3	1650	165	--	--	600	--
$(NH_4)_2SO_4$	--	--	463	400	67	100
KH_2PO_4	170	170	400	170	170	200
$CaCl_2.2H_2O$	400	440	166	600	445	--
$MgSO_4.7H_2O$	370	370	185	300	310	125
$NaH_2PO_4.H_2O$	--	--	--	150	75	--
$FeSO_4.7H_2O$	27.8	--	27.8	--	--	27.8
$FeNa_2$ EDTA	37.3	40	37.3	--	--	37.3
Sesquetrene 330 Fe	--	--	--	40	28	--
KCL	--	--	--	--	150	35
$MnSO_4.4H_2O$	22.3	22.3	4.4	5.0	10	--
$ZnSO4.7H_2O$	8.6	8.6	1.5	2.0	2.0	--
H_3BO_3	6.2	6.2	1.6	5.0	3.0	--
KI	0.83	0.83	0.8	0.8	0.75	--
$NaMoO_4.2H_2O$	0.25	0.25	--	0.25	0.25	--
$CuSO_4.5H_2O$	0.025	0.025	--	0.025	0.025	--
$CoCl2.6H2O$	0.025	0.025	--	0.025	0.025	--
Inositol	100	100	--	2000	100	--
Thiamine.HCl	0.4	0.4	1.0	1.0	2.0	1.0
Pyridoxine.HCl	0.5	--	0.5	0.5	1.0	--
Nicotinic acid	0.5	--	0.5	0.5	1.0	--
Glycine	2.0	--	2.0	--	--	--
Glutamine	--	730	--	--	--	--
Casamino acids	--	--	--	--	250	--
Casein hydro-lysate	--	--	--	300	--	--
Sucrose	--	--	90000	60000	42000	90000
Glucose		--	--	17500	25000	--
Maltose	--	62000	--	--	--	--
Xylose	--	--	--	--	150	--
Ficoll-400	--	200000	--	300000	300000	--
pH		5.6	5.8	6.2	6.0	5.8

[*]   BAC3 medium also contains (mg/l): $KHCO_3$(50); ascorbic a. (1.0); citric a. (10); pyruvic a. (10); $AgNO_3$ (10).

[**] Kao's medium also contains (mg/l): ascorbic a. 1.0); citric a. (10); Na pyruvate (5.0); fumaric a. (10.0); folic a.(0.2); D-Ca pantothenate (0.5); p-Amino benzoic a. (0.01); biotin (0.005); coconut H_2O (10 ml).

[***] Potato II medium (Chen, 1986) includes 10% potato extract.

patent covers the use of sugar concentrations less than 0.03 M of
sucrose, glucose, or oligosaccharides and polysaccharides
containing two or more glucose residues. Subsequent reports
(Hunter, 1988; Hunter et al., 1989) have provided evidence that
high concentrations of sucrose or its breakdown products glucose
and fructose are inhibitory to anther induction and embryo
formation from microspores. Maltose and cellobiose were much
superior to sucrose (Fig. 1). When sucrose was added with maltose
or cellobiose, embryo formation and green plant production was
reduced (Hunter et al., 1989) (Fig. 2). In comparing the
concentrations and breakdown product concentrations of sucrose
(glucose and fructose) and maltose (glucose) Hunter (1988) observed
much higher glucose concentrations from sucrose. He hypothesized
that the more rapid metabolism of sucrose leads to toxic levels of
glucose in the medium. This is consistent with the patent
application data (Hunter, 1987) showing improved anther culture
response with low levels of sucrose and glucose.

     Sorvari (1986) had proposed that replacement of agar by barley
starch led to improved embryo production and plant regeneration
during barley anther culture. These results were further improved
by replacing sucrose with the much less metabolizable sugar
melibiose (Sorvari and Schieder, 1987). The best concentration was
80 to 120 g/l of melibiose with 6% barley starch. They concluded
that the replacement of sucrose by melibiose led to a higher ratio
of green to albino plants and that the main role of melibiose was
as an osmoticum, important for embryo formation. Hunter et al.
(1989) have shown that melibiose by itself was not an improvement.
The medium EDAM (Enzymatically Digestible Agar-free Medium) of
Sorvari and Schieder (1987) is essentially an N6 medium (Table 2)
in which the starch and melibiose have replaced agar and sucrose
and 1.0 mg/l IAA and BAP have been used as hormones. Since the
main breakdown product of barley starch is maltose, it is likely
that starch serves as a slow release sugar source as well as
replacing agar as a medium thickner. A possible role of the less
metabolizable sugars may be to permit better uptake of salt (K^+,
Na^+ etc.), amino acids and organic acids instead of glucose or
sucrose. Both of the above groups would maintain the osmotic
balance required for the cells but the salts and amino acids within
the cells could be important factors for embryogenesis and
regeneration.

     There are strong interactions among media components and of
media components with genotypes. Thus, other media components need
to be re-examined when changes are made to a medium. Hunter (1988)
has reexamined the basal media (salts and minor elements), auxins,
cytokinins and amino acids in conjunction with maltose in deriving
the FHG medium (Table 2). The FHG medium is essentially on MS
medium with lower NH_4NO_3 and high glutamine. In comparing the
auxins IAA, NAA and 2,4-D, he only observed embryo formation with

IAA. In further studies with IAA, the control with no IAA was better than higher concentrations such as 1 mg/l. Only two of five cytokinins had a beneficial effect with the best one being benzylamino-purine (BAP) with Kinetin being the other responsive cytokinin. Both Hunter (1988) and Olsen (1987) have observed beneficial effects of high glutamine concentrations (5 mM.), presumably as a more suitable N source in the medium. Olsen used the same (FHG) medium as Hunter except with sucrose instead of maltose and also observed that either asparagine or glutamine was beneficial in replacing high NH_4 levels.

It is important to note that the results of Hunter and Olsen were obtained with the highly responsive winter barley cultivar Igri and results with other genotypes have not yet been published. It may be noted that results with other genotypes as well as Igri were much lower in the study of Knudsen et al. (1988). Hunter (1988) mentioned that doubled haploids have been produced with other genotypes for breeding programs.

For many years, 2,4-D has been considered by most researchers as essential for induction in cereal tissue culture and its removal important for regeneration. These recent improvements in barley anther culture (Hunter, 1988; Olsen, 1987) were developed from the Foroughi-Wehr medium (essentially MS medium with low NH_4NO_3, Table 2) which contained IAA and no 2,4-D. Kao (1988) has used low concentrations of 2,4-D and varied the concentration (0.5-1.0 mg/l) with genotypes and achieved good results. Our own research utilized high concentrations (8 mg/l) of 2,4-D to induce high frequencies of calli (Marsolais and Kasha, 1985). More recently, working with more genotypes, we have lowered the 2,4-D levels to 2 mg/l in the BAC 3 medium (Table 2).

It is of interest to note that 10-fold improvements in plant production were correlated with embryo formation in two genotypes and that IAA was not a satisfactory replacement for 2,4-D in BAC3 medium (Szarejko and Kasha, 1989). The BAC3 medium also had reduced levels of NH_4^+ and higher KNO_3. Thörn (1988) has used N6 medium with 10% sucrose and observed that 2% polyethylene glycol increased survival and development of embryos from barley microspores. Thus, it appears that it is the combinations of factors that may be important and these requirements may differ with media, environment and genotypes.

B. Wheat

There are various reviews of wheat anther culture (Hu, 1986; Liang and McHughen, 1987) but the most extensive review of the earlier literature is that of Ouyang (1986). Much of the earliest research was carried out in China and France. While the frequencies of green plants produced (Table 1) are not as striking as those in barley, they have been utilized in breeding programs. In China, at least 20 cultivars have been produced in wheat (Hu, 1986). Again, many factors such as media components, plant growth conditions, genotypes, culture temperatures, etc. have been found to be significant. However, the most recent procedures developed for barley (FHG and BAC₃ media) do not appear to work with wheat.

One of the more striking factors in wheat has been the temperature at which anthers are cultured. Ouyang et al. (1987) further demonstrated that the optimum culture temperature for a genotype could vary depending upon the anther donor plant growth conditions. This influence of culture temperature upon anther response has been verified by other researchers (Huang, 1987; Li et al., 1988; Glover and Kasha (unpublished) (Table 3). In barley, higher culture temperatures are also benficial but only up to 28°C (Marsolais et al., 1986).

In contrast to barley and most species, cold pretreatments of spikes have not proven to be effective for wheat (Marsolais et al., 1984; Ouyang, 1986), but can be used to store spikes for short periods. Strong response differences between field and greenhouse grown plants have been noted (Ouyang, 1986; Ouyang et al., 1987) with field grown plants being superior. Bjornstal et al. (1988) compared growth room, phytotron and field grown plants and found that the growth room grown plants gave inferior anther culture response.

Differences in media requirements appear to exist between wheat and barley. Most progress in barley has been with an MS media base while in wheat it has been the N6 medium base or the potato medium (see Table 2). The low sucrose levels of Hunter (1987) have not improved wheat anther culture and most media used for wheat contain between 6 and 10% sucrose. Kelly (1986) observed an improved response when 1.75% glucose was added to media containing 6% sucrose. It may be that wheat microspores can withstand higher levels of glucose and fructose compared to barley, but this remains to be studied.

Nitrogen sources and levels have been studied by Chinese researchers (see Ouyang, 1986) and are one of the main differences between MS and N6 media. Feng and Ouyang (1988) examined the effects of KNO_3 concentrations from 0 to 35 mM on induction and green plant regeneration in wheat. The optimum level for callus induction was 20 mM while green plant ratio increased up to 35 mM KNO_3. Studying the effects of K^+ and NO_3^- separately, they concluded that it was the higher concentrations of NO_3^- that was detrimental to induction, while K^+ was not detrimental. Chu and Hill (1988) observed that the addition of a mixture of amino acids to the N6 media also improved wheat anther culture response.

Table 3. Influence of culture temperature upon the numbers of multinuclate microspores after 10 days of culture on BAC1 medium. Wheat cv. Sinton (Glover and Kasha, unpublished).

No. of	Temperature (°C)					
nuclei	22	24	26	28	30	32
≥ 4	5	14	14	43	90	40
≥ 8	0	3	6	20	49	33
≥ 15	0	2	1	4	18	13
≥ 30	0	1	0	0	9	7
Embryoids	0	1	0	1	11	14
Total	3035	2838	2460	3277	3021	2529

Thus, there appears to be some similarities between wheat and barley anthers relative to N sources and requirements. These similarities may also extend to plant hormones and liquid vs solid culture media.

Most wheat anther culture media have contained low levels of 2,4-D (2 mg/l). However, in studies of direct regeneration, i.e. without transferring calli or embryos to regeneration media, the replacement of 2,4-D by IAA or NAA has been important (Ouyang, 1986; Liang et al., 1987). This approach did not improve plant frequencies but did save time in culturing (Liang et al., 1987). One of the problems with this procedure for breeding may be that most of these progeny (95%) were haploid whereas about 50% are usually spontaneously doubled haploids (Marsolais et al., 1986). This difference may be due to an effect of 2,4-D on mitotic spindles.

Earlier studies (see Ouyang, 1986) on liquid vs solid culture media favored the solid media. However, a number of recent reports have observed about double the induction response in liquid culture media with Ficoll (Chu and Hill, 1988; Jones and Petolino, 1988; Lettre et al., 1989). Regeneration frequencies have tended to be lower and care must be taken to ensure floating or aeration of the anthers and calli. Similar to Hunter (1988) for barley, Chu & Hill (1988) reported a better response in wheat with filter sterilized compared to autoclaved liquid media.

     One research goal should be to develop media, growth
conditions or other treatments to significantly improve response
across all genotypes. To this effect, research with chemicals that
induce male sterility (Picard et al., 1987; Schmid, 1988) or male
sterile mutants (Heberle-Bors and Odenbach, 1985) may provide some
hope. With sterility inducing agents, Picard et al. (1987)
observed an improved induction response of about three times over
the control when compared across genotypes. Their best response
did not exceed the best frequencies reported in the literature
(Table 1) but this could be due to differences in many factors such
as media, plant growth conditions, genotypes or anther culture
temperature. Heberle-Bors and Odenbach (1985) examined anther
culture response of a number of cytoplasmic male sterile lines and
found two with a Triticum timopheevi cytoplasm which had a much
better response. With many of the factors now being sorted out, we
should expect to see rapid improvements in wheat anther culture
success.

C. Rice

     Chen (1986) has provided an excellent summary of the earlier
literature on anther culture in rice, much of it his own research.
The various factors influencing success are very similar to those
on wheat and barley and, in many instances, were first studied in
rice. There are very strong genotype differences and the subsp.
japonica (keng) responds much better than the subsp. indica
(hsien). Chen (1986) reports the average response in keng is over
10 green plants per 100 anthers while with hsien it is less than
two. The best microspore stage for culturing is the mid-
uninucleate stage and cold pretreatments are very beneficial to
response. Low donor plant growth temperatures from the booting
stage on improve the frequency of callus (embryo) induction and
lower the frequencies of albino plants. Plants grown at lower
temperatures have a much thicker tapetum layer in their anther.

     In the development of culture media, the potato extract as a
portion of the medium (20% originally or 10% in Potato II, see
Table 2) gave a substantial increase in response. Analysis for
hormones showed a fairly high ABA level in the potato extract. The
sucrose concentration has been studied quite extensively and its
role as both a carbon source and osmoticum was recognized. Higher
levels of sucrose improved differentiation but tended to lower the
green/albino ratio. Liang et al. (1980) (see Chen, 1986) studied
the effects of 0.24 M and 0.12 M sucrose in medium and noted that
the penetration of sucrose was higher at the lower concentration.
In addition, there was an enhanced rate and larger quantities of
inorganic salts and hormones taken up by the cells, which could
have important secondary effects.

Figs. 3 and 4. High response of microspore cultures of barley cv.
Igri on liquid FHG medium conditional with barley ovules.
Fig. 3. Shed pollen cultures after 21 days, showing shoot
development from embryos. Fig. 4. Mechanically isolated
microspores 21 days after culture initiation showing watery
and organized calli or embryos. Bar indicates 5 mm.

Other media studies have determined the beneficial effect of
high NO_3 and low NH_4 concentrations as well as the value of amino
acid additions such as glutamine. It was also observed that high
levels of 2,4-D (10 mg/l) in media gave higher albino frequencies
than 2 mg/l. Chen (1986) concluded that the concentrations of
hormones required in media was also related to the concentrations
of iron salts, sugars and other compounds in the media.

Hu (1986) reported that over 60 improved lines of rice for
cultivation have been produced in China using anther culture
procedures.

III. MICROSPORE CULTURE

The recent progress in anther culture has renewed interest in
and led to progress in isolated microspore culture in cereals, an
area where progress had been negligible for years. There are two
approaches to microspore isolation. One is to induce the immature
anthers to dehisce the microspores (shed pollen) into a medium or
suitable osmoticum, while the second is to mechanically grind or
disrupt the anthers to remove the microspores at the desired stage
of development. Sunderland and Xu (1982) first reported progress
with shed pollen culture in barley although green plants were not
regenerated. Most earlier studies used either spike pretreatments
or precultured anthers for microspore culture. However, Wei et al.
(1986) isolated into 0.3 M mannitol, freshly harvested barley
pollen at the early binucleate stage and were able to regenerate
plantlets from pollen calli. This procedure has not, apparently,
been repeatable by other researchers but has led to increased
interest in isolated microspore culture. Hunter (1987; 1988),
using cold pretreated anthers of cv. Igri at 4°C for 28 days, has
successfully mechanically isolated barley microspores in liquid FHG
medium and regenerated many plants via direct embryogenesis. This
procedure has essentially been repeated (Olsen, pers. comm.; Jensen
pers. comm.) and earlier stages have responded very well in our lab
so far (Figs. 3 and 4) but plants are still at the early culture
stage. This procedure involves placing microspores in liquid media
drops on a solidified agar medium. The density of these
microspores appears to be important to the frequency of induction.
Hunter (1988) has estimated that up to 27 green plants have been
obtained per anther from isolated microspore culture of Igri.

Köhler and Wenzel (1985) used ovary or anther conditioned
medium and fresh anthers of barley cvs. Dissa and Igri to obtain
plants from shed pollen. The frequency of plants from Dissa was
1.2 per 100 anthers and the ratio of albino to green plants was
13:1. Subsequently, Datta and Wenzel (1987) used a similar shed

pollen procedure to obtain plants from 8 of 13 wheat genotypes tested. In this case, cold pretreated anthers were used and the dehisced anthers were usually not removed. Plant regeneration was low with a high proportion being albino. With this procedure, one question that remains is - what was the stage of development when the embryogenic microspores were shed from the anthers? For some purposes such as mutation and transformation, it would be desirable to have the microspores at the G1 uninucleate stage. This system of direct embryogenesis from shed pollen has now been achieved for barley, wheat, rice and maize (Datta and Potrykus, 1988). Their objective is to obtain microspore derived embryos for gene transfer via microinjection.

The objective of our own research has been to obtain plants from microspores isolated at the uninucleate stage to use as targets for cereal transformation. We have been able to obtain multicellular structures and embryos from both mechanically isolated and shed pollen of a number of North American cultivars of both barley and wheat. We have recently started to work with Igri and the microspores look and behave quite differently than other genotypes, demonstrating the existence of strong genotype differences in response. The first reports on isolated microspore culture of rice were by Chen et al., (1980) (see Chen, 1986) and green plants were reported in 1981. They found that a combination of 10-15 days cold pretreatment plus 3-4 days anther culture prior to isolation was optimum.

Cho and Zapata (1988) have used rice anthers cultured for seven days prior to microspore isolation to obtain plants. In such studies, the debate is whether to call it anther sulture since the induction is during the anther culture phase. In this case responding microspores had undergone 2-4 divisions prior to isolation. Other researchers (Coumans et al., 1989) have obtained plants from isolated microspores of corn.

In summary, while plants have been obtained from isolated microspores of cereals, they almost invariably have developed following cold pretreatments in the anthers, preculturing of anthers before isolation, or from using medium conditioned by anthers or ovules. It is apparent that isolated microspores lack the benefits of the anther substances and further research on the anther factors is critical to obtaining isolated microspore cultures. In our studies with barley and wheat, we have been able to induce divisions in microspores isolated from freshly harvested anthers but have not regenerated plants. We have also been able to induce freshly harvested barley anthers to dehisce the microspores within 24 hours of culture initiation. Through use of conditioned media we hope to be able to obtain populations of uninucleate microspores from freshly harvested anthers that are capable of regeneration.

IV. PROBLEM AREAS IN CEREAL HAPLOIDY

For the utilization of in vitro microspore cultures in cereals
we would like to be able to induce direct embryogenesis from
microspores, to obtain predominantly green plants and to obtain
them in large numbers from any given genotype. To achieve these
objectives, we need more basic information on the process of
embryogenesis and the causes of both albinism and of strong
genotype effects. In this section we shall briefly cover recent
developments and theories on these topics.

A. Pollen Embryogenesis

There has been controversy in the literature over the time
when pollen can be induced to follow an embryogenic pathway. One
group (see Heberle-Bors, 1985) hypothesize that only predetermined
dimorphic pollen (P-grains) are capable of forming calli or embryos
at the time of anther or microspore culture. This predetermination
is considered to have occurred at or prior to meiosis. Others (see
Sunderland and Huang, 1987) believe that pollen embryogenesis can
be also induced at the uninucleate and binucleate stages of
microspores at which they are cultured. The pros and cons of this
debate are very lengthy but there appears to be evidence of
embryogenesis induction at both stages. The application to plants
of chemical sterilants at the meiotic stages have been effective in
increasing the frequency of haploids (Picard et al., 1987; Schmid,
1988). Their potential for a general increase in induction
frequencies across genotypes is important.

The increased frequencies in certain male sterile genotypes
(Heberle-Bors, 1985) is further evidence that events prior to
culturing can influence frequencies. On the other hand, the stage
at which microspores are cultured is also critical and for cereals
this appears to be the mid- to late-uninucleate microspore stage.
Culturing at this stage can greatly increase the frequency of
pollen dimorphism as shown by cytological studies (Sunderland and
Dunwell, 1974; Sun et al., 1983; Chen and Wu, 1983; Chen et al.,
1984; Huang, 1986). Up to 70% of microspores have been found to
undergo the initial divisions leading to sporophytic development.

Based upon cytological examinations, Sunderland and Huang
(1987) have postulated that the microspores do not become fully
committed to gametophytic development until after completion of the
first mitotic division. At any point up until that time, it should
be feasible to block the gametophytic pathway and induce cell
division and potential embryo formation. With the recent higher
frequencies of direct embryogenesis from cereal microspores, it
should now be easier to obtain better ultrastructural and
biochemical knowledge of this process.

Senescence of anthers occurs both in vitro and as a normal in vivo process. Therefore, the question is - why are calli or embryos produced only in in vitro cultures? Senescence can be modified by culture conditions (high or low temperatures), hormones that induce ethylene production, wounding, dehydration and media components such as sugars and salts. Cho (1988) and Cho and Kasha (unpublished) have observed that when anthers are plated on culture medium, there is a strong burst of ethylene production. It is known that auxins induce ethylene production which can, in turn, stimulate the induction of cell divisions. However, when the ethylene concentration becomes too high it can have a detrimental effect on anther culture response. There are various reports that could be related to the ethylene level. In some studies, culture vessel size which could influence ethylene concentration has been shown to be a factor (Dunwell, 1985). Cho (1988) measured the senescence of barley anthers based on changes in fatty acid composition of anthers. Chemicals which influenced senescence also influenced the induction of embryogenic development. Absicic acid is known to induce senescence and Torrizo and Zapata (1986) have demonstrated that increased concentrations of ABA have increased the production of green plants from anther cultures of rice. The most abundant hormone found in potato extract is ABA (Chen, 1986). The use of chemical sterilants (Picard et al., 1987) for induction may also be related to ethylene production as some sterilants like Ethephon (Ethrel) operate by producing ethylene. Thus, it would appear that some senescence may be required to release factors essential for the initial induction response from anther or microspore culture. However, a too rapid or severe senescence could lead to microspore degeneration which can be correlated with high ethylene concentrations.

With so many factors such as donor plant growth, pretreatments, culture media components and genotype influencing the induction of pollen embryogenesis, we are only beginning to understand the process. It would appear that factors in the anther itself are essential for the initial induction of cell division. Köhler and Wenzel (1985) isolated a factor from barley anthers that could improve induction frequencies. In isolated microspore cultures, it appears that cold pretreatments, preculture of anthers, or conditioned media are necessary to ensure induction of initial divisions. Much remains to be learned about induction and further studies of anther extracts for plant hormones, enzymes, etc. are required.

Another major limitation to obtaining high frequencies of plants, in addition to the induction of cell divisions, would appear to be in sustaining the development of induced microspores. Most induced microspores abort later at multicellular stages. To enable further embryos to develop, Hunter (1988) found it necessary to remove the most developed embryos every week for up to six

weeks, at which time the remaining calli and embryos aborted.
Replenishment in liquid culture media would appear to be beneficial
in this respect (Marsolais and Kasha, 1985).

One of the concerns about producing doubled haploids by anther
culture has been the extent of gametoclonal variation induced from
cultures. With much higher frequencies of direct embryo formation,
we are likely to see much less induced variation. Unfortunately,
this may also result in most progeny being haploid rather than
doubled haploids and reduce one of the time saving advantages of
anther culture in breeding.

B. Albinism

Albino plant regenerants encountered in the culture of anthers
and microspores of cereals and other monocots is a problem not
found in dicots. The frequencies of albinos has been close to 70-
80% on average across barley and rice genotypes and has been a
problem to a lesser extent in wheat and other monocots. Huang
(1986) hypothesized that metamorphosis of plastids was a reflection
of the transition from sporophytic to gametophytic development
where plastids no longer have the capacity to develop into
chloroplasts. From ultrastructural studies she observed that
microspores in barley and wheat that have just passed the pollen
tetrad stage have numerous proplastids with dense stroma. However,
by the time the microspores have become vacuolate (late uninucleate
stage) this dense stroma has been lost. In contrast, this plastid
metamorphosis in tobacco, Datura and other dicots that have been
studied does not start until the late bicellular pollen stage
(Huang, 1986).

Day and Ellis (1984, 1985) observed deletions of plastid DNA
(ctDNA) when examining albino plants derived from anther culture of
wheat and barley. This observation could be consistent with the
hypothesis of plastid metamorphosis associated with the transition
to gametophytic development. Presumably, the metamorphosis could
have caused some degradation of ctDNA in some microspores that were
still capable of forming multicellular structures. This hypothesis
is consistent with other observations on albinism frequencies. The
ratio of green to albino plants is fairly consistent for a genotype
but varies with genotypes. Hu and Huang (1987) indicated that
culturing at earlier stages of the uninucleate microspore resulted
in higher proportions of green plants. The use of higher culture
temperatures are thought to result in a higher proportion of albino
plants in wheat (Ouyang et al., 1983), triticale (Bernard, 1980)
and rice (see Chen, 1986). Chen (1986) examined the literature on
anther culture for media factors influencing albino frequencies.
High 2,4-D concentrations enhanced albino frequencies as did
replacing 2,4-D with NAA. Too low a concentration of Fe^{++} and high
sucrose concentrations appeared to decrease green plant formation.

Sorvari and Schieder (1987) found a greatly reduced frequency of albinos in barley on the barley starch-melibiose medium. Feng and Ouyang (1988) observed that the proportion of green plants in wheat increased sharply as the concentration of KNO_3 increased from 0 to 35 mM.

With a reasonable hypothesis of the cause of albinism in cereals and the knowledge that some factors can influence their frequency, it should be feasible to develop procedures to overcome this problem, particularly where high frequencies of regenerants can be produced.

C. Genotype Effects

It is important to note that the recent order of magnitude increase in barley anther culture success has been based mainly upon one responsive genotype Igri. Other genotypes show a wide range of response to these conditions and media components. It is these strong genotype differences that we would like to resolve for breeding and research purposes. At present, very little is known about the sources of genotypic differences in microspore culture response. As mentioned in section II, studies have been conducted to show that the traits (anther response and plant regeneration) are independently heritable. However, the knowledge of what causes these differences to be expressed is lacking.

Microspore culture success appears to be sporophytically determined. The role played by the anther is seen in studies on the anther factor (Köhler and Wenzel, 1985), the use of conditioned media, the need for long cold pretreatments, or the preculture of anthers prior to isolation for microspore culture. Cho (1988) and Cho and Kasha (unpublished) have observed genotype differences in anther senescence and ethylene production in barley. Cho (1988) has demonstrated that anthers of the cv. Klages contain higher levels of the ethylene precursor ACC (1-aminocyclopropane-1-carboxylic acid) than do those of cvs. Elrose or Bruce. Klages produces high frequencies of calli on BAC3 medium whereas Bruce forms more embryoid structures with better regeneration ability (Szarejko and Kasha, 1988).

Surveys have shown that the genotype of the anther donor plant also strongly influences anther culture response in wheat (Marsolais et al., 1986; Lettre et al., 1989). Again, the consensus is that callus induction frequency, regeneration frequency and the proportion of green plants regenerated are independently inherited traits, usually governed by more than one gene (Lazar et al., 1984; Deaton et al., 1987; Agache et al., 1988; Szakacs et al., 1988). The regeneration frequency of plants from F_1 plants tends to show some dominance (Ouyang, 1986; Marsolais et al., 1986). However, this response could be associated with the vigor of the donor plants as Deaton et al. (1987) have observed mostly additive genetic effects for these traits. Szakacs et al. (1988) identified wheat chromosomes involved in the inheritance of anther culture ability by examining the chromosome substitution lines of Cheyenne in Chinese Spring. Lazar et al. (1987) studied the addition lines of the seven rye chromosomes to wheat. They concluded that rye chromosome 4 improved anther response to culture while chromosomes 6 and 7 improved regeneration. Henry and de Buyser (1985) found improved anther culture ability associated with the presence of 1B/1R wheat-rye translocation chromosomes. While the traits of callus induction and plant regeneration are not inherited as single gene traits, it would appear feasible to transfer or select for them.

These studies on donor plant condition and anther components indicate that many genetically determined physiological factors may be involved in the success of anther culture in cereals and other plants. Thus, there are many steps at which genotypes could differ. In order to find procedures that work more uniformly across genotypes, we will need to understand the processes at the biochemical and ultrastructural levels and devise treatments that will provide a more uniform level of response across different genotypes. Treatment applied to plants at earlier stages around meiosis may be one approach.

ACKNOWLEDGEMENTS

We are most grateful to the Natural Sciences and Engineering Research Council of Canada and the Ontario Ministry of Agriculture and Food for financial support of our research. We wish to acknowledge the permission of C.P. Hunter to reproduce information from his publications and valuable personal communications with M. Coumans, A.A. Marsolais, M. Bolic, B. Foroughi-Wehr, G. Wenzel, F.L. Olsen and C.J. Jensen. The excellent technical assistance of Ecaterina Simion and Rosalinda Oro is gratefully acknowledged.

REFERENCES CITED

Agache, S., de Buyser, J., Henry, Y., and Snape, J. W., 1988, Studies of the genetic relationship between anther culture and somatic tissue culture abilities in wheat, Plant Breeding, 100:26.

Baenziger, P. S., Kudirka, D. T., Schaeffer, G. W., and Lazar, M. D., 1984, in: "Gene Manipulation in Plant Improvement," J. P. Gustafson, ed., Plenum, New York (1984), pp. 385-414.

Bernard, S., 1980, In vitro androgenesis in hexaploid triticale - determination of physical conditions increasing embryoid and green plant production, Z. Pflanzenphysiol., 85:308.

Bjørnstad, A., Opsahl-Ferstad, H. G., and Aasmo, M., 1988, Effects of environmental and incubation conditions upon induction and regeneration from wheat anther cultures, Eucarpia Meetings - Elsinore, Denmark, Sep., 1988, Abstract.

Chen, C. C., Howarth, M. J., Peterson, R. L., and Kasha, K. J., 1984, Ultrastructure of androgenic microspores of barley during the earley stages of anther culture, Can. J. Genet. Cytol., 26:484.

Chen, C. C., and Wu, Y. H., 1983, Segmentations in microspores of rice during anther culture, Proc. Natl. Sci. Counc. B. ROC, 7:151.

Chen, Y, 1986, Anther and pollen culture of rice, in: "Haploids of Higher Plants In Vitro," H. Hu, and H. Yang, eds., Springer-Verlag, Berlin Heidelberg New York Tokyo (1986), pp. 3-25.

Cho, U. H., 1988, Senescence of barley (Hordeum vulgare L. cv. Klages) anthers cultured in vitro, MSc Thesis, University of Guelph, Canada.

Cho, M. S., and Zapata, F. J., 1988, Callus formation and plant regeneration in isolated pollen culture of rice (Oryza sativa L. cv. Taipei 309), Plant Science, 58:239.

Choo, T. M., Reinberg, E., and Kasha, K. J., 1985, Use of haploids in breeding barley, Plant Breeding Reviews, 3:219.

Chu, C. C., and Hill, R. D., 1988, An improved anther culture method for obtaining higher frequency of pollen embryoids in Triticum aestivum L., Plant Science, 55:175.

Comeau, A., Plourde, A., St-Pierre, C., and Nadeau, P., 1988, Production of doubled haploid wheat lines by wheat X maize hybridization, Genome, 30 (Suppl.):482.

Coumans, M. P., Sohota, S., and Swanson, E. B., 1989, Plant development from isolated microspores of Zea mays L., Plant Cell Reports, (in press).

Datta, S. K., and Wenzel, G., 1987, Isolated microspore derived plant formation via embryogenesis in Triticum aestivum, Plant Science, 48:49.

Datta, S. K., and Potrykus, I., 1988, Direct embryogenesis and plant regeneration from microspore of barley; maize; rice and wheat, Eucarpia Meetings - Elsinore, Denmark, Sep., 1988, Abstract.

Day, A., and Ellis, T. H. N., 1984, Chloroplast DNA deletions associated with wheat plants regenerated from pollen : possible basis for maternal inheritance of chloroplasts, Cell, 39:359.

Day, A. & Ellis, T. H. N., 1985, Deleted forms of plastid DNA in albino plants from cereal anther culture, Current Genetics, 9:671.

Deaton, W. R., Metz, S. G., Armstrong, T. A., Mascia, P. N., 1987, Genetic analysis of the anther-culture response of three spring wheat crosses, Theor. Appl. Genet., 74:334.

Dunwell, J. M., 1985, Anther and ovary culture, in: "Cereal Tissue

and Cell Culture," S. W. J. Bright, and M. G. K. Jones, eds.,
    Martinus/Nijhoff/Dr. W. Junk Publishers (1985), pp. 1-44.
Feng, G. H., and Ouyang, J., 1988, The effects of KNO$_3$
    concentration in callus induction medium for wheat anther
    culture, Plant Cell Tiss. Org. Cult., 12:3.
Foroughi-Wehr, B., Friedt, W., and Wenzel, G., 1982, On the genetic
    improvement of androgenetic haploid formation in Hordeum
    vulgare L., Theor. Appl. Genet., 62:233.
Hagberg, A., and Hagberg, G., 1987, Production of spontaneously
    doubled haploids in barley using a breeding system with marker
    genes and the "hap"-gene, Biol. Zentralbl., 106:53.
Heberle-Bors, E., 1985, In vitro haploid formation from pollen: a
    critical review, Theor. Appl. Genet., 71:361.
Heberle-Bors, E., and Odenbach, W., 1985, In vitro pollen
    embryogenesis and cytoplasmic male sterility in Triticum
    aestivum, Z. Pflanzenzuchtg., 95:14.
Henry, Y., and de Buyser, J., 1985, Effect of the 1B/1R
    translocation on anther culture ability in wheat (Triticum
    aestivum L.), Plant Cell Reports, 4:307.
Hu, H., 1986, Wheat: Improvement through anther culture, in:
    "Biotechnology in Agriculture and Forestry," Y. P. S. Bajaj,
    ed., Springer-Verlag, Berlin, Heidelberg, New York, (1986),
    pp. 55-72.
Hu, H., and Yang, H, 1986, "Haploids of Higher Plants In Vitro,"
    Springer-Verlag, Berlin, Heidelberg, New York, Tokyo.
Huang, B., 1986, Ultrastructural aspects of pollen embryogenesis in
    Hordeum, Triticum and Paeonia, in: "Haploids of Higher Plants
    In Vitro," H. Hu, and H. Yang, eds., Springer-Verlag, Berlin,
    Heidelberg, New York, Tokyo (1986), pp. 91-117.
Huang, B., 1987, Effects of incubation temperature on microspore
    callus production and plant regeneration in wheat anther
    cultures, Plant Cell Tiss. Org. Cult., 9:45.
Hunter, C. P., 1987, Plant generation method, European Patent
    Application, No. 87200773.7.
Hunter, C. P., 1988, Plant regeneration from microspores of barley,
    Hordeum vulgare L., Ph.D. Thesis, Wye College, University of
    London.
Hunter, C. P., Loose, R. W., Clerk, S. P., and Lyne, R. L., 1989,
    Maltose - the preferred carbon source for barley anther
    culture, Eucarpia Meetings, Elsinore, Denmark, Sep., 1988, in
    press.
Jensen, C. J., 1986, Haploid induction and production in crop
    plants, in: "Genetic Manipulation in Plant Breeding," W. Horn,
    C. J. Jensen, W. Oldenbach, O. Schieder, eds., Walter de
    Gruyter & Co, Berlin, New York (1986), pp. 231-256.
Jones, A. M., and Petolino, J. F., 1987, Effects of donor plant
    genotype and growth environment on anther culture of soft-red
    winter wheat (Triticum aestivum L.), Plant Cell Tiss., Org.
    Cult., 8:215.
Jörgensen, R. B., Jensen, C. J., Anderson, B., and Von Bothmer, R.,

1986, High capacity of plant regeneration from callus of intersepecific by hybrids with cultivated barley (Hordeum vulgare L.), Plant Cell Tiss. Org. Cult., 6:199.

Kao, K. N., 1988, In vitro culture in barley breeding. Proc. FAO/IAEA Res. Coord. Meeting, Casaccia, Rome, Dec. 1985.

Kao, K. N. and Horn, D. C., 1982, A Method for induction of pollen plants in barley. Proc. 5th Intl. Cong. Plant Tissue and Cell Culture. A. Fujiwara, ed., pp. 529-530.

Kasha, K. J., and Seguin-Swartz, G., 1983, Haploidy in crop improvement, in: "Cytogenetics of Crop Plants," M. S. Swaminathan, P. K., Gupta, and U, Sinha, eds., MacMillan India Ltd, New Delhi (1983), pp. 19-68.

Kelly, S. L., 1986, The effect of sucrose, dextran and glucose on anther culture of winter wheat (Triticum aestivum), MSc Thesis, University of Guelph, Canada.

Knudsen, S., Due, I. K., and Anderson, S. B., 1988, Genotypic effects on anther culture in barley, Eucarpia Meetings - Elsinore, Denmark, Sep, Sep., 1988, Abstract.

Köhler, F., and Wenzel, G., 1985, Regeneration of isolated barley microspores in conditioned media and trials to characterize the responsible factor, J. Plant Physiol., 121:181.

Laurie, D. A., and Bennett, M., 1988, Cytological evidence for fertilization in hexaploid wheat X sorghum crosses, Plant Breeding, 100:73.

Lazar, M. D., Schaeffer, G. W., Baenziger, P. S., 1984, Cultivar and cultivar X environment effects on the development of callus and polyhaploid plants from anther cultures of wheat, Theor. Appl. Genet., 67:273.

Lazar, M. D., Chen, T. H. H., Scoles, G. S., and Kartha, K. K., 1987, Immature embryo and anther culture of chromosome addition lines of rye in chinese spring wheat, Plant Science, 51:77.

Lettre, J. J., Kelly, S. L., Marsolais, A. A., and Kasha, K. J., 1989, Wheat anther culture using liquid media, in: "Biotech in Agric. & Forestry" Vol. 8, Y. P. S. Bajaj, ed., Springer-Verlag. (1989), in press.

Li, H., Qureshi, J. A., and Kartha, K. K., 1988, The influence of different temperature treatments on anther culture response of spring wheat (Triticum aestivum L.), Plant Science, 57:55.

Liang, G. H., McHughen, A., 1987, Novel approaches to wheat improvement, in: "Wheat and Wheat Improvement," E. G. Heyne, ed., Agronomy Monograph No. 13. (2nd ed.), ASA (1987) pp. 472-506.

Liang, G. H., Xu, A., and Tang, H., 1987, Direct generation of wheat haploids via anther culture, Crop Sci., 27:336.

Linde-Laursen, I., and Von Bothmer, R., 1986, Preferential loss and gain of specific Hordeum vulgare chromosomes in hybrids with three alien species, in: "Genetic Manipulation in Plant Breeding," W. Horn, C. J. Jensen, W. Odenbach, O. Shieder, eds., Walter de Gruyter & Co., Berlin New York (1986), pp. 179-182.

Marsolais, A. A., Seguin-Swartz, G., and Kasha, K. J., 1984, The
    influence of anther cold pretreatments and donor plant
    genotypes pn in vitro androgenesis in wheat (Triticum aestivum
    L.), Plant cell Tiss. Org. Cult., 3:69.
Marsolais, A. A., and K. J. Kasha, 1985, Callus induction from
    barley microspores. The role of sucrose and auxin in a barley
    anther culture medium, Can. J. Bot., 63:2209.
Marsolais, A. A., Wheatley, W. G., and Kasha, K. J., 1986, Progress
    in wheat and barley haploid induction using anther culture,
    in: "Proceedings of The DSIR Plant Breeding Symposium 1986",
    Agronomy Society of N. Z. Special Publication, 5:340-343.
O'Donoughue, L. S., Laurie, D. A., and Bennett, M. D., 1988, Bread
    wheat X corn and other novel sexual hybrids, Genome, 30
    (Suppl.):467.
Olsen, L., 1987, Induction of microspore embryogenesis in cultured
    anthers of Hordeum vulgare. The effects of ammonium nitrate,
    glutamine and asperagine as nitrogen sources, Carlesberg Res.
    Commun., 52:393.
Ouyang, J. W., Zhou, S. M., and Jia, S. E., 1983, The response of
    anther culture to culture temperature in Triticum aestivum,
    Theor. Appl. Genet., 66:101.
Ouyang, J. W., 1986, Induction of pollen plants in Triticum
    aestivum, in: "Haploids of Higher Plants In Vitro," H. Hu, and
    H. Yang, eds., Springer-Verlag, Berlin Heidelberg New York
    Tokyo (1986), pp. 26-41.
Ouyang, J. W., He, D. G., Feng, G. H., and Jia, S. E.,1987, The
    response of anther culture to culture temperature varies with
    growth conditions of anther-donor plants, Plant Science,
    49:145.
Picard, E., Hours, C., Gregoire, S., Phan, T. H., and Meunier, J.
    P., 1987, Significant improvement of androgenetic haploid and
    doubled haploid induction from wheat plants treated with a
    chemical hybridization agent, Theor. Appl. Genet., 74:289.
Prakashi, J., and Giles, K. L., 1987, Induction and growth of
    androgenic haploids,Intl Rev. Cytol., 107:273.
Sangwan, R. S., and Sangwan-Norreel, B. S., 1987, Biochemical
    cytology of pollen embryogenesis, Intl Rev. Cytol., 107:221.
Schmid, J., 1988, Application of gametocides and different chemical
    agents to wheat anther donor plants and their effects on the
    induction of androgenesis, Eucarpia Meetings - Elsinore,
    Denmark, Sep., 1988, Abstract.
Sorvari, S., 1986, The effect of starch gelatinized nutrient media
    in barley anther cultures, Ann. Agricult. Fenn., 25:127.
Sorvari, S., and Schieder, O., 1987, Influence of sucrose and
    melibiose on barley anther cultures in starch media, Plant
    Breeding, 99:164.
Subrahmanyam, N. C., and Von Bothmer, R., 1987, Interspecific
    hybridization with Hordeum bulbosum and development of hybrids
    and haploids, Hereditas, 106:119.

Sun, C., Chu, C., and Li, H., 1983, Electron microscope observation of microspore division of wheat in vitro, Acta Bot. Sinica, 25:295.

Sunderland, N., and Dunwell, J. M., 1974, Anther and pollen culture, in: "Tissue Culture and Plant Science," H. E. Street, ed., Academic Press, London (1974), pp. 141-167.

Sunderland, N., and Xu, Z. H., 1982, Shed pollen culture in Hordeum vulgare, J. Exp. Bot., 33:1086.

Sunderland, N., and Huang, B., 1987, Ultrastructural aspects of pollen dimorphism, Intl Rev. Cytol., 107:175.

Szakacs, E., Kovacs, G, Pauk, J., and Barnabas, B., 1988, Substitution analysis of callus induction and plant regeneration from anther culture in wheat (Triticum aestivum L.), Plant Cell Reports, 7:127.

Szarejko, I., and Kasha, K. J., 1989, Induction of anther culture derived doubled haploids in barley, Proc. FAO/IAEA Research Coord. Meeting, Katowice, Poland, July 1988, in press.

Thörn, E. C., 1988, Effect of melibiose and polyethylene glycol on anther culture response in barley, Eucarpia Meetings - Elsinore, Denmark, Sep., 1988, Abstract.

Torrizo, L. B., and Zapata, F. J., 1986, Anther culture in rice: IV. The effect of abscisic acid on plant regeneration, Plant Cell Reports, 5:136.

Wei, Z. M., Kyo, M., and Harada, H., 1986, Callus formation and plant regeneration through direct culture of isolated pollen of Hordeum vulgare cv. Sabarlis, Theor. Appl. Genet., 72:252.

Wenzel, G., and Foroughi-Wehr, B., 1984, Anther culture of cereal and grasses, in: "Cell Culture and Somatic Cell Genetics of Plants," Vol. 1., I. K. Vasil, ed., Academic Press (1984), pp. 311-327.

TRANSGENIC PLANTS

R. Dekeyser, D. Inzé, and M. Van Montagu

Laboratorium voor Genetica
Rijksuniversiteit Gent
B-9000 Gent (Belgium)

INTRODUCTION

Currently the term "transgenic organism" is used when referring to an organism which harbors additional genetic information as the result of a genetic engineering step, namely the transfer of purified or cloned DNA. For plants there are two very different approaches for obtaining such DNA transfer. First there is the so called "natural" way of DNA transfer. This method exploits the conjugation-like DNA transfer which can occur when some soil bacteria such as *Agrobacterium tumefaciens* colonize plants (for recent reviews, see Zambryski, 1988; Gheysen et al., 1989). Many gene vectors have been constructed based on this transfer mechanism and these have allowed the engineering of the first transgenic plants expressing selectable marker genes (Herrera-Estrella et al., 1983). *Agrobacterium*-mediated gene transfer has also been the method of choice for introducing new economically important traits into plants such as insect resistance (Vaeck et al., 1987), virus resistance (Abel et al., 1986; Nelson et al., 1988) and also for constructing plants with engineered seed proteins which can be the starting material for producing peptides of importance to mammalian physiology (Vandekerckhove et al., 1989). An appreciated advantage of the *Agrobacterium* system is the fact that the majority of the transformed plants obtained after selection harbour one or two copies of a well defined DNA sequence. However, several important crops such as most leguminous plants and all of the Graminae remain recalcitrant to this type of DNA transfer. Some results have been obtained with such plants by employing the other DNA transfer system which is equivalent to the *in vitro* DNA uptake methods used with other organisms. To introduce the DNA,

Gene Manipulation in Plant Improvement II
Edited by J. P. Gustafson
Plenum Press, New York, 1990

polyethyleneglycol (PEG), electroporation or micro injection can
be used, but the recipient cell has to be a protoplast capable
of regenerating (Lazzeri and Lörz, 1988; Gasser and Fraley, 1989).
This severely limits the usefulness of this approach.
Nevertheless the method has allowed a breakthrough in the
transformation of rice (Shimamoto *et al.*, 1989). Recently
promising results have been obtained with a spectacular new
mechanical method, the particle gun (Klein *et al.*, 1987; McCabe *et
al.*, 1988; Sanford, 1989). This "ballistic" approach can probably
be used with any plant species. It should allow the
transformation of meristematic cells, hence enhancing the chance
of obtaining transgenic plants from those species or cultivars
which as yet cannot be taken through a cell culture step.

The progress in the efficiency with which it is possible to
generate transgenic plants has opened up the field of plant
molecular biology and made plant biotechnology possible. Actually
it is often due to the prospects of the latter that the attention
of funding agencies and industry has been attracted. This has
also generated support for the basic research work.

The early confirmation of the power of plant engineering was
a tremendous stimulus for the fundamental research. It is easy to
predict that this mutual interdependency of fundamental and
applied success will go on and by this plant molecular biology can
become particularly attractive to young scientists.

This progress report on some projects ongoing at the
University lab and at Plant Genetic Systems N.V. (Belgium)
illustrates further the intricate link between research and
development in plant molecular biology.

S-ADENOSYLMETHIONINE SYNTHETASE (SAM) IS ENCODED BY GENES
PREFERENTIALLY EXPRESSED IN VASCULAR TISSUE

SAM is the enzyme which catalyzes the biosynthesis of S-
adenosylmethionine (AdoMet) out of methionine and ATP (Mudd,
1963). AdoMet acts as the methyl group donor in numerous
transmethylation reactions involving acceptor molecules as
proteins, lipids, polysaccharides, and nucleic acids. In plants,
AdoMet is also a precursor molecule in the biosynthesis of
ethylene (Yang and Hoffman, 1984). *Arabidopsis thaliana* contains,
just like yeast, two *sam* genes (Peleman *et al.*, 1989a, 1989b).
Through gene cloning our group could demonstrate that the coding
region of the two genes share 96% homology at the amino-acid-
level.

Northern analysis using gene-specific *sam-1* and *sam-2* probes
showed that both genes are expressed 10 to 20 times better in

stems and roots than in leaves, inflorescences, and seedpods.
This similar expression pattern of both genes might be mediated by
the presence of three highly conserved sequences in the 5' region
of the *sam* genes. Expression specificity at cellular level was
analyzed with a construct consisting of the 750-bp 5' region from
sam-1, the β-glucuronidase gen (*uidA*), and the octopine synthase
gene terminator (P*sam1-gus*). This chimeric gene was introduced
into *Arabidopsis* via *Agrobacterium*-mediated root transformation
(Valvekens *et al.*, 1988). The high number of transformed plants
obtained with this method is closely correlated with the high
shoot-regenerating potential of *Arabidopsis* roots. Root explants
cultured for one week on a callus induction medium containing 2,4-
dichlorophenoxyacetic acid (2,4-D) and then transferred to shoot
induction medium with high cytokinin:auxin ratio, are converted in
a dense growth of shoots within 2 to 3 weeks. The short
preincubation on the 2,4D-containing medium was an essential step
in the regeneration procedure, since immediate incubation on shoot
induction medium leads to shoot formation at the proximal end of
the shoot only. Cocultivation of root explants with *Agrobacterium*
strains yields transformation frequencies of 20% to 80% based on
the initial explant number.

The P*sam1-gus* transformants were subjected to a β-
glucuronidase (GUS) assay and positively stained plant material
was analyzed at histological level. Thin sections through leaves,
stems, and roots demonstrated that the gene is expressed primarily
int he vascular tissues (phloem and parenchyma cells of the xylem)
of these organs. Additionally, high expression was also observed
in sclerenchyma and in root cortex. Since xylem, sclerenchyma
tissue, and to a lesser extent phloem are highly lignified, it
seems that the expression of the *sam-1* gene is correlated with the
lignification tissues undergo. Lignin is methylated prior to
polymerization and the synthesis of each monolignol needs two
AdoMet molecules (Higuchi, 1981). Consequently, high amounts of
AdoMet must be consumed in cells which produce large amounts of
lignin, thus explaining the enhanced SAM levels.

In *Nicotiana tabacum* plants transformed with the P*sam1-gus*
construct, the highest expression level directed by the *sam1*
promoter is detected in the same type of tissues as in
Arabidopsis. This indicates that the regulatory mechanism for
this type of tissue-specific expression is sufficiently conserved
between *Arabidopsis* and tobacco.

Further construction of transgenic plants where the
expression of a selectable marker is regulated by *sam* control and
promoter regions might allow the isolation of mutant cell lines
with altered *sam* expression. It will also be interesting to
construct transgenic plants which overproduce SAM and to study
their physiological characteristics.

TOWARDS THE ENGINEERING OF SUPEROXIDE RADICAL PROTECTION

Superoxide radicals are ubiquitously generated in many
biological oxidations within all compartments of the cell (for
review, see Fridovich, 1978). The toxicity of superoxide radicals
is mainly the result of its interaction with hydrogen peroxide
(Halliwell, 1984). In the presence of trace amounts of iron
salts, the combination of these two products leads rapidly to the
formation of hydroxyl radicals:

$$H_2O_2 + O_2^- \quad \overset{Fe^{2+}}{\underset{Fe^{3+}}{\rightarrow}} \quad OH^- + O_2 + OH\bullet$$

Hydroxyl radicals react with DNA, proteins, lipids, and almost any
other organic constituent of living cells. This can lead to
chromosome deletions, membrane damage, and finally lysis of the
cell.

To prevent the formation of such deleterious oxygen species,
living cells possess a battery of protective enzymes that detoxify
superoxide radicals and hydrogen peroxide. Catalases and
peroxidases neutralize hydrogen peroxide and a class of
metalloproteins, called superoxide dismutases (SOD) catalyze the
dismutation of superoxide radicals to hydrogen peroxide and
oxygen. Three classes of SOD can be distinguished based on their
metallic cofactor: the iron (Fe), the manganese (Mn), and the
copper/zinc (Cu/Zn) forms (for review, see Bannister et al.,
1987). In eukaryotes the nuclear-encoded MnSOD is most often
found in the mitochondrial matrix. In plants, the Cu/Zn form is
often present in several isoforms of which one is present in the
cytosol and the other one in the chloroplasts (Halliwell, 1987).

Recently, a full-length cDNA clone for the MnSOD from
Nicotiana plumbaginifolia has been isolated (Bowler et al.,
1989a). A comparison between the amino acid sequence derived from
the cDNA sequence and the amino acid sequence from the NH_2-
terminus from the mature protein determined by protein sequencing,
revealed that the first amino acid from the cDNA-derived amino
acid sequence. The 24 preceding amino acids have features typical
of a signal peptide for translocation to the mitochondrial matrix
(Schatz, 1987). It is characterized by five arginines,
distributed among uncharged amino acids, devoid of acidic
residues, and rich in hydroxylated amino acids. Several
experiments were designed to determine whether this 24 amino acid
leader sequence can translocate the MnSOD to the mitochondria.
Subcellular fractionation of the N. plumbaginifolia leaf extracts
demonstrated that the MnSOD copurified with the mitochondrial
fraction. Further evidence for the translocation capacity was

obtained by expressing the plant preprotein in a yeast MnSOD-deficient mutant (Bowler *et al.*, 1989b). The yeast mutant still contains a functional Cu/ZnSOD which is located in the cytosol, and this compartmentalization evidently precludes it from replacing the MnSOD. This suggests that introduction of the plant MnSOD will only complement the mutation if the protein is efficiently translocated to the yeast mitochondria. The yeast mutant is unable to utilize nonfermentable carbon sources as ethanol, therefore complementation could be scored by growth on ethanol. The full-length MnSOD cDNA clone was fused with the yeast ethanol-inducible *CYC1* promoter and introduced into the yeast mutant. The complemented mutant grew as well on ethanol as the wild-type yeast strain, suggesting an efficient import of the plant MnSOD into the yeast mitochondria. Protein sequencing of the NH_2-terminus of the processed plant MnSOD in yeast showed that the MnSOD preprotein is cleaved just before the leucine at position 25. This means that the position of cleavage is identical in yeast and *N. plumbaginifolia*.

In a second set of experiments, the expression pattern of the MnSOD was investigated by Northern analysis and SOD activity assays (Bowler *et al.*, 1989a). The expression level in leaves is very low, two to three times higher in roots, and 50 times higher in suspension cultures. The dramatic difference in expression level of MnSOD in leaves and suspension cultures can be due to (i) the difference in cell type or (ii) the difference in culture conditions. Indeed, the expression level of MnSOD in leaf discs incubated for 48 hours in suspension culture medium was 50 times higher than in leaf discs incubated in water for 48 hours. The 0.1-M sucrose in the medium is the crucial factor for induction. A linear dose-response could be observed at different concentrations of sucrose (0.001 M-0.1 M). The same increase in MnSOD expression was also generated by glucose, but not by 0.1 M mannitol. This observation suggests that the induction of MnSOD by sugars is due to a trophic rather than an osmotic effect.

Since the MnSOD is the mitochondrial isozyme, its induction by sugars can be considered as a response against superoxide radicals generated in mitochondria by enhanced respiratory oxidation of sugars. Indeed, there is a clear correlation between the activities of MnSOD and cytochrome oxidase, which is the enzyme involved int he last step of the respiratory chain in mitochondria.

Similar sucrose-mediated induction patterns are observed in SOD activity assays (Beauchamps and Fridovich, 1971) performed directly on nondenaturing protein gels. Remarkably, these assays showed that the Cu/ZnSOD is not induced after incubation of leaf discs on 0.1 M sucrose. Alternatively, the Cu/ZnSOD is induced by treating leaves with the herbicide paraquat, which induces superoxide radical production in chloroplasts, whereas the MnSOD expression does not change (Matters and Scandalios, 1986; Bowler *et al.*, 1989a). A possible interpretation of this observation could be that superoxide radicals produce a specific triggering molecule in each subcellular compartment, which is capable of acting as a signal to induce only those nuclear genes which code for the SOD isoforms belonging to that particular compartment.

The MnSOD expression level is also induced after treatment of leaves with ethylene, salicylic acid, and *Pseudomonas syringae*. All these treatments enhance the plant respiration (Laties, 1982; Raskin *et al.*, 1987) which is demonstrated by increased cytochrome oxidase levels. After *P. syringae* infection, the Cu/ZnSOD is induced in concert with the MnSOD, this being in contrast with the more restricted response to sucrose.

The observation that stress situations leading to increased superoxide radical levels go along with increased SOD levels, suggest a protective role for the SOD. This is further supported by several physiological data demonstrating that plants with enhanced SOD levels are more resistant to chilling (Clare *et al.*, 1984), paraquat (Shaaltiel and Gressel, 1987), phyto-oxidative damage (Rabinowitch *et al.*, 1982), anoxia and hyperoxia (Monk *et al.*, 1987). We are currently analyzing transgenic plant over-expressing the MnSOD isoform int he mitochondria, in the chloroplasts or in the cytosol for increased protection against different stress agents.

PLANT ENGINEERING FOR INSECT RESISTANCE

Modern agriculture spends several billion dollars a year to control insect damage in crop plants. Many of the insecticidal chemicals used are more toxic to other organisms than the pest aimed at. In view of the high concern to find environmentally acceptable pesticides, the insecticidal crystal proteins present in spores from the Gram-positive bacterium *Bacillus thuringiensis* (Bt toxins) are a very interesting alternative to the chemical insecticides. The insecticidal activity of the Bt toxin is highly species-specific and is not toxic for higher animals (Krieg, 1986). Among the more than 3000 natural *B. thuringiensis* isolates screened by Plant Genetic Systems N.V. (Belgium), nearly all were active against a series of lepidopteran larvae and only five against coleopteran larvae (H. Joos, personal communication). Insecticidal crystal proteins (ICPs) produced by different *B. thuringiensis* strains have a different insecticidal spectrum. The ICPs toxic against lepidopteran larvae, for instance, show marked different spectra within this order as illustrated in Table 1. The classification of these *B. thuringiensis* strains according to the insecticidal spectrum of their *B. thuringiensis* toxin is based

Table 1. Toxicity of different ICPs toward four lepidopteran pest insects (the large white butterfly, *Pieris brassicae*; the tobacco hornworm (*Manduca sexta*; the tobacco budworm, *Heliothis virescens*; and the cotton leafworm, *Spodoptera littoralis*)

ICP	*P. brassicae*	*M. sexta*	*H. virescens*	*S. littoralis*
Bt2	0.7	8.6	10.7	>1350
Bt3	0.8	5.2	90.0	>1350
Bt73	0.3	5.3	1.6	>1350
Bt15	38.3	138.8	>625	70.65
Bt4412	2.8	>525	>625	>1350

Toxicities are expressed as 50% lethal doses (LD 50) in ng/cm^2 applied on artificial medium, except for *P. brassicae* where the data are in $\mu g/ml$ and where 5-μl samples are applied on leaf discs (Höfte and Whiteley, 1989).

on a correlation between the insecticidal spectrum and specific reactions of the ICPs in ELISA assays against a set of antibodies that has been generated against different *B. thuringiensis* toxins (Höfte *et al.*, 1988). Based on the reactivity of an ICP to this set of antibodies it is possible to predict its insecticidal spectrum. The technique is now used to sort out new natural isolates. Only those strains that produce interesting new patterns in the ELISA analysis are used in bioassays on various insects.

To understand the factors determining the specificity of the insecticidal spectrum of the different ICPs, we have to look to the fate of the insecticidal crystals. When insecticidal crystals are ingested by insects, they become solubilized in the alkaline environment of the insect's midgut. Most of the so released ICPs are protoxins which are proteolitically activated in the insect midgut to smaller, active toxins. For example, the 130-kDa *B. thuringiensis* 2 protoxin from *B. thuringiensis* sub-species berliner is cleaved into an active 60-kDa toxin (Lilley *et al.*, 1980). The active toxin binds to the membranes of the target cells in the midgut and this finally leads to a membrane disruption. Although this causes no immediate knock-down, it upsets the insect balance in such a way that it stops feeding and dies. That there is a correlation between the insect specificity of ICPs and the presence of specific receptors for these ICPs in the brush border membranes of the target cells in the insect midgut is demonstrated by binding studies performed with two ^{125}I-labeled toxins, Bt2 which is toxic to *Manduca sexta* and *Pieris brassicae* larvae and Bt4412 which is toxic to *Pieris brassicae* only (Hofmann *et al.*, 1988). The Bt2 toxin binds saturably and with high affinity to brush border membrane vesicles from the midgut of both *M. sexta* and *P. brassicae*, whereas the Bt4412 toxin shows high affinity saturable binding to *P. brassicae* but not to *M. sexta* vesicles. Furthermore, other Bt toxins active against *M. sexta* could compete for binding of ^{125}I-labeled Bt2 toxin, whereas toxins active against dipteran or coleopteran larvae do not compete. Other factors such as the solubilization and proteolytic activation of the ICPs in the insect midgut can also account for some of the remarkable species specificity of the ICPs (Haider *et al.*, 1986).

B. thuringiensis strains can be developed as bio-insecticidal sprays for the control of plant pests. Alternatively, chimeric constructs with plant transcriptional signals and the coding region of a specific ICP can be cloned in transformation vectors and introduced into the crop plant of interest (Vaeck *et al.*, 1987).

A chimeric gene with the transcript 2' promoter (Velten *et al.*, 1984) and a truncated *bt2* gene construct where the 3' end of the *bt2* gene, encoding the past of the ICP which is proteolytically digested in the insect midgut, was removed has been introduced into tobacco via *Agrobacterium tumefaciens*-mediated transformation.

The insecticidal activity of transgenic plants on first instar *M. sexta* larvae was demonstrated in leaf discs assays and subsequently in greenhouse and field tests. Typically, plants producing more than 0.01% or their total protein as the toxin caused 100% mortality. In field trails, the transgenic tobacco plants were not only protected against *M. sexta* larvae, but also against the tobacco hornworm (*Heliothis virescens*). Tomato and potato plants transformed with the same chimeric constructs synthesize also high levels of Bt2. Field trials (Plant Genetic Systems N.V., unpublished results) demonstrated that transgenic potatoes are protected against the larvae of the tuber moth (*Phthorimaea oppercullela*).

Whereas first-generation commercial plants are close by, the next step is to engineer plants with different ICPs, to introduce these ICPs in other crops, and to make further use of enhanced and regulated Bt toxin expression. In this concept, the use of the transcript 2' promoter in our chimeric constructs is already an example of applied regulated expression, because this promoter is wound-inducible (Teeri *et al.*, 1989). This means that insect feeding damage leads to enhanced levels of Bt toxin.

PLANT ENGINEERING FOR HERBICIDE RESISTANCE

Many of the cheap herbicides still in use, particularly in
third-world countries, are too toxic and are creating serious
environmental concern. New classes of more acceptable herbicides
are becoming available. They have several, or all, of the
following characteristics: they inhibit enzymes unique to plants,
are rapidly degraded in the soil, and are effective when used at
low application dose. Often it is difficult to add to that the
former requirement of herbicides, namely being active against a
maximum of weeds and not active against the crop plant. Indeed,
most of these new herbicides are total herbicides and their use
for post-emergence applications is very limited. Hence, there is
a high need to engineer some of the major crop plants for
resistance to these herbicides.

To obtain crops resistant to broad-spectrum herbicides by
applying the gene transfer and/or selection techniques, three
approaches can be followed: (i) a mutant form of the target enzyme
is produced which is still biologically active, but less sensitive
to the herbicide, (ii) over production of a sensitive target
protein, and (iii) degradation or detoxification of the
herbicide.

For instance, glyphosate is an herbicide that inhibits 5-
enol-pyruvylshikimate-3-phosphate synthase (EPSP) which is the
key enzyme in the biosynthesis of aromatic acids. Glyphosate-
resistant tobacco plants were obtained by introducing a
glyphosate-tolerant form of EPSP (Comai et al., 1985) and
resistant Petunia plants were obtained by overexpressing the epsp
gene (Shah et al., 1986). At Plant Genetic Systems, Inc.
(Belgium) the engineering of bialaphos-resistant plants was
obtained by introduction of a detoxifying enzyme (De Block et al.,
1987). Bialaphos is a tripeptide antibiotic produced by
Streptomyces hygroscopicus. It consists of phosphinothricin
(PPT), an analog of glutamic acid, and two L-alanines. Upon
removal of the alanine residues by peptidases, PPT is a very
potent inhibitor of glutamine synthetase (GS). Since GS is the
only enzyme in plants that can detoxify ammonia, inhibition of GS
by PPT causes a very rapid accumulation of ammonia which leads to
the death of the plant cell. At present two products are
available as commercial herbicides: glyphosinate ammonia is the
ammonia salt of the chemically synthesized PPT (Basta, Hoechst AG)
and bialaphos is produced by fermentation of S. hygroscopicus
(Herbiace, Meiji Seika). The bialaphos resistance (bar) gene,
whose gene product confers resistance to PPT, was isolated from
the bialaphos biosynthesis pathway of S. hygroscopicus (Murakami
et al., 1986). It encodes the phosphinothricin acetyl transferase
(PAT), which converts PPT with high efficiency to a non-herbicidal
acetylated form. Expression of the bar gene under the control of

the cauliflower mosaic virus 35S promoter in transgenic tobacco, tomato, and potato plants renders the plants resistant to applications of glyphosinate and bialaphos under greenhouse conditions (De Block *et al.*, 1987). In a similar way, herbicide-resistant cabbage, poplar, and sugar beet plants were obtained (Plant Genetic Systems N.V., unpublished results). Spraying of the Fl progeny of the transgenic tobacco plants with Basta demonstrates that the PAT resistance is inherited as a single dominant trait.

Two herbicide resistant transgenic tobacco lines and four herbicide resistant potato lines were analyzed under field conditions (De Greef *et al.*, 1989). Complete resistance to field-dose applications of glyphosinate was observed, although PAT levels in these lines varied by two orders of magnitude. Typically, plants containing 0.01% of their total protein as PAT are fully resistant. The leaf-length increase of transgenic tobacco plants and the tuber yield of transgenic potato lines equals the numbers obtained with untransformed tobacco and potato plants. Thus, the transformants reveal the same agronomic performances as untransformed controls. We can conclude that glyphosinate can be applied as a selective post-emergence herbicide on engineered crops.

ACKNOWLEDGEMENTS

We thank M. De Cock for the preparation of the manuscript. R.D. is a Research Assistant and D.I. a Senior Research Assistant of the National Fund for Scientific Research (Belgium).

REFERENCES

Abel, P. P., Nelson, R. S., De, B., Hoffman, N., Rogers, S. G., Fraley, R. T., and Beachy, R. N., 1986, Delay of disease development in transgenic plants that express the tobacco mosaic virus coat protein gene, Science, 232:738.

Bannister, J. V., Bannister, W. H., and Rotilio, G., 1987, Aspects of the structure, function and applications of superoxide dismutase, CRC Crit. Rev. Biochem., 22:111.

Beauchamps, C., and Fridovich, I., 1971, Superoxide dismutase: improved assays and an assay applicable to acrylamide gels, Anal. Biochem., 44:276.

Bowler, C., Alliotte, T., De Loose, M., Van Montagu, M., and Inzé, D., 1989a, The induction of manganese superoxide dismutase in response to stress in *Nicotiana plumbaginifolia*, EMBO J., 8:31.

Bowler, C., Alliotte, T., Van den Bulcke, M., Bauw, G., Vandekerckhove, J., Van Montagu, M., and Inzé, D., 1989b, plant mitochondrial preprotein is efficiently imported and correctly processed by yeast mitochondria, Proc. Natl. Acad. Sci. USA, 86:3237.

Clare, D. A., Rabinowitch, H. D., and Fridovich, I., 1984, Superoxide dismutase and chilling injury in *Chlorella ellipsoidea*, Arch. Biochem. Biophys., 231:158.

Comai, L., Facciotti, D., Hiatt, W. R., Thompson, G., Rose, R. E., and Stalker, D. M., 1985, Expression in plants of a mutant *aroA* gene from *Salmonella thyphimurium* confers tolerance to glyphosate, Nature (London), 317:741.

De Block, M., Botterman, J., Vandewiele, M., Dockx, J., Thoen, C., Gosselé, V., Movva, R., Thompson, C., Van Montagu, M., and Leemans, J., 1987, Engineering herbicide resistance in plants by expression of a detoxifying enzyme, EMBO J., 6:2513.

De Greef, W., Delon, R., De Block, M., Leemans, J., and Botterman, J., 1989, Evaluation of herbicide resistance in transgenic crops under field conditions, Bio/technology, 7:61.

Fridovich, I., 1978, The biology of oxygen radicals. The superoxide radical is an agent of oxygen toxicity: superoxide dismutases provide an important defense, Science, 201:875.

Gasser, C. S., and Fraley, R. T., 1989, Genetically engineering plants for crop improvement, Science, 244:1293.

Gheysen, G., Herman, L., Breyne, P., Van Montagu, M., and Depicker, A., 1989, Agrobacterium tumefaciens as a tool for the genetic transformation of plants, in "Genetic transformation and expression", L. O. Butler, ed., Intercept, London, in press.

Haider, M. Z., Knowles, B. H., and Ellar, D. J., 1986, Specificity of *Bacillus thuringiensis* var. *colmeri* insecticidal δ-endotoxin is determined by differential proteolytic processing of the protoxin by larval gut proteases, Eur. J. Biochem., 156:531.

Halliwell, B., 1984, "Chloroplast metabolism - The structure and function of chloroplasts in green leaf cells", Clarendon Press, Oxford.

Herrera-Estrella, L., Depicker, A., Van Montagu, M., and Schell, J., 1983, Expression of chimaeric genes transferred into plant cells using a Ti-plasmid-derived vector, Nature (London), 303:209.

Higuchi, T., 1981, Biosynthesis of lignin, in "Plant Carbohydrates II", (Encyclopedia of Plant Physiology, New Series Vol. 12B), W. Tanner, ed., Springer-Verlag, Berlin, pp. 194-224.

Hofmann, C., Vanderbruggen, H., Höfte, H., Van Rie, J., Jansens, S., and Van Mellaert, H., 1988, Specificity of *Bacillus thuringiensis* δ-endotoxins is correlated with the presence of high-affinity binding sites in the brush border membrane of target insect midguts, <u>Proc. Natl. Acad. Sci. USA</u>, 85:7844.

Höfte, H., and Whiteley, H. R., 1989, Insecticidal crystal proteins of *Bacillus thuringiensis*, <u>Microbiol. Rev.</u>, 53:242.

Höfte, H., Van Rie, J., Jansens, S., Van Houtven, A., Vanderbruggen, H., and Vaeck, M., 1988, Monoclonal antibody analysis and insecticidal spectrum of three types of lepidopteran-specific insecticidal crystal proteins of *Bacillus thuringiensis*, <u>Appl. Envir. Microbiol.</u>, 54:2010.

Klein, T. M., Wolf, E. D., Wu, R., and Sanford, J. C., 1987, High-velocity microprojectiles for delivering nucleic acids into living cells, <u>Nature</u> (London), 327:70.

Krieg, A., 1986, *Bacillus thuringiensis* ein mikrobielles Insektizid, <u>Acta Phytomedia</u>, 10:1.

Laties, G. G., 1982, The cyanide-resistant alternative path in higher plant respiration, <u>Ann. Rev. Plant Physiol.</u>, 33:519.

Lazzeri, P., and Lörz, H., 1988, *In vitro* genetic manipulation of cereals and grasses, <u>Adv. Cell Culture</u>, 6:291.

Lilley, M., Ruffell, r. N., and Somerville, H. J., 1980, Purification of the insecticidal toxin in crystals of *Bacillus thuringiensis*, <u>J. Gen. Microbiol.</u>, 118:1.

Matters, G. L., and Scandalios, J. G., 1986, Effect of the free radical-generating herbicide paraquat on the expression of the superoxide dismutase (*Sod*) genes in maize, <u>Biochem. Biophys. Acta</u>, 882:29.

McCabe, D. E., Swain, W. F., Martinell, B. J., and Christou, P., 1988, Stable transformation of soybean (*Glycine max*) by particle acceleration, <u>Bio/technology</u>, 6:923.

Monk, L. S., Fagerstedt, K. V., and Crawford, R.M.M., 1987, Superoxide dismutase as an anaerobic polypeptide. A key factor in recovery from oxygen deprivation in *Iris pseudacorus*? <u>Plant Physiol.</u>, 85:1016.

Mudd, S. H., Finkelstein, J. D., Irreverre, F., and Laster, L., 1965, Transsulfuration in mammals. Microassays and tissue distributions of three enzymes of the pathway, <u>J. Biol. Chem.</u>, 240;4382.

Murakami, T., Anzai, H., Imai, S., Satoh, A., Nagaoka, K., and Thompson, C. J., 1986, Bialaphos biosynthetic genes of *Streptomyces hygroscopicus*: molecular cloning and characterization of the gene cluster, <u>Mol. Gen. Genet.</u>, 205:42.

Nelson, R. S., McCormick, S. M., Delanney, W., Dubé, P., Layton, J., Anderson, E. J., Kaniewska, M., Proksch, R. K., Horsch, R. B., Rogers, S. G., Fraley, R. T., and Beachy, R. N., 1988, Virus tolerance, plant growth, and field performance of transgenic tomato plants expressing coat protein from tobacco mosaic virus, <u>Bio/technology</u>, 6:403.

Peleman, J., Boerjan, W., Engler, G., Seurinck, J., Botterman, J., Alliotte, T., Van Montagu, M., and Inzé, D., 1989, Strong cellular preference in the expression of a housekeeping gene of *Arabidopsis thaliana* encoding S-adenosylmethionine synthetase, The Plant Cell, 1:81.

Peleman, J., Saito, K., Cottyn, B., Engler, G., Seurinck, J., Van Montagu, M., and Inzé, D., 1989, Structure and expression of the S-adenosylmethionine synthetase gene family in *Arabidopsis thaliana*, Gene, in press.

Rabinowitch, H. D., Sklan, D., and Budowski, P., 1982, Photo-oxidative damage in the ripening tomato fruit: protective role of superoxide dismutase, Physiol. Plant., 54:369.

Raskin, I., Ehmann, A., Melander, W. R., and Meeuse, B.J.D., 1987, Salicylic acid: a natural induce of heat production in *Arum* lilies, Science, 237:1601.

Sanford, J. C., 1988, The biolistic process, Trends Biotech., 6:299.

Schatz, G., 1987, Signals guiding proteins to their correct locations in mitochondria, Eur. J. Biochem., 165:1.

Shaaltiel, Y., and Gressel, J., 1987, Kinetic analysis of resistance to paraquat in *Conyza*. Evidence that paraquat transiently inhibits leaf chloroplast reactions in resistant plants, Plant Physiol., 85:869.

Shah, D. M., Horsch, R. b., Klee, H. J., Kishore, G. M., Winter, J. A., Tumer, N. E., Hironaka, C. M., Sanders, P. R., Gasser, C. S., Aykent, S., Siegel, N. R., Rogers, S. G., and Fraley, R. T., 1986, Engineering herbicide tolerance in transgenic plants, Science, 233:478.

Shimamoto, K., Terada, R., Izawa, T., and Fujimoto, H., 1989, Fertile transgenic rice plants regenerated from transformed protoplasts, Nature (London), 338:274.

Teeri, T. H., Lehväshlaiho, H., Franck, M., Uotila, J., Heino, P., Palva, E. T., Van Montagu, M., and Herrera-Estrella, L., 1989, Gene fusions to *lacZ* reveal expression patterns of chimeric genes in transgenic plants, EMBO J., 8:343.

Vaeck, M., Reynaerts, A., Höfte, H., Jansens, S., De Beuckeleer, M., Dean, C., Zabeau, M., Van Montagu, M., and Leemans, J., 1987, Insect resistance in transgenic plants expressing modified *Bacillus thuringiensis* toxin genes, Nature (London), 328:33.

Valvekens, D., Van Montagu, M., and Van Lijsebettens, M., 1988, *Agrobacterium tumefaciens*-mediated transformation of *Arabidopsis* root explants using kanamycin selection, Proc. Natl. Acad. Sci. USA, 85:5536.

Vanderkerckhove, J., Van Damme, J., Van Lijsebettens, J., Botterman, J., De Block, M., Vandewiele, M., De Clercq, A., Leemans, J., Van Montagu, M., and Krebbers, E., 1989, Enkephalins produced in transgenic plants using modified 2S seed storage proteins, Bio/technology, 7:929.

Velten, J., Velten, L., Hain, R., and Schell, J., 1984, Isolation
    of a dual plant promoter fragment from the Ti plasmid of
    Agrobacterium tumefaciens, <u>EMBO J</u>., 3:2723.
Yang, S. F., and Hoffman, N. E., 1984, Ethylene biosynthesis and
    its regulation in higher plants, <u>Ann. Rev. Plant Physiol</u>.,
    35:155.
Zambryski, P., 1988, Basic processes underlying *Agrobacterium*-
    mediated DNA transfer to plant cells, <u>Ann. Rev. Genet</u>.,
    22:1.

TRANSFORMATION AND REGENERATION OF IMPORTANT CROP PLANTS: RICE AS THE MODEL SYSTEM FOR MONOCOTS

RAY WU, ELIZABETH KEMMERER AND
DAVID MCELROY
Section of Biochemistry, Molecular and Cell Biology
Biotechnology Building, Cornell University
Ithaca, N. Y. 14853, U. S. A.

INTRODUCTION

The use of transgenic plants as experimental tools in gene analysis and for the alteration of agronomic characters is fundamental to our attempts to improve crops. There are at least five ways to introduce foreign genes into plants. Here we would like to describe the merits and shortcomings of each method and present our results in producing transgenic rice plants. In addition, we propose a method using transgenic plants to facilitate cloning of genes which are difficult to identify by other methods.

The plant genome can be specifically manipulated once efficient methods for gene transfer and plant regeneration are developed. The major reasons for developing methods to produce transgenic plants are twofold. First, one can use the system for molecular analysis of gene expression *in vivo*. Second, one can introduce specific, agronomically important traits into plants.

Presently, the most convenient method for introducing foreign genes into dicotyledonous plants (dicots) is based on the *Agrobacterium tumefaciens*-Ti plasmid system. Infection by *Agrobacterium* results in the integration of a specific segment of its Ti plasmid, together with any foreign gene joined to it, into the plant genome (Gheysen et al., 1985; Fraley et al., 1986). Using this system, transgenic tobacco (*Nicotiana tobaccum*), tomato (*Lycopersicon esculentum*) and *Brassica napus* plants have been produced routinely. A significant limitation of this method is that important monocotyle-donous plants (monocots) such as maize (*Zea mays*), rice (*Oryza sativa*) and wheat (*Triticum aestivum*) cannot be transformed by the *A. tumefaciens* –Ti plasmid system (Cocking et al., 1987; Goodman et al., 1987, Wu 1989). Consequently, our laboratory and others have been involved in developing alternative methods for introducing foreign genes into monocots.

Gene Manipulation in Plant Improvement II
Edited by J. P. Gustafson
Plenum Press, New York, 1990

Over the past few years, several different methods for transferring foreign genes into protoplasts or cells have been examined. The introduction of DNA into protoplasts by electroporation (Fromm et al., 1986; Ou–Lee et al., 1986) or by polyethylene glycol treatment (Potrykus et al., 1984; Negrutiu et al., 1987) in both dicots and monocots is relatively straightforward. The major problem in the monocots is in the regeneration of fertile transgenic plants from protoplasts.

TRANSFORMATION AND REGENERATION OF IMPORTANT MONOCOTS

In maize, stable integration of foreign genes into protoplasts has been reported (Fromm et al., 1986). Transgenic plants were regenerated (Rhodes et al., 1988), but they were not fertile.

In rice, three laboratories have succeeded in producing transgenic plants using the protoplast method (Zhang and Wu, 1988; Toriyama et al., 1988; Zhang et al., 1988). Only one group has shown that the transgenic plants were fertile by producing viable seeds (Zhang and Wu, 1988).

Two other procedures for producing transgenic plants in monocots are transformation by microinjection and the pollen–tube pathway method. For the first method, transgenic rye (*Secale cereale*) plants were produced by injecting foreign DNA into immature florets (de la Pena et al., 1987). Using the pollen–tube pathway method, transgenic rice plants were obtained by placing a drop of foreign DNA on the excised stigmas of rice florets. The DNA solution flows down along the pollen tube pathway and enters the cells in the embryonic sac. Using this technique, mature seeds were obtained and transgenic plants were germinated (Luo and Wu, 1988). In both cases experiments have not yet been carried out to show that the foreign gene is transmitted to the next generation.

A novel method of transformation, the biolistic method, can deliver foreign DNA into intact cells by high–velocity tungsten particles (Klein et al., 1987). In rice, maize and wheat, bombardment of intact cells (grown as cell suspensions) with DNA–coated tungsten particles resulted in transient expression of the foreign gene (Wang et al., 1988). In maize, bombardment of intact cells resulted in stable transformation (Klein et al., 1988), but no plants have been generated yet.

CRITERIA FOR A USEFUL TRANSGENIC SYSTEM

A useful system for generating transgenic plants must be one that is reproducible. The system must allow stable integration in the transformed plant and transfer of the foreign gene to the subsequent generations. The most convincing evidence for transformation is genetic segregation of the transformed phenotype in progeny. Furthermore, a useful system must enable the scientists to generate hundreds of transgenic plants. So far, of the three methods described in the earlier sections, only the protoplast method in rice comes close to meeting the above criteria. In rice, a fairly large number of transgenic plants (86 in one experiment using 10^7 protoplasts) were regenerated (Zhang and Wu, 1988). The experiment can be efficiently scaled up fivefold to produce over 400 transgenic plants. This system is suitable for the molecular analysis of gene expression *in vivo* in delineating the essential DNA sequences responsible for the specific expression of different genes. A heterologous system can be used in certain cases to study the expression of a foreign gene, e.g. the analysis of a rice or maize gene in the tobacco system. However, it is more desirable that a homologous or near homologous system be used. Now one can study the expression of a rice or maize gene in the rice protoplast transformation system. Moreover, one can use this system to facilitate cloning of genes in a new method to be described later.

One limitation of the rice protoplast system for transformation and regeneration of transgenic rice is that, so far, only a small number of *japonica* varieties can be used to regenerate plants from protoplasts. Thus, the protoplast system is far from ideal for introducing specific agronomically important genes into rice. Since about 80% of the rice cultivars used around the world are *indica* varieties, more work is needed to develop regeneration methods from protoplasts of these varieties.

In principle, the methods based on the pollen–tube pathway and the biolistic process are not limited to specific cultivars of rice. Thus, transgenic plants from the majority of rice cultivars may be produced. So far, these two methods have not yet been sufficiently developed to prove that the introduced foreign genes have been integrated into the chromosomes and transmitted to the following generations. Additional work is being carried out in our laboratory to firmly establish the usefulness of these methods. On the other hand, Duan and Chen (1985) transferred total DNA from a variety of rice with purple leaves and glumes (donor) to a variety of rice with normal green pigmented tissue (recipient) by the pollen–tube pathway method. Seeds which matured from the treated florets were planted and some of the resulting plants showed purple leaves and glumes. These characters were transmitted to the second and third generation as well. However, Duan and Chen (1985) did not present any molecular evidence to confirm that these phenotypic changes were the result of the transferred donor DNA, as opposed to some sort of mutational event in some of the recipient plants.

Once efficient methods for producing transgenic rice plants are established, one needs to decide which gene or genes to introduce into rice to improve agronomically important traits. So far, only a limited number of useful genes have been isolated, all from other sources, and they have been applied to dicots only. These include (1) herbicide resistance genes (Botterman and Leemans, 1988) such as glyphosate resistance and sulfonylurea resistance genes (Haughn et al., 1988), (2) genes that produce toxic protein products which kill insects such as a cowpea trypsin inhibitor gene (Hilder et al., 1987) and a *bacillus thurengiensis* toxin gene (Schnepf et al., 1985), and (3) genes that produce protein products that cross–protect plants against virus infection (Abel et al., 1986).

In most instances, the identity of agronomically important genes is lacking, thus deterring isolation of the DNA encoding the gene. An interesting approach has been developed but has been applied only to isolating genes which are readily detectable but do not have agronomic value. It is known as transposon

tagging, which traces genes inactivated by transposon insertions. This method
has been successfully used to clone approximately ten different genes from maize
(Federoff et al., 1984; O'Reilley et al., 1986). Although several families of
transposons have been identified in maize, none was found in rice. Very recently,
Yong Xie in our laboratory identified a Mu–like transposon which has all the
structural characteristics of the Mu transposon from maize (Lillis and Freeling,
1986). Once an active Mu–like transposon in rice is characterized, the transposon
tagging approach can be used to isolate agronomically useful genes from rice.

We now propose a new method for cloning unknown genes. This method
involves restriction fragment length polymorphism (RFLP) analysis followed by
cloning of the DNA fragments in between two DNA markers and identifying the
desired gene after transformation and regeneration of transgenic plants. This new
method will be described more fully since it has not yet been used to isolate a
single gene.

A PROPOSED NEW METHOD FOR GENE CLONING THROUGH RFLP ANALYSIS, PULSE FIELD GEL ELECTROPHORESIS AND TRANSFORMATION

First, we will briefly describe the principle of the method. Later on, we
will use a hypothetical example to illustrate the method in more detail. The
method can be divided into six steps as shown in Fig. 1.

Step I: a specific gene or phenotypic trait is mapped by RFLP analysis
using cloned DNA markers as probes (Botstein et al., 1980).

Step II: the rice DNA is specifically digested to sizes between 3 and 10
megabase pairs (mbp) and the DNA fragments fractionated by pulse field gel
electrophoresis (PFGE). Through hybridization analysis, the DNA fragment that
hybridized to both M1 and M2 probes (Fig. 1, Step I and Fig. 2b) is identified.

Step III: the desired DNA from Step II is digested to 15 kbp size and
cloned into a λ vector system. Forty single–copy rice DNA sequences in these
clones are used to further map the DR gene by RFLP so that the distance between
M3 and M4 probes is in the range of 0.3–0.6 mbp.

Step IV: total rice DNA is first digested with a restriction enzyme and
fractionated as in Step II above and the desired DNA fragment (Fig. 2b) is cut out
from the gel and digested with NotI (or SfiI) enzyme to give fragments in the
range of 0.3–0.6 mbp. The gel band is placed in the slot at the origin of another
gel (Fig. 2c) and PFGE performed. After hybridization analysis, the DNA band
that hybridized to both M3 and M4 probes is excised (Fig. 2d).

Step V: the DNA from Step IV above is digested to an average size of 40
kbp and the fragments cloned in a cosmid vector system.

Step VI: the desired gene is identified by transforming rice protoplasts
with DNA prepared from the cosmid library and screening for the trait in
transgenic plants. Even though the proposed method has not yet been carried out
fully, each individual part of the method has been tested by others or in our
laboratory.

I. Isolate DNA from a DR rice plant and map the DR gene by RFLP analysis.

3 to 6 megabase pairs (mbp)

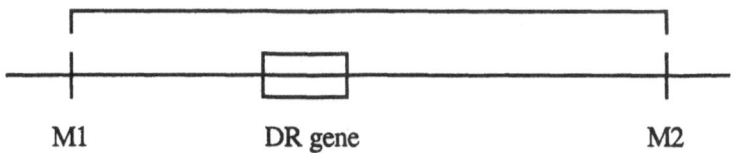

II. Digest the chromosomal DNA to 3–10 mbp and fractionate the DNA by PFGE. Identify the fragment that hybridizes with both M1 and M2 probes.

III. Digest the 3–10 mbp DNA to 15 kbp fragments and clone them in a λ vector system to produce more probes. Additional RFLP analysis to place the DR gene between M3 and M4.

0.3 – 0.6 mbp

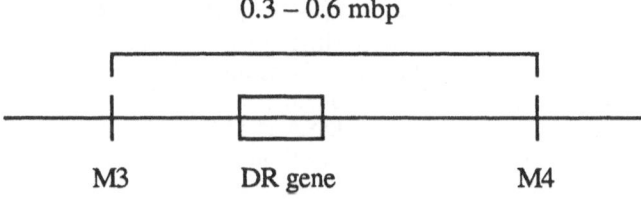

IV. Repeat Step II above followed by digesting the desired fragment to 0.3–0.6 mbp and PFGE. Identify and excise the fragment that hybridizes with both M3 and M4 probes.

V. Digest the 0.3–0.6 mbp fragment from Step IV to 40 kbp and clone them in a cosmid vector system.

VI. Identify the DR gene by transforming rice protoplasts with the cosmid clones and screening the DR trait in transgenic plantlets.

Fig. 1. A proposed new method for cloning an unknown gene, such as a disease resistance (DR) gene.

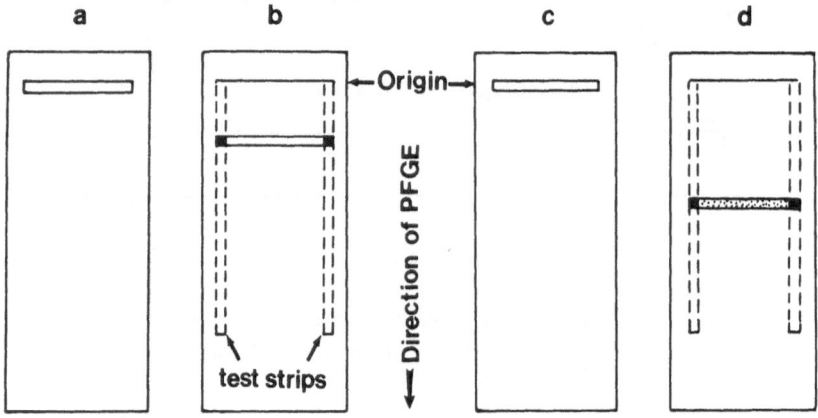

Fig. 2.   A scheme for the sequential fractionation and manipulation of large
          DNAs in an agarose gel.
          (a) An agarose gel for PFGE. The open bar at the origin represents the
          location where the agarose is removed and replaced by a gel band that
          contains purified full length chromosomal rice DNA. The total amount
          of DNA in this band is approximately 1 mg. The apparatus used for
          PFGE is based on the hexagonal design (Orbach et al., 1988).
          (b) After PFGE, the DNA molecules are distributed throughout the gel.
          Since the average size of the DNA is 16,700 kb, one expects
          approximately 40 fragments based on the rice genome size of 600,000
          kb. The two vertical bars within the dotted lines are regions of the gel to
          be cut out as test strips to find out by hybridization analysis which part
          of the gel has the DNA molecule that includes the BDR gene. The open
          bar ends with black squares represent the DNA that hybridized to both
          RFLP probes and thus includes the BDR gene. Within this region there
          may be up to 50 μg of DNA including several different fragments of
          DNA of similar length.
          (c) An agarose gel for PFGE. The open bar at the origin represents the
          location where the agarose is removed and replaced by a gel band
          excised in step (b) above.
          (d) After PFGE, the DNA is distributed in different parts of the gel.
          The average size of the DNA is 40 kb. The two vertical bars within the
          dotted lines are regions of the gel to be cut out as test strips. The open
          bar ends with black squares represents the DNA that hybridized to both
          RFLP probes. Within this region, there may be 5 μg of DNA which is
          likely to be of high purity.

In order to illustrate the method in more detail, we will describe the cloning of a gene that confers blast disease resistance (BDR) in rice. There are a dozen different BDR genes, each specific for a given race of blast fungus. The BDR characteristic behaves like a single gene and the BDR trait can be introduced from a BDR rice cultivar into a sensitive one by traditional plant breeding. Genetic analysis indicates that the BDR trait behaves like a dominant gene. Several BDR genes have been mapped by traditional genetic methods. For example, Pi–z, *Pyricularia oryzae* resistance–1, belongs to linkage group I (*wx* group) at gene locus 68 (Oka and Khush, 1984). Thus, in principle, it is possible to isolate the BDR gene from a blast resistant cultivar and use it to transform a sensitive rice plant. The BDR gene can be identified if it codes for a protein that makes the transgenic plant resistant to the fungus. This characteristic is used as an assay to identify a cloned rice DNA segment that includes the BDR gene. Since circumstantial evidence indicates that the difference between the BDR gene and its counterpart in sensitive plants may be due to simple mutations, the common method of differential screening of a cDNA library is not likely to work.

Step I: Mapping the BDR Trait by RFLP Analysis

A RFLP map in rice has been developed by McCouch et al. (1988). There are now close to 200 DNA markers placed on the 12 rice chromosomes. The genome size of rice is 600,000 kb (Bennett and Smith, 1976). This number divided by 200 gives 3,000 kb, which is the average distance between any two adjacent DNA markers. The BDR gene can be mapped using the RFLP method by looking for co–segregation of the BDR trait with specific DNA markers.

Step II: Identifying and Isolating a Very Large DNA Fragment Containing the BDR Gene

A rice variety harboring the mapped BDR gene is chosen as the source of donor DNA. Rice leaves can be used to prepare DNA but a better material is cells grown in tissue culture. The rice cells or segments of finely chopped leaves or cultured tissue are immobilized in agarose (Smith and Cantor, 1987) and treated with cellulase and other enzymes to remove the cell wall (Cocking et al., 1986). Proteinase K is added to release the DNA from the cells into the agarose (Luo and Wu, unpublished data).

(A) The rice DNA (ca. 2 mg) immobilized in the agarose block is subjected to PFGE to separate intact chromosomal DNA from broken DNA as well as from mitochondrial and chloroplast DNA.

(B) The region of the gel that includes the full–sized chromosomal DNA is cut out and treated *in situ* with a XbaI methylase to produce a 12–base–pair recognition sequence which can be digested by the restriction enzyme DpnI to produce DNA fragments with an average size of 16,700 kb (Weil and McClelland, 1989), perhaps covering a range of 4,000 to 40,000 kb. After DpnI digestion, the agarose gel band is placed in a gel slot (Fig. 2a, at the origin of the gel) and subjected to PFGE. Next, two narrow test strips of the gel are cut out from the edges of the gel (Fig. 2b, areas shown by the dotted lines) and transferred to nitrocellulose paper for hybridization with the two DNA probes that flank the BDR gene. The gel band (shown as the open bar in Fig. 2b) in between the hybridizing regions (black square) of the test strips that hybridized to both probes is cut out from the gel. If the desired gel band is much larger than 6 mb, the rice DNA in step (B) above will be treated with ClaI methylase to produce a 10–base–pair recognition sequence which can be digested with DpnI to produce DNA fragments with an average size of 10,000 kb.

Step III: Further Mapping the BDR Gene by RFLP Analysis

The 3–10 mbp DNA from Step II is digested to 15 kbp size by partial Sau3A digestion and the fragments are ligated to a λ phage vector to make a small DNA library. Between 920 and 3070 clones are needed to cover the entire region with a probability (P) of 99% that every region is included. The calculation is based on the formula by Clarke and Carbon (1979): $N = \ln(1–P)/\ln(1–f)$, where N represents the number of clones to be screened in order to be 99% certain of encountering at least one copy of the desired clone, and f represents the fractional abundance of relevant clones (or a fraction of the entire region). This formula is also used in subsequent sections for calculating the number of clones needed to reach 99% probability. However, only approximately 200 overlapping clones or 40 non–overlapping clones are needed for further mapping the BDR gene by RFLP analysis. Before selecting the 40 non–overlapping clones, 200 clones are tested by dot blot hybridization to find the clones that contain only single copy genes, and do not cross hybridize with one another. To make dot hybridization easier, the 200 clones are tagged by the non–radioactive method using a biotinylated probe (Gebeyechu et al., 1987). Also, the clones are screened by dot blot hybridization to DNA isolated from a rice variety which is trisomic for the chromosome containing the M1 and M2 probes. Additional RFLP analysis with the 40 clones using 40–80 F2 plants should allow mapping the DR gene within two markers (M3 and M4) which are within 0.3–0.6 mbp apart.

Step IV: Isolating and Purifying the 3–10 mbp and then the 0.3–0.6 mbp Fragments by Consecutive PFGE

Digesting the total rice DNA to an average size of 3–10 mbp should give between 60–200 fragments. After PFGE, there may be 4–10 fragments which are of similar size as the desired one that hybridized to both M1 and M2 probes. However, digestion of these fragments with NotI (or SfiI) followed by PFGE should give almost pure 0.3–0.6 mbp DNA that hybridizes to both M3 and M4 probes.

Step V: Digesting the Large DNA From Step (IV) for Cloning in a Cosmid Vector

The NotI band is eluted from the agarose and the DNA (ca. 2 µg) is partially digested with a restriction enzyme (such as Sau3A) to an average size of 40 kb. The DNA fragments are ligated into a cosmid vector (DeLella and Woo, 1987) which also contains a selective marker, such as a sulfonylurea resistance gene (Haughn et al., 1988). If the large DNA fragment from Step IV above is 300–600 kbp, in principle only 67 cosmid clones are needed to cover this region. In practice, around 70–100 clones are to be analyzed to find a clone that includes the BDR gene.

Step VI: Identifying the BDR Gene in Transgenic Rice Plantlets

(A) Seventy cosmid clones are randomly selected for transformation of protoplasts isolated from a plant which is sensitive to blast disease. The presence of the BDR gene in each DNA sample is searched for after transforming 10^6 rice protoplasts and growing them to the callus stage (ca 2 weeks). Calli containing cosmid DNA are selected by exposure to sulfonylurea. Resistant calli are transferred to regeneration medium for plantlet formation. Approximately 5 plantlets can be regenerated in each transformation experiment, based on the efficiency obtained in our laboratory (Zhang and Wu, 1988). In principle, if DNA from only one cosmid clone were integrated and expressed in each plantlet, only one plantlet is needed for each experiment. Thus, 5 plantlets provides a safety factor of five. A total number of 350 plantlets need to be regenerated in the 70 experiments, which can be readily managed.

(B) Direct selection for plantlets containing an expressed BDR gene occurs by spraying 8–10 cm tall plantlets with a solution containing spores of the fungus in a high–humidity chamber. One week later, the plantlets are examined for fungus infection on leaves. The plantlets expressing the BDR genes would be resistant and show no (or much decreased) symptoms of infection. If a plantlet in a given experiment shows no sign of infection, DNA from that cosmid used in the transformation experiment is likely to include the BDR gene. Assuming that suspension cells are available for making protoplasts, steps (A) and (B) require approximately 10–12 weeks.

(C) The cosmid that contains the BDR gene identified in step (B) above is partially digested with a restriction enzyme to an average size of 4 kb. The DNA fragments are ligated to a plasmid (such as pUC13) to construct a small plasmid library. In principle, only 45 clones are needed. Plasmid DNA is isolated from each of the 45 clones and separate transformation experiments are carried out. The plantlets are analyzed for resistance to the blast fungus, as in Step VI (B). If one or several BDR plantlets are found, the 4 kb plasmid DNA used in that transformation experiment is likely to include the BDR gene. This gene will be sequenced and the gene can be transferred to sensitive plants by transformation to render the plants disease resistant.

Should the number of transgenic plants to be regenerated in (A) prove too numerous, for easy management a two–step selection procedure can be employed. The cosmid clones can be pooled into groups of 7 to 10 members. Initial selection in (B) would be at this group level prior to individual selection of those members of the group which is found to contain the desired gene. With 70 cosmid clones to assay in the above example, and an initial pool size of 7 clones, one would only have to analyze 85 transgenic plants (10x5 + 7x5) in such a two–step selection procedure.

In the example given, it is assumed that the BDR gene codes for a protein that makes the plants, including the transgenic plants, resistant to the blast fungus. However, it is also possible that disease resistance is due to the lack of a protein in the resistant rice plant. According to this possibility, the rice plant is resistant because it fails to produce the protein that the fungus recognizes prior to infection. If this were the case, we can use the following approach to achieve our goals.

Assume that the resistant plant has a nonfunctional "recognition protein" gene (rp) and does not produce the recognition protein. The sensitive plant has a functional "recognition protein" gene (RP) and produces the recognition protein. In Step II, we will start with the sensitive plant and use the RFLP and PFGE methods to isolate the sensitivity gene RP. We then transform resistant plants with the 70 cosmid clones (Step VI), grow to callus stage, select for transformants with sulfonylurea, regenerate transformants to plantlets, spray with the fungus spores and select the plants which show signs of infection (that is, resistant plants transformed to the sensitive phenotype).

Once the RP gene from sensitive rice plants has been identified and sequenced, we can use this information to confer resistance to sensitive plants. For example, we can construct a transformation vector containing a DNA sequence which would produce anti–sense RNA in transformed plants. It would be even better if a ribozyme sequence (Walbot and Bruening, 1988) is included as part of the synthetic DNA so that the construct can cleave and destroy the mRNA coded for by the RP gene. Theoretically, either approach would result in the removal of the recognition protein mRNA from the translatable mRNA pool. The recognition protein would not be made and the sensitive plant will have been transformed to the resistant phenotype. Alternatively, we may be able to identify an enzyme which interferes with the biosynthesis of the mature recognition protein. The introduction of the gene for this enzyme into sensitive plants would also prevent biosynthesis of the recognition protein and render the transformed sensitive plant resistant.

The same approach for cloning the disease resistance gene can be used to clone any gene as long as there is an assay system for identifying positive transgenic plants. For example, genetic studies indicated that a single gene, Pl, is responsible for producing rice plants with purple leaves (Nagao et al., 1968; Oka and Khush, 1984). One can clone the Pl gene by the same procedures described for the cloning of a BDR gene. However, it is much easier to identify the Pl gene transformed plants. In Step III (C), for screening the cosmid clones for the one that include the Pl gene, all one needs to do is to look for the regenerated plantlets with purple leaves. In fact, it would be a good idea to first use this as a model system before trying to clone the BDR gene.

ACKNOWLEDGEMENTS

I thank R. Grumbles and N. Ayres for careful reading of this manuscript and for making helpful suggestions. This work was supported by research grants RF 84066, Allocation No. 3, from the Rockefeller Foundation, and GM29279 from NIH, U. S. Public Health Service. E. Kemmerer was supported by a fellowship from the Plant Science Center, Cornell University. D. McElroy was supported by a U. K. SERC/NATO Studentship, a Fulbright Fellowship and scholarships from the British Universities North America Club and the St. Andrew's Society of Washington, D. C., USA.

REFERENCES

Abel, P. P., Nelson, R. S., De, B., Hoffmann, N., Rogers, S. G., Fraley, R. T., and Beachy, R. N., 1986, Delay of disease development in transgenic plants that express the tobacco mosaic virus coat protein gene, Science, 232:738–743.

Abdullah, R., Cocking, E. C., and Thompson, J. A., 1986, Efficient plant regeneration from rice protoplasts through somatic embryogenesis, Bio/Technology, 4:1087–1097.

Bennett, M. D., and Smith, J. B., 1976, Nuclear DNA amounts in angiosperms, Philos. Trans. Royal Soc. (London), 274:227–274.

Botstein, D., White, R. L., Skolnick, M., and Davis, R. W., 1980, Construction of a genetic linkage map in man using restriction fragment length polymorphisms, Am. J. Human Genet., 32:314–331.

Botterman, J., and Leemans, J., 1988, Engineering herbicide resistance in plants, Trends in Genen., 4:219–222.

Carle, G. F., and Olson, M. V., 1987, Orthogonal–field–alteration gel electrophoresis, Methods in Enzymol., 155:468–482.

Clarke, L., and Carbon I., 1979, Selection of specific clones from colony banks by suppression or complementation tests, Methods in Enzymol., 68:346–408.

Cocking, E. C., and Davey, M. R., 1987, Gene transfer in cereals, Science, 236:1259–1262.

de la Pena, A., Lorz, H., and Schell, J., 1987, Transgenic rye plants obtained by injecting DNA into young floral tillers, Nature, 325:274–276.

DeLella, A. G., and Woo, S. L. C., 1987, Cloning large segments of genomic DNA using cosmid vectors, Methods in Enzymol., 152:199–212.

Duan, X., and Chen, S., 1985, Variation of the characters in rice (Oryza sativa) induced by foreign DNA uptake, China Agricul. Sci., 3:6–9.

Fedoroff, N., Furtek, D., and Nelson, O., 1984, Cloning of the bronze locus in maize by a simple and generalizable procedure using the transposable controlling element, Ac, Proc. Natl. Acad. Sci. U. S. A., 81:3825.

Fraley, R. T., Rogers, S. G., and Horsch, R. B., 1986, Genetics transformation in higher plants, CRC Crit. Rev. Plant Sci., 4:1–46.

Fromm, M. E., Taylor, L. P., and Walbot, V., 1986, Stable transformation of maize after gene transfer by electroporation, Nature, 319:791–793.

Gebeyechu, G., Rao, P. Y., Soo Chan, P., Simms, D. A., and Klevan, L., 1987, Novel biotinylated nucleotide–analogs for labeling and colorimetric detection of DNA, Nucleic Acids Res. 15:4513–4534.

Goodman, R. M., Hauptli, H., Crossway, A., and Knauf, V. C., 1987, Gene transfer in crop improvement, Science, 236:48–54.

Gheysen, G., Dhaese, P., van Montagu, M., and Schell, J., 1985, Advances in plant gene research, in: "Genetic Flux in Plants," B. Hohn, and E. S. Dennis, eds., Springer, New York (1985), Vol. 2, pp. 11–47.

Haughn, G. W., Smith, J., Mazur, B., and Somerville, C., 1988, Transformation with a mutant Arabidopsis acetolactate synthase gene renders tobacco resistant to sulfonylurea herbicides, Mol. Gen. Genet., 211:266–271.

Hilder, V. A., Gatehouse, A. M. R., Sheerman, S. E., Barker, R. F., and Boulter, D., 1987, A novel mechanism of insect resistance engineered in tobacco, Nature, 330:160–163.

Klein, T. M., Wolf, E. D., Wu, R., and Sanford, J. C., 1987, High–velocity microprojectiles for delivering nucleic acids into living cells, Nature, 327:70–73.

Klein, T. M., Gradziel, T., Fromm, M. E., and Sanford, J. C., 1988, Factors influencing gene delivery into Zea mays by high–velocity microprojectiles, Bio/Technology, 6:559–563.

Lillis, M., and Freeling, M., 1986, Mu transposons in maize, TIG, July:183–187.

Luo, Z. X., and Wu, R., 1988, A simple method for the transformation of rice via the pollen–tube pathway, Plant Molec. Biol. Reporter, 6:165–174.

McCouch, S. R., Kochert, G., Yu, Z. H., Wang, Z. Y., Khush, G. S., Coffman, W. R., and Tanksley, S. D., 1988, Molecular mapping of rice chromosomes, Theor. Appl. Genet., 76:815–829.

Nagao, S., Takahashi, M., and Kinoshita, T., 1968, Heterotic effect of alleles at Pl–locus in rice plant, Genetic studies on rice plant, XXX, J. Fac. Agr. Hokkaido Univ., 56 (1):45–56.

Negrutiu, I., Shillito, R., Potrykus, I., Biasini, G., and Sala, F., 1987, Hybrid genes in the analysis of transformation conditions. I. Setting up a simple method for direct gene transfer in plant protoplasts, Plant Mol. Biol., 8:363–373.

Oka, H. I., and Khush, G. S., eds., 1984, Rice Genetics Newsletter, Vol. 1, Japanese Rice Genetics Information Committee, Japan (1984).

Orbach, M. J., Vollrath, D., Davis, R. W., and Yanofsky, C., 1988, An electrophoretic karyotype of Neurospora crassa, Mol. Cell Biol., 8:1469–1473.

O'Reilley, C., Shepherd, N. S., Pereira, A., Schwarz–Sommers, Z., Bertram, I., and Peterson, P. A., 1985, Molecular cloning of the al locus of Zea mays using the transposable elements En and Mu1, EMBO J., 4:877–882.

Ou–Lee, T. M., Turgeon, R., and Wu, R., 1986, Expression of a foreign gene linked to either a plant virus or a Drosophila promoter, after electroporation of protoplasts of rice, wheat, and sorghum, Proc. Natl. Acad. Sci. U. S. A., 83:6815–6819.

Potrykus, I., Shillito, R. D., Saul, M. W., and Paszkawski, J., 1985, Direct gene transfer—state of the art and future potential, Plant Mol. Biol. Reporter, 3:117–128.

Rhodes, C. A., Pierce, D. A., Mettler, I. J., Mascarenhas, D., and Detmer, J. J., 1988, Genetically transformed maize plants from protoplasts, Science, 240:204–207.

Schnepf, H. E., Wong, H. C., and Whiteley, H. R., 1985, The amino acid sequence of a crystal protein from Bacillus thuringiensis deduced from the DNA base sequence, J. Biol. Chem., 260:6264–6272.

Schwartz, D., and Cantor, C. R., 1984, Separation of yeast chromosome–sized DNA's by pulsed field gradient gel electrophoresis, Cell, 37:67–75.

Smith, C. L., and Cantor, C. R., 1987, Purification, specific fractionation, and separation of large DNA molecules, Methods in Enzymol., 155:449–467.

Toriyama, K., Arimoto, Y., Uchimiya, H., and Hinata, K., 1988, Transgenic rice plants after direct gene transfer into protoplasts, Bio/Technology, 6:1072–1074.

Walbot, V., and Bruening, G., 1988, Plant development and ribozymes for pathogens, Nature, 334:196–197.

Wang, Y. C., Klein, T. M., Fromm, M., Cao, J., Sanford, J. C., and Wu, R., 1988, Transient expression of foreign genes in rice, wheat and soybean cells following particle bombardment, Plant Molec. Biol., 11:433–439.

Weil, M. D., and McClelland, M., 1989, Enzymatic cleavage of a bacterial genome at a 10–base–pair recognition site, Proc. Natl. Acad. Sci. U. S. A., 86:51–55.

Wu, R., 1988, Methods for transforming plant cells, in: "Plant Biotechnology," S.D. Kung, and C.J. Arntzen, eds., Butterworth Publishers (1988), pp. 35–51.

Zhang, H. M., Yang, H., Rech, E. L., Golds, T. J., Davis, A. S., Mulligan, B. J., Cocking, E. C., and Davey, M. R., 1988, Transgenic rice plants produced by electroporation mediated plasmid uptake into protoplasts, Plant Cell Reports, 7:379–384.

Zhang, W. G., and Wu, R., 1988, Efficient regeneration of transgenic plants from rice protoplasts and correctly regulated expression of the foreign gene in the plants, Theor. Appl. Genet., 76:835–840.

GENETIC TRANSFORMATION OF MAIZE CELLS BY PARTICLE BOMBARDMENT AND THE INFLUENCE OF METHYLATION ON FOREIGN-GENE EXPRESSION

T. M. Klein, L. Kornstein and M. E. Fromm

Plant Gene Expression Center, USDA-ARS,

Albany, CA, 94710

INTRODUCTION

Much of the progress in understanding plant gene regulation has been due to the availability of gene transfer systems for many plant species. The results from these studies of the structure of plant gene regulatory regions have been reviewed recently (Schell, 1987; Willmitser, 1988). Additionally, the development of techniques for transferring genes into crop plants allows new agronomic traits to be introduced (Goodman et al., 1987). Recently, resistance to insects (Fischhoff et al., 1987; Hilder et al., 1987; Vaeck et al., 1987), viruses (Abel et al., 1986; Gerlach et al., 1987; Harrison et al., 1987; Cuozzo et al., 1987; Nelson et al., 1987), and herbicides (Shah et al., 1986; della-Cioppa et al., 1987; Fillatti et al., 1987; Cheung et al., 1988) have been introduced into crop plants via gene transfer techniques. Unfortunately much of this progress has not benefited cereals, as general techniques for transferring genes into cereal plants, such as rice (*Oryza sativa*), wheat (*Triticum aestivum*) and maize (*Zea mays*), are not as advanced as for many of the dicot plants. The first instance of a transgenic rice plant has recently been reported (Toriyama et al., 1988). A mature but sterile

Gene Manipulation in Plant Improvement II
Edited by J. P. Gustafson
Plenum Press, New York, 1990

transgenic maize plant has also been obtained (Rhodes et al., 1988). Both of these cases utilized protoplasts as the recipient cell for DNA transfer. The major difficulty with using cereal protoplasts is that they are generally very difficult to regenerate into fertile plants (Cocking and Davey, 1987). Alternative gene transfer techniques have sought to use intact cells as DNA recipients to avoid the difficulties of protoplasts. One of the more promising and dramatic techniques for introducing DNA into intact cells is the use of high-velocity microprojectiles.

The high-velocity microprojectile gene transfer technique consists of coating small metal particles, up to several microns in size, with the DNA to be transferred and accelerating these DNA-coated microprojectiles into intact plant cells. To date, two acceleration devices have been described. One uses a gun powder explosion to accelerate a 22 caliber plastic bullet carrying the microprojectiles (Sanford et al., 1987; Klein et al., 1987; Klein et al., 1988a; Klein et al., 1988b; Klein et al., 1988c; Wang et al., 1988), while the second design uses a high-voltage explosion of a water droplet to propel a thin plastic square carrying the microprojectiles (McCabe et al., 1988; Christou et al., 1988). Both techniques use a stopping plate or screen to stop the large plastic projectile while allowing the microprojectiles to pass through and impact on the target cells. In order to reduce the friction the particles encounter travelling through the air, the process is carried out in a partial vacuum. This microprojectile bombardment process has resulted in gene transfer into cells of tobacco (*Nicotiana tabacum*; Klein et al., 1988c), onion (*Allium cepa*; Klein et al., 1987), maize (*Zea mays*; Klein et al., 1988a; 1988b;), wheat (*Triticum aestivum*), rice (*Oryza sativa*; Wang et al., 1988), and soybeans (Glycine max; McCabe et al., 1988; Christou et al., 1988) as measured by transient gene expression (short term gene expression that does not require integration of the plasmid into one of the cell's chromosomes). Stable transformation of tobacco (Klein et al., 1988c) and soybean (McCabe et al., 1988) plants has been demonstrated recently. High-velocity microprojectiles have not been successfully used to obtain transgenic cereal cells or plants as yet. In this article we described obtaining stably transformed maize cells after bombardment of intact maize tissue culture cells.

Additionally, the transferred plasmid contained a beta-glucuronidase (GUS) gene. The presence of the GUS enyzme in the transformed cells causes the cells to turn blue (Jefferson et al., 1987) in the presence of the histochemical substrate 5-bromo-4-chloro-3-indoyl-ß-D-Glucuronic acid (X-Gluc). The transformed maize calli obtained by gene transfer with high-velocity microprojectiles showed variable levels of Gus activity. Upon staining with X-Gluc, the variation in Gus expression was apparent in different parts of the same callus. Since DNA methylation has been observed to influence gene expression in plants (Hepburn et al., 1987), we investigated the methylation of the GUS gene in different transformed calli. We observed that the calli that express low levels of Gus enzyme contained Gus genes that were hyper-methylated, while calli that express high levels of Gus enzyme contained Gus genes that were hypo-methylated.

RESULTS

Structure of the Transforming Plasmid pNGI

The plasmid pNGI (Fig. 1) contains two genes that are expressed in the plant cells. A neomycin phosphotransferase II (NPT II) gene is used to confer resistance to kanamycin, to which the maize cells are normally sensitive. A ß-glucuronidase (GUS) gene is used to visualize the transformed cells, as cells expressing the GUS enzyme turn blue in the presence of X-gluc (Jefferson et al., 1987). The NPT II gene in pNGI was derived from pCaMVNEO (Fromm et al., 1986) and is composed of the 35S promoter from cauliflower mosaic virus, the NPT II coding region, and the nopaline synthase 3' end. The GUS gene in pNGI is comprised of the promoter and intron 1 (nucleotides 1 to 1775) from the alcohol dehydrogenase 1 (Adh1) gene of maize (Dennis et al., 1984), a GUS coding region consisting of the *Pst* I fragment from pRAJ260 (Jefferson et al., 1986), and a modified 3' end from the nopaline synthase gene (Fromm et al., 1985).

Microprojectile Gene Transfer into Maize Cells

A suspension culture of *Z. mays* cv. Black Mexican Sweet (BMS) was maintained in MS media as previously described (Fromm et al., 1985). The bombardment (Klein

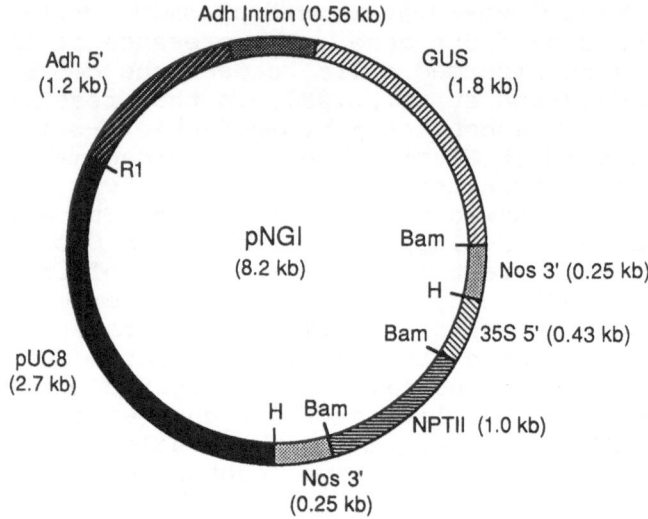

Fig. 1.   Structure of pNGI. Adh, Alcohol Dehydrogenase
          1; GUS, ß-Glucuronidase; Nos3', nopaline
          synthase polyadenylation region; H, *Hind*III;
          Bam, *Bam*HI; R1, *Eco*RI.

et al., 1988a, 1988b) and subsequent selection (Fromm et
al., 1986) of kanamycin resistant clones was performed
as previously described. Briefly, 2 ml of suspension
culture (about 2×10^5 cells) of BMS cells was distributed
over the surface of a filter paper (Whatmann #4, 5.5 cm
in diameter). The filter paper bearing the cells was
placed over 3 layers of filter paper to which 2.5 ml of
MS media had been added. The cells were then bombarded
under a partial vacuum with tungsten particles
(average diameter 1.2 µm) to which DNA was previously
adsorbed. The DNA was absorbed to the particles by
mixing a solution of 8 ug of DNA, 25 ul of 1 M calcium
chloride and 10 ul of 0.1M spermidine, in the presence
of 25 ul of tungsten particles (50 mg/ml), which causes
the DNA to precipitate onto the particles (Klein et al.,

1988a, 1988b).   A 2.5 ul aliquot of the DNA-particle slurry was placed on the surface of the 22 caliber plastic bullet, and fired into the stopping plate of the microprojectile gun, causing the microprojectiles to pass through the small hole in the stopping plate and to impact on the BMS cells on the other side.

Following bombardment, the cells were washed from the filter with 6 ml of MS medium into a 10 cm petri dish.   The petri dish was sealed and then incubated in the dark at 26 C. After 2 days, 6 ml of fresh medium containing 300 µg of kanamycin per ml was added to each Petri dish and half of the volume was transferred to a fresh dish.   After 1 week the culture was again diluted 2 fold with fresh medium containing 150 µg of kanamycin per ml and half the culture transferred to a new petri dish.   Subsequent additions of fresh medium containing kanamycin were made at 2 week intervals until the majority of cells in the culture clearly ceased to grow. This generally occurred after 4 to 6 weeks of selection. The cells were then transferred from the liquid medium to a membrane filter supported on a cellulose adsorbent pad (Gelman) that was on agarose-solidified MS medium containing 100 µg/ml kanamycin.   Calli that developed on the filter were transferred to agarose-solidified medium supplemented with kanamycin after they had reached a diameter of about 0.5 cm.   Up to 117 kanamycin resistant calli were recovered from a single bombardment of about 2×10^5 cells.   For 8 independent bombardments the average number of kanamycin resistant calli that was recovered was 56.   Kanamycin resistant calli were not recovered from unbombarded samples.

In one set of experiments a comparison was made between the number of kanamycin resistant colonies recovered following bombardment with pCaMVNEO and pCaMVINEO, which is identical to pCaMVNEO but contains an Adh1 intron between the CaMV 35S promoter and the NPT II coding region.   The presence of the Adh1 intron between the promoter and coding region has been shown to increase transient expression of the NPT II gene in maize (Callis et al., 1987).   Similar numbers of kanamycin-resistant calli were recovered following bombardment with pCaMVINEO, pCaMVNEO, and pNGI.   The increase in NPT II expression as a result of the presence of the Adh1 intron did not result in an increase in transformation efficiency.   Apparently the

level of expression of the gene that lacks the intron is
sufficient to confer resistance to kanamycin. pCaMVINEO
and pCaMVNEO were linearized by digestion with *Eco*RI
prior to their adsorption to microprojectiles and
acceleration into maize cells. We found that
linearization did not increase the frequency of
transformation in relation to supercoiled plasmid DNA.

Verification of Transformation

GUS Expression in Transformed Calli. A small
amount of each kanamycin-resistant callus was placed
into the well of a microtiter dish containing 200 μl of
GUS assay solution (5-bromo-4-chloro-3-indoyl-ß-D-
glucuronic acid [X-gluc; Research Organics], 2 mM;
potassium ferricyanide, 0.05 mM; potassium ferrocyanide,
0.05 mM; Triton X-100, 0.1% [v/v]).   Triton X-100 was
present in the substrate mixture to permit X-gluc to
pass through the cell membrane.   After 12 hours the
calli that were expressing GUS had developed the blue
color indicative of GUS expression (Jefferson et al.,
1987).   All kanamycin-resistant isolates had at least
some GUS-expressing cells (Fig. 2). Blue cells were not
observed in unbombarded samples (data not shown). The
staining pattern of the calli ranged from virtually all
of the cells turning blue in some samples (Fig. 2A) to
those samples with only a few blue cells (Fig. 2B).
Samples that strongly expressed GUS often exhibited a
rather 'patchy' distribution of GUS expression with
densely stained aggregates of cells interspersed with
aggregates that were only faintly blue.   The levels of
GUS expression in the kanamycin-resistant calli as
determined with X-gluc are expressed qualitatively in
Table 1.

The variation in GUS expression may have been the
result of differences in the ability of the substrate to
diffuse into the detergent-treated calli.   Therefore,
GUS expression in extracts from kanamycin-resistant
calli was analyzed quantitatively using the fluorometric
substrate, 4-methylumbelliferyl-ß-D-glucuronide (MUG).
The buffer used during preparation of the plant cell
extract was as described by Jefferson (1987) except the
concentration of Triton X-100 was reduced to 0.01%.
Methyl umbelliferyl levels in the reaction mixture were
determined after 15, 45, and 75 min of incubation.
Protein in tissue extracts was quantified according to

Bradford (1976). Levels of expression ranged from 5 to 172 pM of MU produced per min per μg of protein (Table 1). Levels of GUS expression as assayed with the fluorometric substrate generally corresponded with the degree of staining observed with the histochemical substrate indicating that the low levels of GUS expression as indicated by the histochemical assay was not caused by a lack of penetration of the substrate into the cells.

Southern Analysis of Transformants. To confirm the transformed nature of the kanamycin-resistant calli, hybridization analyses were performed to probe for the

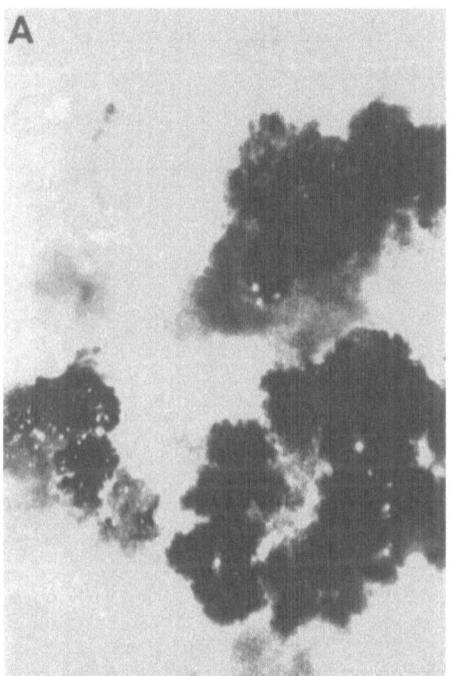 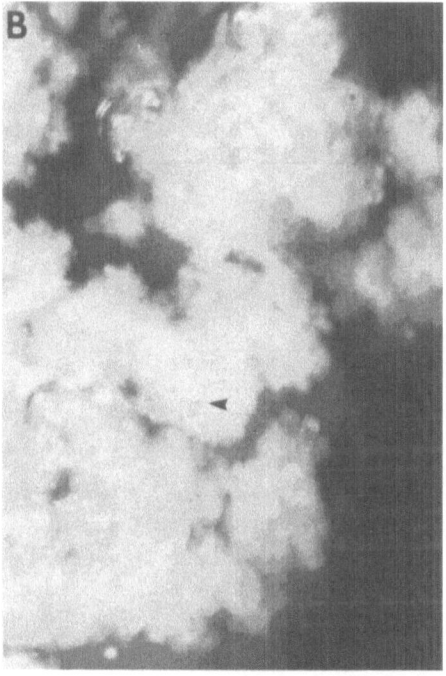

Fig. 2.   GUS expression in kanamycin-resistant calli that were recovered from cells bombarded with pNGI. The tissue was treated with the histochemical substrate X-gluc. A) An isolate that expressed high levels of GUS. B) An isolate that expressed little GUS activity. The arrow points to one of the few blue cells in this isolate.

presence of the NPT II region of the pNGI plasmid used
to transform the BMS cells. Genomic DNA was isolated
from the various kanamycin-resistant calli using a
published method (Chilton et al., 1982). The isolated
DNA (8 μg) was digested for 3 hours using a 3- to 4-fold
excess of restriction enzyme and electrophoresed in 0.8%
agarose gels. The DNA was transferred (Southern, 1975)

Table 1. GUS expression in kanamycin-resistant
calli as determined qualitatively by
histochemical staining using X-gluc or
quantitatively by an enzyme assay using
MUG as the substrate. Also given is the
number of intact copies of the GUS gene
present in the various isolates.

Sample	GUS Staining Blue Color[a]	GUS Activity x 10^{-3} (pmol MU/μg protein/min)	GUS Copy Number
1-1	3	104	8
1-4	4	172	2
1-5	2	18	2
1-9	2	50	8
1-14	3	98	1
1-15	2	17	10
2-1	1	5	10
2-2	1	8	5
2-3	2	14	1
2-5	2	12	1
2-8	4	94	8

[a] Calli were qualitatively evaluated for their
level of GUS expression following treatment with
X-gluc with 4 representing isolates that
uniformly and densely stained blue while a value
of 1 representing weak expression with only a
few blue cells present in the isolate. Calli
that had GUS staining values of 2 or 3 had some
densely stained aggregates of cells interspersed
among aggregates that exhibited little or no
staining.

to nylon membranes (Hybond N, Amersham) which were then illuminated with UV light (254 nm) for 5 min. The Southern blot containing the transferred DNA was probed with a radioactive *Bam*HI NPT II fragment from pNGI. Radioactive probes were prepared by incorporation of ^{32}P-dCTP by the random-hexamer primer method (Feinberg and Vogelstein, 1983).

Bam HI digestion of genomic DNA from transformed cells should release a 1.0 kb NPT II fragment from the pNGI DNA present in the transformed calli (for the structure of pNGI, see Fig. 1). All of the kanamycin resistant calli tested yielded a 1.0 kb band that co-migrates with the 1.0 kb NPT II fragment from *Bam*HI-digested pNGI DNA (Fig. 3a). Copy number reconstructions indicated that the transformants analyzed contained from 1 to 8 intact copies of the NPT II gene per diploid genome. Most of the calli analyzed had additional rearranged copies integrated into the genome. Unbombarded tissue did not contain DNA that hybridized to the NPT II probe (Fig. 3a, Lane pNGI-0).

Southern blot analyses were also performed to determine the copy number and structure of the GUS gene in the kanamycin-resistant calli. Digestion of pNGI with *Eco*RI and *Hind*III releases the 3.8 kb GUS gene (see Fig. 1). Southern blots of genomic DNA digested with *Eco*RI and *Hind*III were hybridized to a 1.8 kb fragment of the GUS coding region. All of the kanamycin-resistant calli tested contained both intact and rearranged copies of the GUS gene (Fig. 3b) with the number of rearranged sequences often exceeding the number of intact copies. The number of intact copies of the GUS gene varied from 1 to 10 in the transformed calli. Unbombarded calli did not contain DNA that hybridized to the GUS probe.

Variation of GUS Expression and Methylation of the GUS Gene.

Adh Expression of Transformants. The number of intact copies of the GUS gene was not correlated with the level of GUS expression (Table 1). One potential explanation for the variation in GUS activity is that the anaerobically inducible Adh1 promoter (Sachs et al., 1980) that is used to express the GUS gene in pNGI may have been induced to different degrees between transformants. Levels of endogenous Adh1 activity were

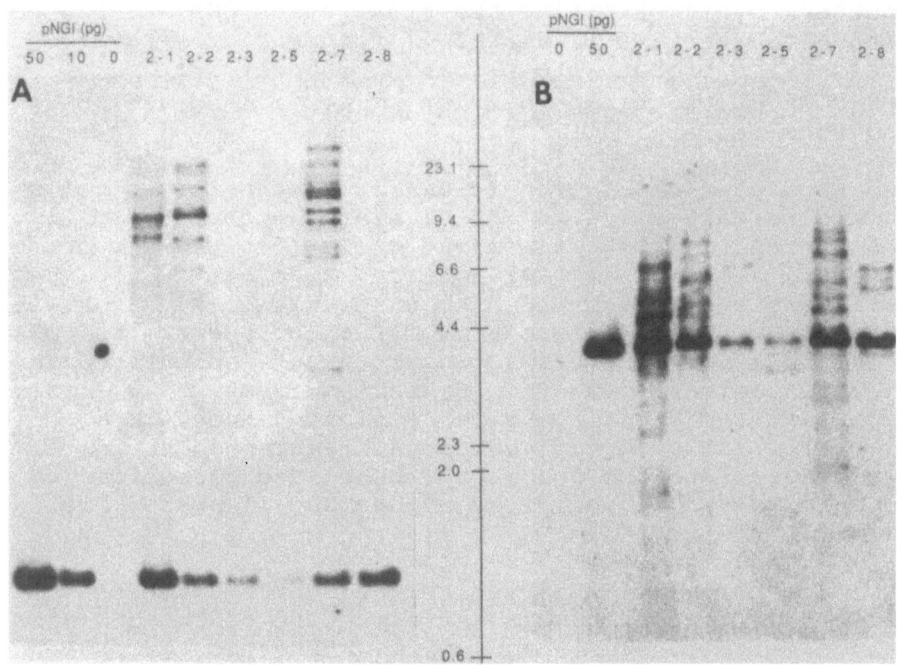

Fig. 3.   Analysis of the NPT II and GUS gene structures
          in maize calli transformed with pNGI DNA. DNA
          was isolated from kanamycin-resistant calli 12
          weeks after bombardment. The DNA was digested
          with either *Bam*HI (A) or *Eco*RI and *Hin*dIII (B),
          separated by electrophoresis in a 0.8% agarose
          gel and transferred to nylon membranes. A
          radioactive NPT II (A) or GUS (B) probe was
          hybridized to the DNA on the membrane and the
          hybridizing sequences visualize
          autoradiography. The position of the size
          markers (kb) is indicated between the two
          panels. The lanes designated pNGI contain the
          indicated amount (pg) of pNGI DNA digested with
          either *Bam*HI (A) or *Eco*RI and *Hin*dIII (B) as
          well as 10 µg of similarly digested DNA from
          untransformed maize callus. Digestion of pNGI
          with *Bam*HI releases a 1.0 kb hybridizing
          fragment while digestion with *Eco*RI and *Hin*dIII
          releases a 3.8 kb GUS gene hybridizing
          fragment.

therefore measured to determine the variation in the
levels of Adh1 induction between the different
transformed calli. Cell extracts were prepared from
calli, run on non-denaturing polyacrylamide gels and Adh
enzyme activity was visualized in the gel by incubating
it in a solution containing the appropriate substrates
(Schwartz and Endo, 1966; Sachs et al., 1980). Adh
enzyme levels were found to be similar between all calli
(data not shown). This demonstatrates that the level of
induction of the Adh1 genes is similar in the different
calli. This indicates that the Adh1 promoter of the GUS
construct should be induced to similar levels in
different calli. Thus the variation in GUS expression
is larger that the variation in Adh expression and
larger than the variation in GUS gene copy number.
Therefore, GUS expression from the pNGI construct, which
uses the Adh1 promoter, is being regulated differently
then the endogenous Adh1 gene.

        <u>Methylation of the GUS Gene</u>. Since methylation can
influence the expression of plant genes, the methylation
of the introduced GUS gene was investigated. This was
done by comparing the ability of methylation-sensitive
and -insensitive restriction enzymes to digest GUS
sequences in DNA isolated from strongly and weakly
expressing calli. The calli chosen for this comparison
contained similar numbers of the intact GUS gene (about
5 to 10). Calli 1-1 and 2-8 had high levels of GUS
expression while 2-1 and 2-2 had low levels of
expression (Table 1). These calli were derived from
different bombardments. Genomic DNA was digested with
EcoRII or BstNI which both recognize the sequence
CC(AT)GG. EcoRII will not cut this sequence if the
internal C of the recognition site is methylated while
BstNI will cut regardless of the methylation state of
this base. The chimeric GUS gene has one EcoRII-BstNI
site in the promoter region, one in the intron, and
several in the coding region (Fig. 4, top). The
Southern blots were hybridized to a probe derived from a
PstI to EcoRV fragment from the first 0.6 kb of the GUS
coding region.

        The restriction pattern produced by digestion with
BstNI was similar for calli that express GUS at either
high or low levels (Fig. 4, lanes labeled B). The two
predominant hybridizing bands observed correspond to the
expected 0.9 and 0.4 kb fragments produced upon
digestion of the pNGI-GUS gene by BstNI. In contrast,

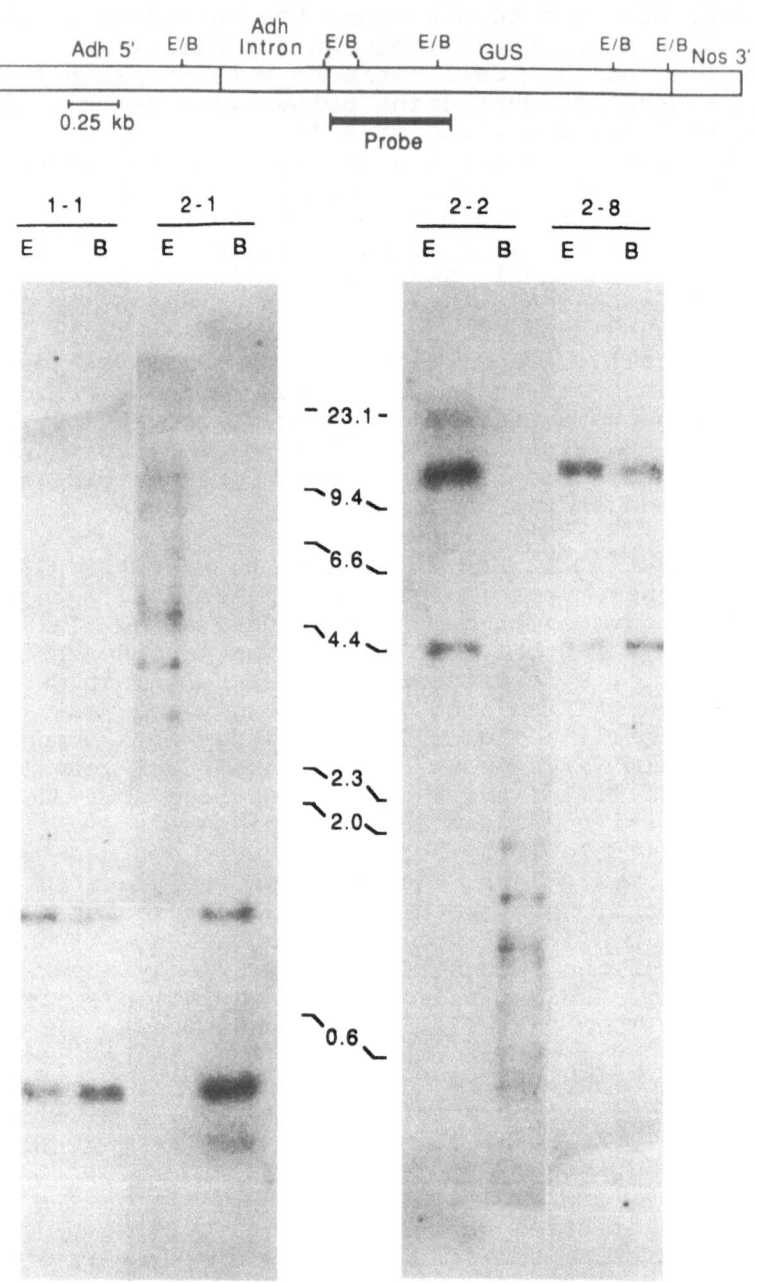

Fig. 4.   Restriction analysis of the methylation status
          of GUS gene DNA from calli expressing low or
          high levels of GUS. Genomic DNA was isolated

EcoRII digestion of DNA from calli that expressed GUS at
low levels produced a different restriction pattern than
DNA from calli that expressed GUS at high levels. Only
bands greater than 2.5 kb were observed in EcoRII-
digested DNA from calli with low levels of GUS (lanes 2-
1 and 2-2 E). The hybridization profile of EcoRII-
digested DNA from calli that express high levels of GUS
produced 0.9 and 0.4 kb hybridizing fragments similar to
those produced by BstNI digestion. Therefore the sites
recognized by EcoRII and BstNI are methylated in calli
that possess low levels of GUS activity. Complete
EcoRII digestion in each sample was verified by
rehybridizing the Southern blot with an Adh1 probe that
recognizes the endogenous Adh1 gene.

    Similar results were obtained when genomic DNA was
digested with AvaII or PvuII, both of which are 5-
methylcytosine-sensitive restriction enzymes. The
restriction patterns expected from complete digestion
with AvaII or PvuII were observed for DNA from calli
that possess high levels of GUS expression. In
contrast, higher molecular weight fragments were
produced when DNA from low expressing calli were used.
This indicates that methylation of these sites
interfered with the ability of these enzymes to cut.
Taken together, these results indicate that methylation
of the GUS gene is correlated with low levels of its
expression.

from pNGI-transformed calli that expressed GUS
at low (calli 2-1 and 2-2) and high (calli1-1
and 2-8) levels. DNA (10 µg) was digested with
either EcoRII or BstNI and analyzed by Southern
blots. EcoRII will not restrict methylated
recognition sites while BstNI will digest the
recognition site regardless of its methylation
status (see text). E or B at the top of a lane
denotes digestion with EcoRII or BstNI,
respectively. The structure of the GUS gene in
pNGI, the position of EcoRII and BstNI (E/B)
sites in this gene, and the region from which
the radioactive probe was derived are shown at
the top of the figure. Complete digestion of
the GUS gene produces 0.9 and 0.4 kb
hybridizing fragments (lanes 2-1E and 2-2E).
The position of the size markers (kb) is
indicated between the two panels.

  5-Azacytidine Treatment. To further investigate
the association of low levels of GUS expression with
methylation of the GUS gene, calli were exposed to 5-
azacytidine (AZC). This compound can be incorporated
into DNA in place of cytosine and often results in the
appearance of unmethylated sites after DNA replication
(Jones and Taylor, 1980). Kanamycin-resistant maize
calli were treated with AZC by placing about 100 mg of
tissue into a well of a microtiter plate with 500 µl of
MS medium supplemented with kanamycin (100 µg/ml) and
AZC at either 0, 25, 50 or 100 mM. The tissue was
incubated at 27 C for 48 hours and then transferred from
the liquid medium to agarose-solidified MS medium
containing kanamycin where the calli were allowed to
grow for 11 days on medium lacking AZC. After
this time GUS expression was visualized with X-gluc.
Calli that had strongly expressed GUS prior to AZC
treatment continued to do so following treatment. Many
GUS expressing cells (about 1% of the total number of
cells) were observed following AZC treatment of calli 2-
1 or 2-2. These calli showed no or very few (<0.001%)
GUS expressing cells prior to exposure to AZC (Fig. 5).

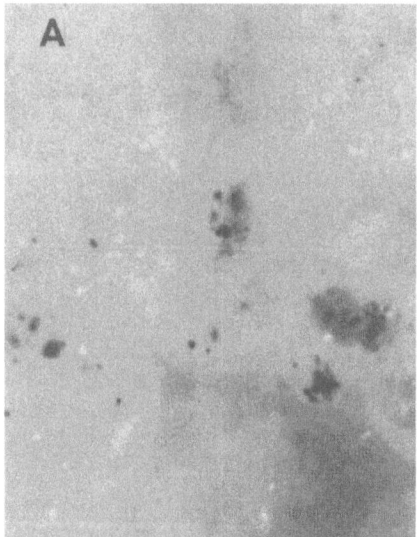

Fig. 5. GUS expression in callus 2-1 cells following
         treatment with AZC. A) Cells exposed to 50 mM
         AZC prior to addition of X-gluc. b) Control
         tissue treated with X-gluc but not with AZC.

GUS expression was stimulated by AZC treatment at any of
the 3 concentrations of tested (25, 50 or 100 mM). GUS
activity was not induced by AZC treatment in control
calli that did not receive the GUS gene. These results
provide additional evidence of the influence of
methylation on GUS expression.

DISCUSSION

Stably transformed calli of maize can readily be
recovered following bombardment of intact cells with
DNA-bearing microprojectiles. The frequency of
transformation found in this study (about 1 in 5000 of
the cells treated with microprojectiles) was similar to
that previously found for microprojectile-mediated
transformation of intact tobacco (Klein et al., 1988c)
and soybean (Christou et al., 1988) cells. The frequency
of transformation by microprojectiles is comparable to
that observed for methods used for the transformation of
plant protoplasts such as electroporation (Fromm et al.,
1986) or PEG-mediated delivery (Lörz, 1985). Recently,
as a result of improved methods for the culture of
protoplasts, transformed plants of several cereal
species have been produced by gene delivery to
protoplasts. For example, transgenic rice plants have
been obtained by electroporation-mediated delivery of
DNA to protoplasts (Toriyama et al., 1988). Transgenic
maize plants have also been recovered by gene transfer
to protoplasts, but the regenerated plants were sterile
(Rhodes et al., 1988). In spite of this progress,
regeneration of most cereal species from protoplasts is
still difficult and therefore the pursuit of alternative
approaches for plant transformation is necessary.
Microprojectile-mediated transfer represents a means for
delivering DNA into intact cells within tissues.
Therefore, with the appropriate selectable or screenable
markers, it should be possible to recover transformed
calli from bombarded embryogenic maize tissues that have
the potential to develop into whole plants. Improvements
in transformation efficiencies as a result of further
refinements in the design of microprojectiles and
acceleration devices will aid in the recovery of
transformed plants of recalcitrant species.

Histochemical analysis of GUS expression provided a
rapid means for visually screening the kanamycin-
resistant calli to verify their transformed nature.

This visual screen is far more rapid and simple than screens utilizing enzymatic assays of extracts or that utilize Southern hybridizations for the detection of transforming DNA. The only drawback of the GUS screen is that the substrate does not readily enter living maize cells. Therefore, a small portion of the callus must be treated with a membrane-disrupting agent, which often kills the cells. The development of GUS assays that do not kill cells would greatly facilitate the recovery of rare transformants from large numbers of untransformed cells.

All kanamycin-resistant calli tested carried intact copies of the NPT II and GUS genes and expressed GUS to some degree. However, GUS expression varied greatly between isolates. This variation was not related to the number of intact copies of the GUS gene. In some calli low levels of GUS activity were observed despite the fact that many intact copies of the gene were present. Variation in expression of the chimeric GUS gene could have been caused by different levels of induction of the Adh1 promoter, which can be induced by several environmental factors (Sachs et al., 1980). However, levels of endogenous Adh1 activity were consistent between the transformants. Therefore, the factors that cause the variation in GUS expression in different transformants are not directly related to the factors that control Adh1 gene expression.

A possible explanation for the variation in GUS expression between isolates could be differences in the methylation of the introduced gene. The methylation of cytosine residues is known to influence transcriptional activity of genes in plants (Hepburn et al., 1987) and , animals (Razin and Cedar, 1984) and is thought to play a role in the control of tissue-specific gene expression. The methylation of sites within the promoter region often inhibits gene expression (Bussingler et al., 1983), although methylation in the coding region or 3' end may also influence expression (Keshet et al., 1985). In plants, methylation of transposable elements has been shown to influence their activity (Chandler and Walbot, 1986; Chomet et al., 1987). Methylation of ribosomal genes (Blundy et al., 1987; Watson et al., 1987) and T-DNA genes involved with gall formation (Gelvin et al., 1983; Hepburn et al., 1983; Amasino et al., 1984; van Slogteren et al., 1984) is also correlated to reduced transcription. About 30% of the total cytosine residues

in the DNA of higher plants are methylated and these
modified residues occur in the sequences CG and CXG
where X represents any of the four nucleotides. This is
in contrast with animal cells where only about 8% of the
cytosines are methylated and these methylated sites are
restricted to the sequence CG (Shapiro, 1976).
Therefore, methylation may play a greater role in
influencing gene expression in plants than it does in
animal cells.

Our results with methylation-sensitive restriction
enzymes indicated a correlation between methylation of
the GUS gene and its expression. Each of the three
methylation-sensitive restriction enzymes tested did not
restrict the GUS gene to completion in DNA from two
calli that expressed low levels of GUS activity. In
contrast, these enzymes readily digested GUS sequences
from two calli that expressed GUS at relatively high
levels. Therefore, methylation of the gene appears to
influence its inactivation. Because of the distribution
of these restriction sites and the probe used, only the
methylation of the GUS coding region could readily be
discerned. Treatment with AZC increased levels of GUS
expression in transformants carrying methylated copies
of the GUS gene. This result reinforces the conclusion
that methylation influences GUS expression since AZC
blocks methylation of cytosine residues and has been
used previously to reactivate the expression of
methylated T-DNA genes in plant cells (Hepburn et al.,
1983). Methylation of the GUS gene may explain the
irregular pattern of expression that was observed
following treatment of transformed calli with the
histochemical substrate. Within a GUS-expressing calli,
sectors of cells that do not express GUS may arise from
the growth of cells with methylated GUS genes.
Progressive methylation is also a likely explanation for
the almost total lose of GUS activity over time noted
for some of the transformants.

In contrast to the variability observed with our
Adh1/GUS gene, the cloned maize Adh1S gene exhibited a
consistent correlation between copy number and
expression when stably introduced into maize cells
(Callis et al., 1987). A comparison between the
endogenous Adh activity (Adh1F) and Adh activity from
the introduced gene (Adh1S) was possible because the
products of these two genes possess different
electrophoretic mobilities. Each transferred Adh1S

(slow electrophoretic allele) gene was expressed at
levels equivalent to that produced by one endogenous
Adh1F gene (fast electrophoretic allele). This suggests
that the chimeric Adh1/GUS gene is regulated differently
from introduced copies of the intact Adh1S gene, even
though the promoter and 5' flanking sequences are
identical. It is important to note that methylation of
the Adh1 gene does not appear to play a role in its
regulation. Nick et al. (1986) found that changes in
the expression of the endogenous Adh1 gene were not
caused by changes in its methylation. They observed
that in maize leaves, where the Adh1 gene is not
expressed, the 900 base pairs 5' to the Adh1 coding
region remain unmethylated. However, the introduction of
the GUS coding region appears to create a gene that is
susceptible to suppression by methylation. This may be
because the GUS gene is of bacterial origin and has more
CG and CXG sequences than the Adh1 gene. Computer
analysis showed that the Adh1 coding region contains 32
CG and 51 CXG sequences per kb while the GUS coding
region has 82 CG and 88 CXG sequences per kb.
Alternatively, other more subtle differences between the
GUS and Adh1 coding regions may be responsible for the
differences in susceptibility to methylation between the
two genes.

     The differences in methylation of the GUS gene
between isolates may in part be due to the site of
integration of the introduced gene. Integration into
methylated regions of the genome could result in the
subsequent methylation of the foreign gene. Methylation
may account for the frequent lack of correlation between
copy number and expression found when genes are
transferred into plant cells (Jones et al., 1985; An,
1986). Differences in expression of transferred genes
are often ascribed to poorly defined position effects
(Weintraub, 1985) which may be due to the susceptibility
of the gene to methylation as a result of integration in
a methylated region of the genome (Bird et al., 1979).
Position effects may also be the result of integration
into inactive chromatin structure or integration into
sites adjacent to sequences which in some way interfere
or promote expression.

     In rare cases a good correlation between copy
number and expression exists. Besides the direct
correlation found for the maize Adh1 gene already
mentioned (Callis et al., 1987), Stockhaus and coworkers

(1987) found that a light-dependent gene from potato (*Solanum tuberosum*), which was tagged by an exon modification, was expressed in a dosage dependent fashion when reintroduced into potato or tobacco. Both of these examples involve transfer of nonchimeric genes into homologous hosts. It is therefore interesting to speculate that chimeric genes are more susceptible to inactivation by methylation or other undefined position effects than are non-chimeric genes. Taken together, these results suggest that the properties of the particular transferred gene may determine the likelihood of its becoming inactivated. Apparently, methylation can be one of the significant factors involved in the suppression of expression of genes introduced into plant cells. Therefore, a better understanding of the factors that control the methylation of DNA transferred into plant cells may have an impact on the design of chimeric genes.

ACKNOWLEDGEMENTS

A Biolistics, Inc (Geneva, N.Y., USA) high-velocity microprojectile gun was used in this research. The use of this product does not constitute an endorsement or advertisement by the USDA. This work was supported by Pioneer Hi-Bred Inc. and the US Department of Agriculture-Agricultural Research Service.

REFERENCES

Abel, P.P, Nelson, R.S., De, B., Hoffmann, N., Rogers, S.G., Fraley, R.T. and Beachy, R.N., 1986, Delay of disease development in transgenic plants that express the tobacco mosaic virus coat protein gene, Science, 232:738-743.

Amasino, R. M., Powell, A. L. T., and Gordon, M. P., 1984, Changes in T-DNA methylation and expression are associated with phenotypic variation and plant regeneration in a crown gall tumor line, Mol. Gen. Genet., 197:437-446.

An, G., 1986, Development of plant promoter expression vectors ant their use for analysis of differential activity of naopaline synthase promoter in transformed tobacco cells, Plant Physiol., 81:86-91.

Bird, A. P., Taggart, M. H., and Smith, B. A., 1979,
    Methylated and unmethylated DNA compartments in the
    sea urchin genome, Cell, 17:889-901.
Blundy, K. S., Cullis, C. A., and Hepburn, A. G., 1987,
    Ribosomal DNA methylation in flax genotroph and a
    crown gall tumour, Plant Mol. Biol., 8:217-225.
Bradford, M. M., 1976, A rapid and sensitive method for
    quantitation of microgram quantities of protein
    utilizing the principle of protein-dye binding,
    Analyt. Biochem., 72:248-254.
Busslinger, M., Hurst, J., and Flavell, R., 1983, DNA
    methylation and the regulation of globin gene
    expression, Cell, 34:197-206.
Callis, J., Fromm, M., and Walbot, V., 1987, Introns
    increase chimeric gene expression in maize, Genes
    and Development, 1:1183-1200.
Chandler, V. L., and Walbot, V., 1986, DNA modification
    of maize transposable element correlates with the
    loss of activity, Proc. Natl. Acad. Sci. USA
    83:1767-1771.
Cheung, A.Y., Bogorad, L., van Montagu, M., and Schell,
    J., 1988, Relocating a gene for herbicide
    tolerance:A chloroplast gene is converted into a
    nuclear gene. Proc. Nat. Acad. Sci. USA, 85:391-
    395
Chilton, M. D., Tepfer, D., Petit, A., David, C., Casse-
    Delbart, F., and Tempe, J., 1982, Agrobacterium
    rhizogenes inserts T-DNA into genomes of the host
    plant root cells, Nature, 295:432-434.
Chomet, P. S., Wessler, S., and Dellaporta, S. L., 1987,
    Inactivation of the transposable element *Activator
    (Ac)* is associated with its DNA modification, EMBO
    J., 6:295-302.
Christou P., McCabe, D. E., and Swain, W. F., 1988,
    Stable transformation of soybean callus by DNA-
    coated gold particles, Plant Physiol., 87:671-674.
Cocking, E. C., and Davey, M.R., 1987, Gene transfer in
    cereals, Science, 236:1259-1262.
Cuozzo. M., O'Connell,, K.M., Kaniewski. W, . Fang.
    R.X., Chua, N.H., and Tumer, N.E., 1988, Viral
    protection in transgenic tobacco plants expressing
    the cucumber mosaic virus coat protein or its
    antisense RNA, Bio/Technology, 6:549-558.
della-Cioppa, G., Bauer, S.C., Taylor, M.L., Rochester,
    D.E., Klein, B.K., Shah, D.M., Fraley, R.R., and
    Kishore, G.M., 1987, Targeting a herbicide-
    resistant enzyme from Escheriachia coli to
    chloroplasts of higher plants, Bio/Technology,
    5:579-584.

Dennis, E. S., Gerlach, W. L., Pryor, A. J., Bennetzen, J. L., Inglis, A., Llewellyn, D., Sachs, M. M., Ferl, R. J., and Peacock, W. J., 1984, Molecular analysis of the alcohol dehydrogenase (Adh1) gene of maize, Nucleic Acid Res., 12:3983-4000.

Feinberg, A. P., and Vogelstein, B., 1983, A technique for radiolabeling DNA restriction endonuclease fragments to high specific activity, Anal. Biochem., 132:6-13.

Fillatti, J.J., Kiser, J., Rose, R., and Comai, L., 1987, Efficient transfer of a glyphosate tolerance gene into tomato using a binary Agrobacterium tumefaciens vector, Bio/Technology, 5:726-730.

Fischhoff, D.A., Bowdish, K.S., Perlak, F.J., Marrone, P.G., McCormick, S.M., Niedermeyer, J.G., Dean, D.A., KusanoKretzmer, K., Mayer, E.J., Rochester, D.E., Rogers, S.G., and Fraley, R.T., 1987, Insect tolerant transgenic tomato plants, Bio/Technology, 5:807-813.

Fromm, M. E., Taylor, L. P., and Walbot, V., 1985, Expression of genes transferred into monocot and dicot plant cells by electroporation, Proc. Natl. Acad. Sci. USA, 82:5824-5828.

Fromm, M. E., Taylor, L. P., and Walbot, V., 1986, Stable transformation of maize after gene transfer by electroporation, Nature, 319:791-793.

Gelvin, S. B., Karcher, S. J., and DiRita, V. J., 1983, Methylation of the T-DNA in Agrobacterium tumefaciens and in several crown gall tumors, Nucleic Acid Research 11:159-174.

Gerlach, W.L., Llewellyn, D., and Haseloff, J., 1987, Construction of a plant disease resistance gene from the satellite RNA of tobacco ringspot virus, Nature, 328:802-805.

Goodman, R.M., Hauptli, H., Crossway, A., and Knauf, V.C., 1987, Gene transfer in crop improvement, Science, 236:48-54.

Harrison, B.D., Mayo, A., and Baulcombe, D.C., 1987, Virus resistance in transgenic plants that express cucumber mosiac virus satellite RNA, Nature, 328:799-802.

Hepburn, A. G., Clarke, L. E., Pearson, L., and White, J., 1983, The role of cytosine methylation in the control of nopaline synthase gene expression in a plant tumor, J. Molecular and Applied Genetics, 2:315-329.

Hepburn, A. G., Belanger, F. C., and Mattheis J. R., 1987, DNA methylation in plants, Developmental Genetics, 8:475-493.

Hilder, V.A., Gatehouse, A.N.R., Sheerman, S.E., Barker,
    R.F., and Boulter, D., 1987, A novel mechanism of
    insect resistance engineered into tobacco, Nature,
    330:160-163.
Jones, P. A., and Taylor, S. M., 1980, Cellular
    differentiation, cytidine analogues and DNA, Cell,
    20:85-93.
Jones, J. D. G., Dunsmuir, P. and Bedbrook, J., 1985,
    High levels of expression of introduced chimaeric
    genes in regenerated transformed plants, EMBO J.,
    4:2411-2418.
Jefferson, R. A., Burgess, S. M., and Hirsh, D., 1986,
    ß-Glucuronidase from Escherichia coli as a gene-
    fusion marker, Proc. Natl. Acad. Sci. USA, 83:8447-
    8451.
Jefferson, R. A., Kavanagh, T. A., and Bevan, M. W.,
    1987, GUS fusions: ß-glucuronidase as a sensitive
    and versatile gene fusion marker in higher plants,
    EMBO J., 6:3901-3907.
Keshet, I., Yisraeli, J., and Cedar, H., 1985, Effect of
    regional DNA methylation on gene expression, Proc.
    Natl. Acad. Sci. USA, 82:2560-2564.
Klee H, Horsch, R., and Rogers, S., 1987, Agrobacterium-
    mediated plant transformation and its further
    applications to plant biology, Ann. Rev. Plant
    Physiol., 38:467-486.
Klein, T. M., Wolf, E. D., Wu, R., and Sanford, J. C.,
    1987, High-velocity microprojectiles for delivery
    of nucleic acids into living cells, Nature, 327:70-
    73.
Klein, T. M., Fromm, M. E., Weissinger, A., Tomes, D.,
    Schaaf, S., Sletten, M., and Sanford, J. C., 1988a,
    Transfer of foreign genes into intact maize cells
    using high-velocity microprojectiles, Proc. Natl.
    Acad. Sci. USA, 85:4305-4309.
Klein, T. M., Gradziel, T., Fromm, M. E., and Sanford,
    J. C., 1988b, Factors influencing gene delivery
    into Zea maize cells by high-velocity
    microprojectiles, Bio/Technology, 6:559-563.
Klein, T.M., Harper, E. C., Svab, Z., Sanford, J. C.,
    Fromm, M. E., and Maliga, P., 1988c, Stable genetic
    transformation of intact Nicotiana cells by the
    particle bombardment process, Proc Natl Acad Sci
    USA, 85:8502-8505.
Lörz, H., Baker, B., and Schell, J., 1985, Gene transfer
    to cereal cells mediated by protoplast
    transformation, Molec. Gen. Genetic, 199:178-182.

McCabe, D. E., Swain, W. F., Marinell, B. J., and
    Christou, P., 1988, Stable transformation of
    soybean (*Glycine max*) by particle acceleration,
    Bio/Technology, 6:923-926.

Nick, H., Bowen, B., Ferl, R. J., and Gilbert, W., 1986,
    Detection of cytosine methylation in the maize
    alcohol dehydrogenase gene by geomoic sequencing,
    Nature, 319:243-246.

Nelson, R.S., McCormick, S.M., Delannay, X., Dube, P.,
    Layton, J., Anderson, E.J., Kaniewska, M., Proksch,
    R.K., Horsh, R.N., Rogers, S.G., Fraley, R.T., and
    Beachy, R.N., 1988, Virus tolerance, plant growth
    and field performance of transgenic tomato plants
    expressing coat protein from tobacco mosaic virus,
    Bio/Technology, 6:403-410.

Razin, A., and Cedar, H., 1984, DNA methylation in
    eucaryotic cells, Int. Rev. Cytol., 92:159-185.

Rhodes, C. A., Pierce, D. A., Mettler, I. J.,
    Mascarenhas, D., and Detmer, J., 1988, Genetically
    transformed maize plants from protoplasts, Science,
    240:204-207.

Sachs, M. M., Freeling, M., and Okimoto, R., 1980, The
    anaerobic protiens of maizew, Cell, 20:761-767.

Sanford, J. C., Klein, T. M., Wolf, E. D., and Allen,
    N., 1987, Delivery of substances into cells and
    tissues using a particle bombardment process, J.
    Part. Sci. Technol., 5:27-37.

Schell,J. St., 1987, Transgenic plants as tools to study
    the molecular organization of plant genes,
    Science, 237:1176-1183.

Schwartz, D., and Endo, T., 1966, Alcohol dehydrogenase
    polymorphism in maize simple and compound loci,
    Genetics, 53:709-715.

Shah, D.M., Horsch, R.B., Klee, H.J., Kishore, G.M.,
    Winter, J.A., Tumer, N.E., Jironaka, C.M., Sanders,
    P. R., Gasser, C.S., Aykent, S., Siegel, N.R.,
    Rogers, S.G., and Fraley, R. T., 1986, Engineering
    herbicide tolerance in transgenic plants, Science,
    233:478-481.

Shapiro, H. S., 1976, Distribution of purines and
    pyrimidines in deoxyribonucleic acids, CRC Handbook
    of Biochemistry and Molecular Biology, Vol. 2, p.
    259.

Southern, E. M., 1975, Detection of specific sequences
    among DNA fragments separated by gel
    electrophoresis, J. Mol. Biol., 98:503-517.

Stockhaus, J., Eckes, P., Blau, A., Schell, J., and
    Willmitzer, L., 1987, Organ-specific and dosage-
    dependent expression of a leaf/stem specific gene
    from potato after taggina and transfer into potato
    and tobacco plants, Nucl. Acids Res., 15:3479-3491.
Toriyama, K., Arimoto, Y., Uchimiya, H., and Hinata, K.,
    1988, Transgenic rice plants after direct gene
    transfer into protoplasts, Bio/Technology, 6:1072-
    1074.
van Slogteren, G. M. S., Hooykaas, P. J. J., and
    Schilperoort, R. A., 1984, Silent T-DNA genes in
    plant lines transformed by Agrobacterium
    tumefaciens are activated by grafting and 5-
    azacytidine treatment, Plant Mol. Biol., 3:333-336.
Vaeck, M., Reynaerts, A., Hofte, H., Jansens, S., de
    Beuckeleer, M., Dean, C., Zabeau, M., van Montagu,
    M., and Leemans, J., 1987, Transgenic plants
    protected from insect attack, Nature, 328:33-37.
Wang, Y.-C., Klein, T. M., Fromm, M. E., Cao, J.,
    Sanford, J. C., and Wu, R., 1988, Transient
    expression of foreign genes in rice, wheat and
    soybean cells following particle bombardment, Plant
    Molec. Biol., 11:433-439.
Watson, J. C., Kaufman, L. S., and Thompson, W. F.,
    1987, Developmental regulation of cytosine
    methylation in the nuclear ribosomal genes of
    Pisium sativum, J. Mol. Biol., 193:15-26.
Weintraub, H., 1985, Assembly and propagation of
    repressed and deprepressed chromosomal states,
    Cell, 42:705-711.
Willmitser, L., 1988, The use of transgenic plants to
    study plant gene expression, Trends Genet., 4:13-
    18

NON-CONVENTIONAL RESISTANCE TO VIRUSES IN PLANTS - CONCEPTS

AND RISKS

Roger Hull

John Innes Institute and AFRC Institute of Plant
Science Research, Colney Lane
Norwich, NR4 7UH, UK

INTRODUCTION

Viruses cause large yield losses in many crop species and, in
some situations, can limit the use of certain crops in some
areas. There are three basic approaches to preventing these crop
losses. Firstly, the sources of virus infection can be removed,
say, by use of virus-free planting stock or by eradication
schemes. Secondly, attempts can be made to prevent virus spread
by cultural techniques such as killing their vectors. Thirdly,
virus-resistant varieties of crops can be used. This latter
approach has several major advantages: it is considered to be the
most economical for farmers; it reduces the ecological problems
caused by the widespread use of pesticides; and it is thought to
be the most effective control measure in the long term.

The breeding programmes for many crops include attempts to
incorporate virus resistance into new cultivars. The targets for
such resistance can be grouped into those which give resistance
to transmission (eg. resistance to vectors, vector non-
preference, resistance to seed transmission), those which give
resistance to disease establishment (eg. prevention of virus
replication, prevention of cell-to-cell spread, localization of
infection with or without necrosis) and those which give
resistance to symptom development (eg. tolerance). These forms
of resistance are often conferred by single genes which, in some
cases, are dominant (see Fraser, 1986; 1987 for reviews).
However, sources of these genes can be very difficult to locate
and the resistance they confer frequently breaks down. For
instance, resistance to tungro disease of rice (Oryza sativa),
the most important virus disease of rice, has been obtained

Gene Manipulation in Plant Improvement II
Edited by J. P. Gustafson
Plenum Press, New York, 1990

through resistance to the leafhopper vector (Heinrichs and
Rapusas, 1983; Hibino et al., 1987; Rapusas and Heinrichs, 1985);
however, this can be overcome by new biotypes of the vector
(Anon, 1988). Another example is the breakdown of resistance to
tobacco mosaic virus (TMV) conferred by various genes in tomatoes
(Lycopersicon esculentum) (Table 1); the strains of TMV which
overcome the resistance genes often differ from wild type by only
one or two amino acid changes in the relevant gene (Meshi et al.,
1988; Hull, 1989).

Table 1.   Interactions between TMV-resistant
           tomato plants and strains of the virus

Host	Virus genotype				Function
genotype	0	1	2	2^2	
+	S	S	S	S	
Tm-1	R	S	R	R	Virus multiplication
Tm-2	R	R	S	R	Cell-to-cell spread
Tm-2^2	R	R	R	S	? cell-to-cell spread

S = susceptible
R = resistant
From Fraser (1986)

    Sanford and Johnston (1985) introduced a new concept for
genetically engineering resistance into organisms. They
suggested that modified pathogen genes could confer resistance by
interfering with the functioning of the pathogen. This
conceptual breakthrough has already yielded non-conventional
forms of resistance to viruses in plants and has opened up the
possibilities of other sources of resistance (for recent reviews
see Baulcombe, 1989; Wilson, 1989). One purpose of this chapter
is to discuss aspects of these forms of resistance not covered in
the above reviews and especially the, as yet, untested forms.
However, the use of viral sequences for conferring protection and
additionally their use in gene vectors poses potential risks in
the field release of genetically engineered crops. Thus, I will
also attempt to identify some of these potential risks which have
to be assessed before the widespread use of these sequences in
transgenic plants can be applied.

NON-CONVENTIONAL RESISTANCE

Current systems

Three forms of non-conventional resistance have already been tested, those conferred by viral coat protein genes, by satellite sequences and by antisense sequences. The viral coat protein-induced resistance will be discussed in detail by another contributor but I will raise just two or three points on it. Firstly, the evidence is that it operates early in the virus replication cycle (Register and Beachy, 1988) and hence prevents significant virus replication. This is an advantage as there is less chance of mutants arising which could overcome the resistance. Secondly, for each virus, resistance is limited to that virus and it is not known what range of strains of the virus is prevented from replicating by this form of resistance. Furthermore, resistance has been obtained only to viruses which, according to the available evidence have a co-translational mechanism of virus disassembly. It is possible that some viruses, e.g. como- or nepo-viruses, might have other disassembly mechanisms (Brisco et al., 1986) and it is not known if coat protein resistance would operate against these.

The expression of satellite sequences in transgenic plants reduces symptom expression and, in some cases, also reduces virus replication (for review see Baulcombe, 1989). However, satellite protection is specific for the 'parent' virus and relatively few viruses have satellites. Future research may result in 'broad spectrum' satellite-like sequences or defective interfering sequences but these would not overcome another basic problem. It is necessary for the 'parent' virus to replicate for the satellite system to be 'activated'. Thus, protected plants will be a source of infection for unprotected ones, and virus and satellite replication would allow for the possibility of mutations which could alter the protection.

Thus far there have been no reports of the successful use of antisense RNA as a protection strategy (see Baulcombe et al., 1987; Cuozzo et al 1988: Hemenway et al., 1988). However, antisense RNA has been used successfully in controlling the expression of nuclear genes in plants (Ecker and Davis, 1986; Delauney et al., 1988; Smith et al., 1988; van der Krol et al., 1988). There are several possible explanations of the failure of antisense RNA to control virus replication. Most of the constructs so far used have been derived from coat protein genes and neighbouring sequences. It may be that other sequences could be more effective, especially those involved in the initiation of replication and of subgenomic RNA formation. The combination of antisense RNA and ribozyme sequences (Haseloff and Gerlach, 1988) which would cleave the viral RNA might increase the efficiency; this will be discussed by another contributor. Every replicating viral sequence would have to be complexed with antisense RNA or else the virus would 'escape' the inhibition. As the antisense RNA is produced in the nucleus and RNA viruses replicate in the cytoplasm there might be compartmentalization which would limit the potential for hybrid formation. Furthermore, it is likely that viral RNA exists in the cytoplasm as nucleoprotein complexes, thus further limiting the chances of hybrid formation. The antisense strategy may work against DNA viruses, such as caulimoviruses and geminiviruses, which have their nucleic acid transcribed in the nucleus.

Potential systems

The simplicity of viral genomes coupled with molecular biological techniques and the ability to transform plants opens up the possibilities of other methods for conferring non-conventional resistance. There are two basic approaches to this. Firstly, if one can understand in detail the interactions involved in viral functions it should be possible to block them. This could be done by designing molecules which, when transformed into and expressed by plants, would either 'decoy' the targetted viral gene product or block its interaction with its 'substrate'. This is basically the manner in which the three forms of protection described above operate but it should be possible to create even more sophisticated forms of resistance. Viral genes or functions which might be amenable to this approach include those which are involved in the spread of viruses within or between plants, those involved in the processing of viral genes and those involved in viral replication. The second approach is to determine the molecular basis of forms of normal host resistance and to transfer such functions to susceptible plants. Genes involved here include those which confer natural or acquired resistance and those which are involved in non-host resistance.

Functions involved in virus spread within and between plants. There is accumulating evidence that the spread of most viruses from the initially infected cell(s) is via plasmodesmata and that this is facilitated by virus-encoded functions (movement proteins) (for review see Hull, 1989). These proteins are thought to interact with cellular components in such a manner that the plasmodesmata are 'gated open' to allow infection to move to adjacent cells. Present knowledge would indicate that there are, at least, two different mechanisms. In one, shown for instance by TMV, the viral gene product appears to interact with the plasmodesmata in a transient manner and enables non-capsid nucleoprotein forms of the virus to pass from cell to cell. In the other, virus particles [eg. those of cauliflower mosaic virus

(CaMV) and cowpea mosaic virus (CPMV)] pass through tubular structures extending from the plasmodesmata into the cytoplasm. Interference with either of these two mechanisms would prevent the virus from spreading from the initially infected cell. This, termed subliminal infection, most likely occurs in many so-called non-hosts of viruses (Zaitlin and Hull, 1987).

Viruses spread from plant to plant by a variety of mechanisms, one of the most common being by arthropod vectors. In most, if not all, cases the virus-vector interaction is very specific. This would suggest that there are specific sites on the coat of virus particles and in the arthropod vector at which these interactions take place (for reviews see Harrison and Murant, 1984; Harrison, 1987). With certain viruses, eg. caulimoviruses and potyviruses, there are non-capsid proteins (transmission or helper proteins) involved in these interactions. The transmission proteins of caulimoviruses are about 18,000 mol. wt. (P18) and have an unusual predicted secondary structure (Fig 1). As was shown firstly for CaMV (Modjahedji et al., 1985) and subsequently derived from other caulimovirus sequences (Hull, R. unpublished data), the N-terminal two-thirds of the transmission protein is nearly all β-sheet and turns whereas the C-terminal one-third is predominantly α-helix.

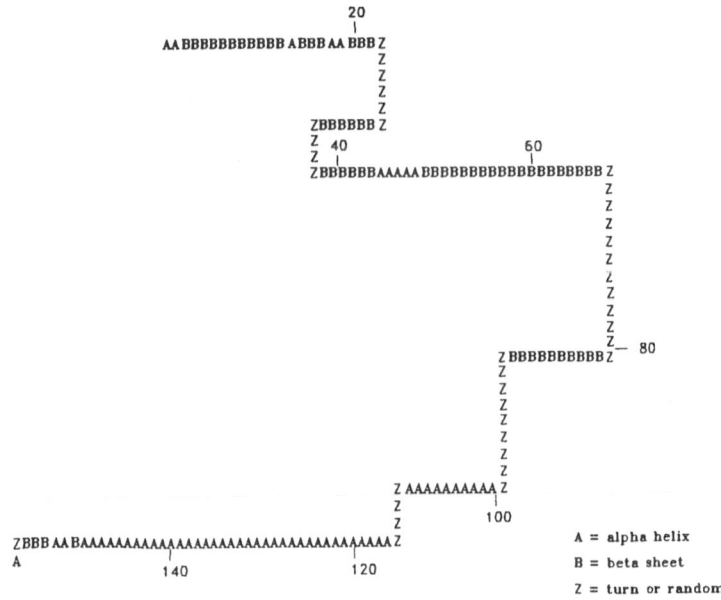

Fig. 1. Prediction of secondary structure of cauliflower mosaic virus gene 2 product (prediction by Joint method of Eliopoulos et al., 1982)

This is indicitative of there being two domains and it is tempting to suggest that one interacts with the virus capsid and the other with a specific region of the mouth parts or anterior alimentary canal of the aphid vector. The Campbell isolate of CaMV encodes a P18 which is non-functional and sequencing has shown that its P18 has a single amino acid difference at position 94 (Woolston et al., 1987) when compared with the amino acid sequence of the P18 of an aphid-transmissible isolate; this substitution is in the interdomain region. Thus, the position of these domains relative to each other might be important.

When sufficient information on the molecular biology of such virus-vector interactions is available it should be possible to devise systems which interfere with these interactions. The major drawbacks on attacking these genes are that the resistance will, in most cases be rather limited in the range of viruses controlled and that the viruses would not be prevented from replicating.

Factors involved in viral expression. The majority of important viral pathogens of crops have plus-strand RNA genomes (see Zaitlin and Hull, 1987). The expression of these, and in fact all other viruses of eukaryotes, is by translation of mRNAs using host ribosomes, a step which obviously could not be blocked. The constraints of translation by eukaryotic ribosomes dictate that, in a polycistronic RNA, only the 5' cistron is translated; those downstream of the 5' cistron are effectively closed. Plant viruses (and other eukaryotic viruses) have two strategies to overcome this problem. They either form subgenomic RNAs which are effectively monocistronic or they translate the whole RNA as a single cistron into a polyprotein which is cleaved into the functional proteins. Comoviruses and potyviruses follow the latter strategy and the evidence is that they encode their own proteases which cleave at specific sites (Goldbach and van Kammen, 1985; Carrington and Dougherty, 1988; Hellmann et al., 1988). If these proteases were either blocked or decoyed the virus would not be able to function. Similarly if the factors which control the formation of subgenomic mRNA are understood, control measures could be based on either blocking them or using them in trans to 'open up', say a toxin gene the product of which would kill the initially infected cell. It is likely that this approach of blocking viral expression functions would be virus-specific and not be usable as a broad spectrum form of resistance.

Viral replicases. The replication of plus-strand RNA viruses is via minus-strand RNA, and thus requires RNA-dependent RNA polymerases. RNA-RNA replication is unusual in eukaryotic cells and the evidence that is available indicates that most, if not all of this function is encoded by the virus (see Hall et al.,

1982; van Kammen and Eggen, 1986). Comparisons of amino acid sequences of putative replicase proteins of the 20 or so RNA plant viruses for which data are available indicate there are two groups, one based on TMV (which has similarities to the animal alphaviruses) and the other on CPMV (which has similarities to the animal picornaviruses) (see Goldbach, 1986; 1987, for reviews). What little is known about the structure of the replication complexes indicates that they are composed of various subunits (Horikoshi et al., 1988; Dorssers et al., 1984) which must interact with each other and with the replicating nucleic acid in specific ways. These should be amenable to molecular blockers.

A strategy of non-conventional resistance which attacked the replicase function of viruses would have several advantages. It is likely that a single non-conventional 'gene' could give resistance to a wide range of viruses. As RNA-RNA replication is rare in the host it should not interfere with normal host functions. Additionally, it would prevent virus multiplication at an early stage in the initially infected cell and thus would minimize the chances of mutations which could overcome the resistance. In view of the conservation of sequences across a wide range of viruses, it would seem that these are interactions which are critically important to virus function and that defects or blockages could not be overcome by simple mutations.

Natural and acquired resistance. As noted above, natural resistance to viruses is sometimes controlled by a single dominant gene. In many cases resistance can not be transferred to crops because of the species barrier. If the resistance gene(s) could be identified it might be possible to bypass this barrier. Information on the molecular mechanisms is available for only one form of resistance, that in Arlington cowpea to CPMV. The resistance is due to an inhibitor of the protease which divides up the polyprotein translated from the viral genome (Ponz et al., 1988). When the gene has been isolated it should be possible to move it to the few other hosts of CPMV; however, because of the specificity of the CPMV protease, it is unlikely that it would function against other viruses.

There are various other forms of resistance which are expressed on infection of plants. These include pathogenesis-related proteins (see Bol et al., 1987 for review), various antiviral factors (see Sela et al., 1987; Loebenstein and Gera, 1981) and hypersensitive responses (see Fritig et al., 1987). One, or several of these forms may provide sources of resistance when their mechanisms are understood. However, all of these induced forms of resistance depend upon an initial virus infection before they operate.

Non-host resistance. A survey of the Commonwealth Myco-
logical Institute/Association of Applied Biologists Descriptions
of Plant Viruses (Table 2) shows that most plant viruses have
host ranges (plants which give, at least, some systemic
expression of infection) of 50 or less species.

Table 2. Host ranges of plant viruses

Number of plant species susceptible	Number of viruses in category
≤ 6	95
5 - 25	75
26 - 50	49
≥ 51	39

While the numbers of species used in a host range determination is usually very limited, experience also shows that any one
virus will systemically infect only a small proportion of higher
plant species. This raises the question as to whether the vast
numbers of non-host species could provide sources of resistance.
As was suggested by Hull (1986) and Zaitlin and Hull (1987)
susceptibility and/or immunity can be considered as operating at
various levels, total susceptibility, hypersensitive response,
subliminal infection and total immunity. The use of the
subliminal infection response as a form of resistance was
discussed above under cell-to-cell spread. It has been noted
that subliminal infection may be a much more common non-host
response than previously recognized (Zaitlin and Hull, 1987). In
spite of this it is likely that much of non-host resistance is
due to total immunity. There is now sufficient information and
technology to determine at what stage in the virus replication
cycle total immunity operates. It would seem most likely that
there are not individual mechanisms for each virus/non-host
combination but that there are just one or a few mechanisms. In
this case their use would give broad spectrum resistance.

RISKS OF USE OF VIRAL SEQUENCES

The above discussion has focused on the use of plant viral
sequences for conferring resistance. There are also developments
in the use of viral sequences in gene vector systems. Both these
areas of research are advancing rapidly and there will shortly be
pressure for the commercial release of plants transgenic in viral
sequences into the field. What has been lacking is a concomitant
assessment of the potential risks in the use of viral sequences.
This section is to draw attention to some possible risks on which

further information is required for the regulating authorities to
make their recommendations.

The basic question that needs to be answered in risk assessment of plants transgenic in viral sequences is:- What is the
risk of viral genes or sequences which are used as vectors, or to
confer resistance, having significant effects on infection by
other viruses? I will consider this question in relation to the
viral genes and sequences discussed above and to existing and
possible vector systems.

Coat protein. There are two main situations that could arise
from the infection of a plant expressing a viral coat protein
gene by another virus. The main one is the possibility of the
infecting viral genome being encapsidated in the expressed viral
coat protein. Reconstitution studies have shown that the coat
protein of one virus can form virus-like particles around the
nucleic acid of other viruses (see Hull, 1970). Coat protein is
involved in the specificity of vector transmission of a virus,
say by insects, nematodes or fungi (see Harrison and Murant,
1984). This raises the possibility of the vector specificity of
the virus infecting the transgenic plant being changed by transcapsidation. Although this would be only for one transmission
passage (in the non-transgenic host it would be encapsidated by
its own coat protein and revert to its normal means of
transmission) the virus might be taken to ecological situations
to which its normal vector would not normally introduce it. This
could change its epidemiology.

A potentially more serious risk is that of making the
infecting virus seed-transmitted in the transgenic host.
Although there are no instances of viral genomes integrating into
host genomes in an infectious form, a high rate of seed
transmission effectively takes a virus into the germ line. For
many seed-transmitted viruses there is variation between strains
in the efficiency of transmission (Carroll, 1972; Haber and
Hamilton, 1980; Hamada and Harrison, 1977; Harrison and Hamada,
1976; Harrison et al., 1974). This indicates that, at least in
part, the efficiency of seed transmission is controlled by viral
gene product(s) but there are no data as to which. Thus, until
further information is available, great care has to be taken in
the selection of the strain of a virus from which the coat
protein gene is used for transformation.

The second situation that could arise from the infection of a
coat protein transgenic plant is that of synergism (Holmes,
1956). Several combinations of viruses, eg. tomato mosaic virus
and potato virus X in tomatoes, cause much more severe symptoms
than those induced by the individual viruses. What controls the

synergistic response is unknown and until it is, care will have
to be taken.

Satellites. Some of the risks associated with the use of
satellite protection have been mentioned above. Perhaps the
greatest is the effects that single base changes in the satellite
sequence might have on the expression of the 'parent' virus. It
might be possible to overcome this problem once there is further
information on the molecular interactions that lead to reduction
or enhancement of symptom expression.

Cell-to-cell spread proteins and replicases. It is unlikely
that cell-to-cell spread proteins or replicases would be used
directly to confer resistance; it is modifications of these
proteins which would be used as decoys. Such molecules are
unlikely to pose any risks.

Among the gene vector systems that are being considered are
ones which contain origin of replication sequences with
replication and cell-to-cell spread functions being supplied by
viral genes transformed into the plant. Thus, the risks of
plants which are transgenic in functional viral replicase and
possibly cell-to-cell spread protein genes would have to be
assessed. From the discussions above it would seem most likely
that the host ranges of viruses are controlled by these genes.
Thus, there could be a major risk that plants containing these
genes would become susceptible to viruses which did not normally
infect them.

Even if the expression of these viral genes did not increase
host range there are the questions of synergism and seed
transmission noted above.

Gene vectors. Viral sequences, especially the promoter
sequence for the 35S RNA of CaMV, are used extensively in vector
constructs for transforming plants. The recent report on
targetting genes in plants (Paszkowski et al., 1988) raises the
possibility that the presence of viral sequences in a plant
chromosome could allow the integration of whole viral genome.
This would only occur with those viruses (eg. caulimoviruses,
geminiviruses) which had part of their replication cycle in the
nucleus.

CONCLUSIONS

The conceptual breakthrough of Sanford and Johnston (1985),
coupled with advances in technology, are leading to some very
exciting approaches to the control of plant viruses. The first
generation of non-conventional forms of resistance are already

being tested and information is rapidly being accrued which should lead to further forms of resistance, including ones conferred by a single 'gene' to a wide range of viruses.

However, this excitement should be tempered by the need for a close assessment of the risks involved in the field release of these plants. Care will have to be taken that these developments do not leave us with even more serious agronomic problems derived from inadequate appreciation of the risks. Even if these perceived risks are not to be real, the experimentation needed to prove this will not be wasted as it would add considerably to the understanding of viral function and thereby could contribute even further forms of non-conventional resistance.

ACKNOWLEDGEMENTS

I thank Drs D. Baulcombe and T.M.A Wilson for kindly giving me preprints of their papers and Drs J.W. Davies and T.M.A. Wilson for helpful comments on the manuscript.

REFERENCES

Anon, 1988, "Top" seed hits snags, International Agricultural Development March/April 1988, p. 21.

Baulcombe, D., 1989, Strategies for virus resistance, Trends in Genetics, in press.

Baulcombe, D.G., Hamilton,W.D.O., Mayo, M.A. and Harrison, B.D., 1987, Resistance to viral disease through expression of viral genetic material from the plant genome, in "Plant Resistance to Viruses", CIBA Foundation Symposium 133, D. Evered and S. Harnett eds., John Wiley and Sons, Chichester, pp. 170-184.

Bol, J.F., van Huijsduijen, H., Cornelissen, B.J.C. and van Kan, J.A.L., 1987, Characterization of pathogenesis-related proteins and genes, in "Plant Resistance to Viruses", CIBA Foundation Symposium 133, D. Evered and S. Harnett eds, John Wiley and Sons: Chichester, pp. 72-91.

Brisco, M., Hull, R. and Wilson, T.M.A., 1986, Swelling of isometric and of bacilliform plant virus nucleocapsids is required for virus-specific protein synthesis in vitro, Virology, 148:210.

Carrington, J.C. and Dougherty, W.G., 1988, A viral cleavage site cassette: identification of amino acid sequences required for tobacco etch virus polyprotein processing, Proc. Natl. Acad. Sci. USA, 85:3391.

Carroll, T.C., 1972, Seed transmissibility of two strains of barley stripe mosaic virus, Virology, 48:323.

Cuozzo,M., O'Connell, K.M., Kaniewski, W., Fang, R.-X., Chua, N.-
H. and Tumer, N.E., 1988, Viral protection in transgenic
tobacco plants expressing the cucumber mosaic virus coat
protein or its antisense RNA, Biotechnology, 6:549.

Delauney, A.J., Tabaeizadeh, Z. and Verma, D.P.S., 1988, A stable
bifunctional antisense transcript inhibiting gene
expression in transgenic plants, Proc. Natl. Acad. Sci.
USA, 85:4300.

Dorssers.L., van der Krol, S., van der Meer, J., van Kammen, A.
and Zabel, P., 1984, Purification of cowpea mosaic virus
RNA replication complex: identification of a virus-encoded
110,000 dalton polypeptide responsible for RNA chain
elongation, Proc. Natl. Acad. Sci. USA, 81:1951.

Ecker, J.R. and Davis, R.W., 1986, Inhibition of gene expression
in plant cells by expression of antisense RNA, Proc. Natl.
Acad. Sci. USA, 83:5372.

Eliopoulos, E.E., Geddes, A.J., Brett, M., Pappin, D.J.C. and
Findlay, J.B.C., 1982, A structural model for the
chromophore binding domain of rhodopsin, Int. J. Biol.
Macromol., 4:263.

Fraser, R.S.S., 1986, Genes for resistance to plant viruses, CRC
Crit. Rev. Plant Sci., 3:257.

Fraser, R.S.S., 1987, Genetics of plant resistance to viruses, in
"Plant Resistance to Viruses", Ciba Foundation Symposium
133, D. Evered and S. Harnett eds., John Wiley and Sons,
Chichester, pp. 6-22.

Fritig, B., Kauffmann, S., Dumas, B., Geoffroy, P., Kopp, M. and
Legrand, M., 1987, Mechanism of the hypersensitivity
reaction of plants, "in Plant Resistance to Viruses", CIBA
Foundation Symposium 133, D. Evered and S. Harnett eds.,
John Wiley and Sons: Chichester, pp. 92-108.

Goldbach, R.W. 1986, Molecular evolution of plant RNA viruses,
Ann. Rev. Phytopath., 24:289.

Goldbach, R., 1987, Genome similarities between plant and animal
RNA viruses, Microbiol. Sci., 4:197.

Goldbach, R. and van Kammen, A. 1985, Structure, replication and
expression of the bipartite genome of cowpea mosaic virus,
in "Molecular Plant Virology", J.W. Davies, ed., CRC
Press: Boca Raton, Florida, Vol. 2. pp. 83-120.

Haber, S. and Hamilton, R.I., 1980, Distribution of determinants
for symptom production, nucleoprotein component
distribution and antigenicity of coat protein between the
two RNA components of cherry leaf roll virus, J. gen.
Virol., 50:377.

Hall, T.C., Miller, W.A. and Bujarski, J.J., 1982, Enzymes
involved in the replication of plant viral RNAs, Adv.
Plant Pathol., 1:179.

Hanada, K. and Harrison, B.D., 1977, Effects of virus genotype
and temperature on seed transmission of nepoviruses, Ann.
appl. Biol., 85:70.

Harrison, B.D. 1987, Plant virus transmission by vectors: Mechanisms and consequences, in "Molecular Basis of Virus Disease", W.C. Russell and J.W. Almond, ed., Cambridge University Press: Cambridge, pp.319-344.

Harrison, B.D. and Hanada, K., 1976, Competitiveness between genotypes of raspberry ringspot virus is mainly determined by RNA-1, J. gen. Virol., 31:455.

Harrison, B.D. and Murant, A.F., 1984, Involvement of virus-coded proteins in transmission of plant viruses by vectors, in "Vectors in Virus Biology", M.A. Mayo and K.A. Harrap, ed., Academic Press: London, pp. 1-36.

Harrison, B.D., Murant, A.F., Mayo, M.A. and Roberts, I.M., 1974, Distribution of determinants for symptom production, host range and nematode transmissibility between the two RNA components of raspberry ringspot virus, J. gen. Virol., 22:233.

Haseloff, J. and Gerlach, W., 1988, Simple RNA enzymes with new and highly specific endonuclease activities, Nature, 334: 585.

Heinrichs, E.A. and Rapusas, H., 1983, Correlation of resistance of the green leafhopper Nephotettix virescens (Hompoptera: Cicadellidae) with tungro virus infection in rice varieties having different genes for resistance, Environ. Entomol., 12:201.

Hellmann, G.M., Shaw, J.G. and Rhoads, R.E., 1988, In vitro analysis of tobacco vein mottling virus NIa cistron: evidence for a virus-encoded protease, Virology, 163:554.

Hemenway, C., Fang, R.-X., Kaniewski, W.K., Chua, N.-H. and Tumer, N.E., 1988, Analysis of the mechanism of protection in transgenic plants expressing the potato virus X coat protein or its antisense RNA, EMBO Jour., 7:1273.

Hibino, H., Tiongco, E.R., Cabunagan, R.C. and Flores, Z.M., 1987, Resistance to rice tungro-associated viruses in rice under experimental and natural conditions, Phytopathology, 77:871.

Holmes, F.O., 1956, A simultaneous-infection test for viral inter-relationships as applied to aspermy and other viruses, Virology, 2:611.

Horikoshi, M., Mise, K., Furusawa, I. and Shishiyama, J., 1988, Immunological analysis of brome mosaic virus replicase, J. gen. Virol., 69:3081.

Hull, R., 1970, Studies on alfalfa mosaic virus. III. Reversible dissociation and reconstitution studies, Virology, 40:34.

Hull, R., 1986, The pathogenesis of cauliflower mosaic virus, in "Genetics and Plant Pathogenesis", P.R. Day and G.J. Jellis ed., Blackwell: Oxford, pp. 25-32.

Hull, R., 1989, The movement of viruses in plants, Ann. Rev. Phytopath., in press.

Loebenstein, G. and Gera, A., 1981, Inhibitor of virus replication released from tobacco mosaic virus-infected protoplasts of a local lesion-responding tobacco cultivar, Virology, 114:132.

Meshi, T., Motoyoshi, F., Adachi, A., Watanabe, Y., Takamatsu, N. and Okada, Y., 1988, Two concomitant base substitutions in the putative replicase genes of tobacco mosaic virus confer the ability to overcome the effects of a tomato resistance gene, Tm-1, EMBO Jour., 7:1575.

Modjtahedi, N., Volovitch,M., Mazzolini, L. and Yot, P., 1985, Comparison of the predicted secondary structure of aphid transmission factor for transmissible and non-transmissible cauliflower mosaic virus strains, FEBS Letters, 181: 223.

Paszkowski, J., Baur, M., Bogucki, A. and Potrykus, I., 1988, Gene targetting in plants, EMBO Jour., 7:4021.

Ponz, F., Glascoch, C.B. and Bruening, G., 1988, An inhibitor of polyprotein processing with the characteristics of a natural virus resistance factor, Molecular Plant-Microbe Interactions, 1:25.

Rapusas, H.R. and Heinrichs, E.A., 1982, Plant age and levels of resistance to green leafhopper, Nephotettix virescens (Distant), and tungro virus in rice cultivars, Crop Prot., 1:91.

Register, J.C. and Beachy, R.N., 1988, Resistance to TMV in transgenic plants results from interference with an early event in infection, Virology, 166:524.

Sanford, J.C. and Johnston, S.A., 1985, The concept of parasite-derived resistance – deriving resistance genes from the parasite's own genome, J. theor. Biol., 113:395.

Sela, I., Grafi, G., Sher, N., Edelbaum, O., Yagev, H. and Gerassi, E., 1987, Resistance systems related to the N gene and their comparison with interferon, in "Plant Resistance to Viruses", CIBA Foundation Symposium 133, D. Evered and S. Harnett, eds., John Wiley and Sons, Chichester, pp. 109-119.

Smith, C.J.S., Watson, C.F., Ray, J., Bird, C.R., Morris, P.C., Schuch, W. and Grierson, D., 1988, Antisense RNA inhibition of polygalacturonidase gene expression in transgenic tomatoes, Nature, 334:724.

van der Krol, A.R., Lenting, P.E., Veenstra, J., van der Meer, I.M., Koes, R.E., Gerats, A.G.M., Mol, J.N.M. and Stuitje, A.R., 1988, An anti-sense chalcone synthase gene in transgenic plants inhibits flower pigmentation, Nature, 333:866.

van Kammen, A. and Eggen, H.I.L., 1986, The replication of cowpea mosaic virus, BioEssays, 5:261.

Wilson, T.M.A., 1989, Plant viruses: A tool-box for genetic engineering and crop protection, BioEssays, in press.

Woolston, C.J., Czaplewski, L.G., Markham, P.G., Goad, A.S., Hull, R. and Davies, J.W., 1987, Location and sequence of a region of cauliflower mosaic virus gene 2 responsible for aphid transmissibility, Virology, 160:246.
Zaitlin, M. and Hull, R., 1987, Plant virus-host interactions, Ann. Rev. Plant Physiol., 38:291.

PLANT TRANSFORMATION TO CONFER RESISTANCE

AGAINST VIRUS INFECTION

Roger N. Beachy

Department of Biology
Washington University
St. Louis, Missouri 63130

INTRODUCTION

Genetic transformation of crop plants to improve production and value are targets for many scientists in both university and private company laboratories. Several years ago the results of experiments was described which demonstrated that it was possible to produce virus resistant transgenic plants by expressing a gene that encoded virus capsid protein (Powell-Abel et al., 1986). Although seeds containing resistance genes have not yet reached the market place, the potential impact on crop production may be substantial.

A number of crop plants are regularly exposed to one or more virus pathogens which can cause little to no yield losses or extremely high losses, depending upon a variety of environmental and vector-related factors that control viral epidemiology. Plant breeders have little defense against virus diseases and their epidemics. Resistance genes have been found in wild relative species for some virus diseases, but the lack of sexual compatibility or other factors made the resistance gene essentially inaccessible to production agriculture. Thus, many crops carry little or no effective resistance against viral pathogens.

Here we summarize the results of experiments which demonstrate that transgenic plants expressing genes encoding viral capsid or coat proteins are resistant to the virus from which the gene was taken and to related viruses. We refer to such resistance as "coat protein-mediated protection". Although this method does not produce immunity, the level of disease resistance may be

Gene Manipulation in Plant Improvement II
Edited by J. P. Gustafson
Plenum Press, New York, 1990

sufficient to be effective under field conditions and will aid
plant breeders in producing virus resistant plant lines.

PRODUCING AND TESTING VIRUS RESISTANT PLANTS

The general method taken to produce plants that are resistant
to virus infection is to create a chimeric gene that, when
expressed in transgenic plants, leads to accumulation of capsid
protein (CP) of the virus for which resistance is desired. Thus,
if the target is resistance against potato virus X (PVX) in potato
(Solanum tuberosum) plants, a gene encoding the CP of PVX is
transferred to potato cells that are regenerated to whole plants.
Regenerated plants and their progeny are then tested for
resistance to infection. In the examples of virus resistance
reported to-date in transgenic plants it has been important to use
chimeric genes that lead to high levels of CP in all or most plant
cells (reviewed by Beachy et al., 1988; Hemenway et al., 1989).
This has been accomplished in most cases with the promoter from
the 35S (p35S) transcript of cauliflower mosaic virus (CaMV).
Under the control of this promoter the levels of CP detected in
transgenic plant are up to 0.2% (w/w) of the extracted plant
protein. This is not to imply that the degree of resistance is
governed solely by the level of gene expression or that high
levels of gene expression are required for resistance. While the
highest levels of resistance against TMV correlated with high
levels of accumulation of TMV-CP (Powell et al., 1990), this was
not the case with protection against the potyviruses, tobacco etch
virus and potato virus Y (Stark and Beachy, 1989). On the other
hand, high levels of resistance against cucumber mosaic virus
(CMV) was apparently accomplished by the accumulation of
relatively low levels of CMV CP (Cuozzo et al., 1988).

Resistance to infection in transgenic plant lines has been
evaluated in vegetatively propagated plant parts, as in the case
of TMV in tobacco (Nicotiana tabacum) (Powell-Abel et al., 1986),
potato viruses X and Y in potato (Hoekema et al., 1989; Lawson et
al., 1990), and in the progeny derived from seeds. Evaluating
resistance in rooted cuttings can be less faithful than in
seedlings, but if sufficient care is taken to produce propagules
of similar size and growth rate, an accurate evaluation of disease
resistance can be achieved. When possible, however, seedling
progeny are preferred in experiments to evaluate disease
resistance.

In general the degree of resistance has been assessed by
inoculating plants with increasing concentrations of virus. By
comparing the minimum level of inoculum required to cause disease
in 100% of the plants that do not express the CP gene [CP(-)] with
the level required to cause disease in 100% of plants that express
the CP gene [CP(+)] resistance can be quantitated. In general 10^2

to 10^4 higher concentrations of virus inoculum are required to overcome resistance in CP(+) plant lines that bear high levels of resistance than are required to cause disease in CP(-) plant lines.

While disease resistance can be scored as a +/- response, other methods to evaluate resistance are also useful. For example, the numbers of necrotic local lesions (Nelson et al., 1987) or starch lesions (Hemenway et al., 1988), each of which quantitate the numbers of sites of infection on inoculated leaves, can be compared in CP(+) vs. CP(-) plant lines. CP(+) plant lines produce many fewer sites of infection than CP(-) plant lines. ELISAs have also been used to quantitate the rate of virus accumulation and to compare the final amount of virus that accumulates in CP(+) vs. CP(-) plant lines. CP(+) plant lines generally have either or both reduced rates of virus accumulation and lower final amounts of virus than CP(-) plants (Nelson et al., 1987; Hemenway et al., 1988; Hoekema et al., 1989). Finally, a subjective comparison of the degree of disease severity in CP(+) and CP(-) plant lines can be used to evaluate disease resistance. Using this approach it was found that CP(+) plant lines that became infected developed disease symptoms that were significantly less severe than the symptoms on CP(-) plant lines (Stark and Beachy, 1989; Nejidat, Nelson, Holt, and Beachy, unpublished data).

In summary, CP(+) plant lines are less likely to become infected by virus than are CP(-) plant lines. Furthermore, if CP(+) plants do become infected, they develop less severe symptoms and accumulate lower levels of virus than CP(-) plants.

To avoid confusion with regard to CP-mediated resistance it should be emphasized that resistance is relatively specific. Thus, while expression of the TMV-CP gene confers resistance against TMV and other tobamoviruses, including tomato mosaic virus, pepper mild mottle virus, and tobacco green mottle virus (Nejidat and Beachy, unpublished data), this gene provides very little protection against other tobamoviruses (Nelson et al., 1987) or viruses in other groups (Anderson et al., 1989). In another study Stark and Beachy (1989) found that a gene encoding the CP of the potyvirus SMV protected transgenic tobacco plants against the potyviruses TEV and PVY. As anticipated there was no protection against TMV in these plant lines. Likewise, Loesch-Fries et al. (1987) demonstrated that the CP of alfalfa mosaic virus provided resistance against AlMV, but not against TMV.

CHARACTERISTICS OF CP-MEDIATED RESISTANCE TO TMV

Since the first report of CP-mediated resistance to TMV in transgenic CP(+) tobacco plants (Powell-Abel et al.. 1986) we have spent considerable effort to elucidate the molecular and cellular mechanisms that are responsible for resistance. In a recent report Powell et al. (1990) demonstrated unequivocally that resistance against TMV was due to accumulation of the CP itself, rather than the RNA transcript. In these studies site-directed mutagenesis was used to prevent translation of the transcript of the CP gene. Although large amounts of gene transcript accumulated in transgenic plants that expressed the mutant gene, these plants were not resistant to TMV infection. In the same study it was demonstrated that transgenic lines that accumulated low levels of CP were less resistant to infection than plant lines with higher levels of CP (Powell et al., 1990).

As described above, CP-mediated resistance is characterized by reduced numbers of sites of infection upon inoculation (Nelson et al., 1987). They also developed a protoplast system to determine whether inhibition of infection was reflected at the protoplast level. Register and Beachy (1988) reported that protoplasts derived from CP(+) tobacco plants were highly resistant to high concentrations of TMV. As indicated above, plant lines that accumulated low levels of CP were less resistant to TMV than other lines. On the other hand, protoplasts isolated from lines that accumulated ultra high or low levels of CP were resistant to infection by TMV (Register, unpublished data). By contrast, protoplasts from CP(+) plant lines were much less resistant to infection by TMV-RNA than to TMV (Register and Beachy, 1988). This study also demonstrated that TMV, briefly treated at pH 8.0 to expose the 5' end of TMV RNA, overcame resistance in protoplasts and in whole plants. These results indicate that CP-mediated resistance acts via a mechanism that blocks an early step in infection, a step which causes disassembly of the virus. Wilson (1984) identified conditions that cause TMV to disassociate in vitro, but it remains to be determined whether the same or different mechanisms are responsible for disassembly in vivo. Likewise it is not known how disassembly is blocked in CP-mediated resistance.

As described above, resistance may be manifested as a reduced rate of disease development in plants that are CP(+) compared to CP(-) plant lines. TMV CP(+) tobacco plants inoculated with high levels of TMV-RNA (to overcome the initial step in resistance) developed systemic infection significantly later than CP(-) plants (Wisniewski, Powell, Nelson and Beachy, unpublished). In this study, which included plant grafting experiments and tissue print studies, it was concluded that while virus replicates to relatively high levels in inoculated leaves, the infection was less likely to enter and pass through the vascular system in CP(+) plants than in CP(-) plants. This lead Wisniewski et al. to suggest that virus was unable to "load" into the phloem and/or companion cells in CP(+) plants. This process, like that of uncoating of virus, is poorly understood. Likewise the exit of

virus from the vascular system into other leaves may be blocked in
CP(+) plants. While this process is also poorly understood it
represents another potential site of resistance in CP(+) plant
lines.

BREADTH OF PROTECTION PROVIDED THROUGH CP-MEDIATED RESISTANCE

    In most examples of disease resistance a specific gene
provides protection against one pathogen or a single strain of the
pathogen. Experiments in several laboratories have demonstrated
that the CP gene derived from one virus provides resistance
against several different viruses. In one example the CP gene of
the "D"-strain of cucumber mosaic virus (CMV) protected tobacco
plants against infection by the "C" strain of CMV (Cuozzo et al.,
1988). Strains of most viruses, including CMV, are distinguished
by their host range, vector specificity, symptoms, and/or
antigenic differences. While the differences in the amino acid
sequences of the C and D strains of CMV are minor, the
experimental results imply that minor changes or drift in virus
sequences are unlikely to overcome CP-mediated resistance.

    This hypothesis is supported by work with the tobamoviruses.
Nelson et al. (1988) reported that the TMV-CP gene in tomato
plants gave protection against TMV as well as the tomato mosaic
virus strains L, ToMV-1, ToMV-2, and ToMV-2[2]. In a subsequent
study Nejidat and Beachy (unpublished data) showed that tobacco
plant lines that express the TMV CP gene were highly resistant to
the tobamoviruses pepper mild mottle virus, tobacco mild green
mosaic virus, less resistant to ondontoglassum ringspot virus.
These plants were much less resistant to the tobamovirus sunn hemp
mosaic virus (Anderson et al., 1989). The CP amino acid sequences
of these tobamoviruses range from approximately 82% to 40%
homologous to the CP sequence of TMV. As anticipated there was
less resistance against the most distantly related strains than to
closely related strains.

    A similar type of study was done with the potyviruses. In
this study a CP gene encoding a strain of soybean mosaic virus
(SMV) was introduced into transgenic tobacco plants (Stark and
Beachy, 1989). SMV is not a pathogen on tobacco; therefore,
plants were challenged by inoculation with tobacco etch virus
(TEV) and potato virus Y, both of which are pathogenic on tobacco.
Although the amino acid sequences of SMV, TEV and PVY CPs differ
substantially and are antigenically distinct from each other, the
SMV-CP(+) tobacco plants are resistant to infection by TEV and
PVY. While some plant lines had low levels of resistance others
were highly resistant to infection (Stark and Beachy, 1989). In
some resistant plant lines inoculum concentrations up to 50 μg/ml
failed to induce disease symptoms on CP(+) plants. However, while
many transgenic lines contained high levels of SMV-CP, not all

lines were protected against infection, and a single line was highly resistant (Stark and Beachy, unpublished data). As expected, plants resistant to the potyviruses were susceptible to infection by TMV. It has yet to be determined why some CP(+) plant CP lines, but not others, were highly resistant to infection.

As a final example of broad protection van Dun and Bol (1988) reported results of experiments with tobacco plants that express the CP gene of the TCM strain of the tobravirus tobacco rattle virus (TRV). The CP(+) plants were resistant to infection by the TCM strain of TRV, but not the PLB strain. The amino acid sequences of the TCM and PLB strains differ by about 60%. On the other hand the TRV CP(+) plant lines were resistant to pea early browning virus, a tobravirus with approximately 60% amino acid sequence homology with TRV-TCM.

Taken together, these examples indicate that the expression of a CP gene can provide broad resistance against viruses and virus strains other than those from which the gene was isolated. This broad resistance undoubtedly is reflective of details relating to the basic mechanisms of resistance. For example, resistance may rely upon similarities not only of primary amino acid sequence, but of protein structure. However, much remains to be done to identify the structural and/or sequence features of coat proteins that govern resistance.

CONCLUSIONS

Coat protein mediated resistance has been demonstrated effective against a number of different viruses. The phenotype of resistant plants is simple -- plants either escape infection, or if infected, develop more mild symptoms with a substantial delay compared with susceptible CP(-) plants. Under field conditions this should result in less infection, lower inoculum potential, and reduce the likelihood of disease epidemics. Although a limited number of approved field experiments have been successfully completed, full evaluation of CP-mediated resistance will take several additional years. Only then will the full potential of CP-mediated resistance be realized.

ACKNOWLEDGEMENTS

The NIH (AI27161) and the Monsanto Company are acknowledged for support of this research. I am grateful to Ms. Nancy Burkhart for preparation of the manuscript.

REFERENCES

Anderson, E. J., Stark, D. M., Nelson, R. S., Tumer N. E., and
    Beachy, R. N., 1989, Transgenic plants that express the coat
    protein gene of TMV or AlMV interfere with disease,
    development of non-related viruses, Phytopathology,
    12:1284-1290.
Beachy, R. N., 1988, "Virus cross-protection in transgenic
    plants," in: "Plant Gene Research. Temporal and Spatial
    Regulation of Plant Genes," D. P. S. Verma and R. B.
    Goldberg, eds., Springer-Verlag, New York., pp. 313-327.
Cuozzo, M., O'Connell, K. M., Kaniewski, W., Fang, R.-X., Chua,
    N.-H., and Tumer, N.E., 1988, Viral protection in transgenic
    plants expressing the cucumber mosaic virus coat protein or
    its antisense RNA, Bio/Technology, 6:549-557.
Hemenway, C., Fang, R.-X., Kaniewski, W. K., Chua, N.-H., and
    Tumer, N. E., 1988, Analysis of the mechanism of protection
    in transgenic plants expressing the potato virus X coat
    protein or its antisense RNA, EMBO J., 7:1273-1280.
Hemenway, C., Tumer, N. E., Powell, P. A., and Beachy, R. N.,
    1989, Genetic engineering of plants for Viral Disease
    Resistance, in: "Cell Culture and Somatic Cell Genetics of
    Plants. Molecular Biology of Plant Nuclear Genes," J. Schell
    and I. K. Vasil, eds., Academic Press, Inc., California., pp.
    406-423.
Hoekema, A., Huisman, M. J., Molendijk, L., van den Elzen, P. J.
    M., and Cornelissen, B. J. C., 1989, The genetic engineering
    of two commercial potato cultivars for resistance to potato
    virus X, Bio/Technology, 7:273-278.
Lawson, C., Kaniewski, W., Haley, L., Rozman, R., Newell, C.,
    Sanders, P., and Tumer N., 1990, Engineering resistance to
    mixed virus infection in a commercial potato cultivar:
    reistance to potato virus X and potato virus Y in transgenic
    Russet Burbank, Bio/Technology, in press.
Loesch-Fries, L. S., Merlo, D., Zinnen, T., Burhop L., Hill, K.,
    Krahn, K., Jarvis, N., Nelson, S., and Halk, E., 1987,
    Expression of alfalfa mosaic virus RNA4 in transgenic plants
    confers virus resistance, EMBO J., 6:1845-1851.
Nelson, R. S., Powell-Abel, P., and Beachy, R. N., 1987, Lesions
    and virus accumulation in inoculated transgenic tobacco
    plants expressing the coat protein gene of tobacco mosaic
    virus, Virology, 158:126-132.
Nelson, R. S., McCormick, S. M., Delannay, X., Dubé, P., Layton,
    J., Anderson, E. J., Kaniewska, M., Proksch, R. K., Horsch,
    R. B., Rogers, R. G., Fraley R. T., and Beachy, R. N., 1988,
    Virus tolerance, plant growth, and field performance of
    transgenic tomato plants expressing coat protein from tobacco
    mosaic virus, Bio/Technology, 6:403-409.
Powell-Abel, P. A., Nelson, R. S., De, B., Hoffmann, N., Rogers,
    S. G., Fraley R. T., and Beachy, R. N., 1986, Delay of
    disease development in transgenic plants that express the

tobacco mosaic virus coat protein gene, <u>Science</u>, 232:738-743.

Powell, P. A., Stark, D. M., Sanders, P., and Beachy, R. N., 1989, Protection against tobacco mosaic virus in transgenic plants that express TMV antisense RNA, <u>Proc. Natl. Acad. Sci. U.S.A.</u>, 86:6949-6952.

Powell, P. A., Sanders, P. R., Tumer N., and Beachy, R. N., 1990, Protection against tobacco mosaic virus infection in transgenic plants requires accumulation of capsid protein rather than coat protein RNA sequences, <u>Virology</u>, in press.

Register, J. C. III, and Beachy. R. N., 1988, Resistance to TMV in transgenic plants results from interference with an early event in infection, <u>Virology</u>, 166:524-532.

Register, J. C. III, and Beachy, R. N., 1989. A transient protoplast assay for capsid protein-mediated protection: Effect of capsid protein aggregation state on protection against tobacco mosaic virus, <u>Virology</u>, in press.

Stark, D. M., and Beachy, R. N., 1989, Protection against potyvirus infection in transgenic plants: Evidence for broad spectrum resistance, <u>Bio/Technology</u>, 7:1257-1262.

van Dun, C. M. P., and Bol, J. F., 1988, Transgenic tobacco plants accumulating tobacco rattle virus coat protein resist infection with tobacco rattle virus and pea early browning virus, <u>Virology</u>, 167:649-652.

Wilson, T. M. A., 1984, Cotranslational disassembly of tobacco mosaic virus <u>in vitro</u>, <u>Virology</u>, 137:255-265.

USING PLANT VIRUS AND RELATED RNA SEQUENCES TO CONTROL

GENE EXPRESSION

Mark Young and Wayne Gerlach

CSIRO Division of Plant Industry
Canberra, ACT 2601 Australia

INTRODUCTION

Plant virus replication relies on the biochemistry
of its host cell, strongly suggesting that viral gene
expression closely mimics the host's own mechanism of
gene expression. This fundamental principle has been a
driving force behind research directed at understanding
the molecular biology of viruses, using them as a tool
for studying both viral and host gene expression. Many
recent advances have been made possible by recombinant
DNA technology, which has allowed a more thorough
analysis of basic   viral genome structure and
function. Armed with this basic knowledge, virologists
have been successful in manipulating and using viral
genomes and related sequences to alter gene expression
and thereby altering phenotypes. Bacterial, animal,
and plant viruses have all been subjected to this
approach.

One concept to emerge in recent years is that RNA
viruses have developed through modular evolution
(reviewed by Strauss and Strauss, 1988). This concept
holds that the virus genome can be viewed as being
comprised of modules of functional activity. These
functional modules cover viral activities such as
replication, assembly, and movement. An examination of
nucleotide sequence data from diverse RNA viruses has
suggested that these various functional modules have

Gene Manipulation in Plant Improvement II
Edited by J. P. Gustafson
Plenum Press, New York, 1990

313

evolved quasi-independently. The forces of mutation and
RNA recombination have allowed reassortment of modular
units to form new viruses. Each of these modular units
may therefore provide their own unique mechanisms for
gene expression. Can we dissect these functional units,
gain insight into their mode of action, and manipulate
them in such a way as to alter host and viral gene
expression?

     The purpose of this paper is to examine how this
reductionist approach has allowed us to influence viral
and host gene expression. It is not our intention to
cover all aspects of plant viral gene expression and
regulation, but to concentrate on an analysis of the
non-coding segments of plant viral genomes and related
genomes which allow for altered expression of both host
and viral genes. These non-coding segments of plant
viral genomes provide a natural source of signal
sequences to be further manipulated for control of gene
expression. In particular, we bias our views to
illustrate how the use of such non-protein coding
sequences of plant viral genomes and related viral
associated elements have been, or potentially could be,
employed to develop novel mechanisms of resistance to
plant viral infection. Where possible we will
illustrate these points using examples from the plant
luteovirus and nepovirus groups which are the subject of
study in our laboratory.

VIRAL PROMOTERS, ENHANCERS AND TRANSLATIONAL CONTROL
SEQUENCES

     Viruses are very efficient transforming agents of
plants. They have evolved mechanisms to introduce and
highly express their genes, altering the gene expression
of the host during infection. The signals by which
plant viruses control their gene expression are only now
beginning to be unravelled. Two classes of control
sequences can be defined; those that affect
transcription and/or replication and those that affect
translation and/or protein stability.

     Transcriptional promoters have already been used to
express heterologous sequences. Certainly, the most
widely used promoter is the constitutive 35S promoter

derived from the ds DNA virus, cauliflower mosaic virus (CaMV)(Odell et al., 1985; Fang et al., 1989; for review see Kuhlemeier et al., 1987). This strong constitutive promoter has been used to drive heterologus gene transcription in both transient expression and stable transformation systems. For example, it has been used to drive virus coat protein expression in transgenic plants in order to confer resistance to viral infection (Abel et al., 1986; Tumer et al., 1987; van Dun et al., 1987). We and others have also used the 35S promoter to express the satellite RNAs of viruses in plants (Gerlach et al., 1987; Harrison et al., 1987). These transgenic plants showed protection when infected with virus. It is interesting to note that the 35S promoter functions well in many dicots and monocots which are not hosts for CaMV. This suggests that many other functions derived from viruses with limited host ranges may be useful in a broad range of plants. Continued research on other promoters of DNA viruses will undoubtedly lead to additional promoters with useful characteristics e.g. tissue specificity or developmentally regulated expression.

Transcriptional promoters found associated with DNA viruses represent only one type of plant viral promoter. RNA viruses and related agents also have promoter sequences required for replication. The majority of plant viruses possess single stranded (+) sense RNA genomes which efficiently utilize their RNA sequences to both express their gene products and as sites for replication of progeny genomes. It is reasonable to expect the development of replication signals derived from RNA viruses and related agents as the basis for RNA episomal vectors. In studies with many plant viruses it has been demonstrated that the 3' end of RNA contains the replication sequences necessary to direct (-) strand synthesis from the (+) strand genome. The divergent structures found at the 3' termini of plant viruses argues that there are many different replication sequences operating in plant viruses. Characterization of these RNA replication sequences is just beginning. Mutations in the 3' tRNA-like terminus of brome mosaic virus (BMV) either prevents polymerization or alters the specificity of the polymerization start site (Bujarski et al., 1985 and 1986). Viroids and satellite RNAs must also possess signals for RNA replication, and in the case of viroids must possess sequences which are recognized by the replication machinery of the host plant.

Characterization and manipulation of RNA replication sequences should provide the basis for construction of RNA episomal elements which would be capable of expressing foreign genes in plants without the obligatory plant regeneration systems now required. Already, foreign genes have been expressed using RNA viral vectors. Chloramphenicol acetyl transferase (CAT) has been expressed at high levels using either tobacco mosaic virus (TMV; W. O. Dawson, pers. comm.) or BMV (French et al., 1986) as RNA vectors. Viral based RNA episomal vectors have the advantage of being capable of rapidly transforming whole plants and directing high levels of gene expression. Thus, expression of viral resistance factors from episomal vectors could ultimately provide a rapid method for introduction of resistance into a field crop.

Other control elements associated with plant virus genomes will also be exploited to control gene expression. Viral enhancers which increase gene transcription by promoting the binding of transcription factors could be manipulated to alter the binding affinity of transcription factors and thereby altering gene transcription. The characterization and manipulation of the CaMV 35S promoter enhancer has already led to increases in transcription of genes in transgenic plants (Odell et al., 1988; Fang et al., 1989). Translational enhancer sequences have also been described in plant viruses which appear to increase the translatability of a mRNA. For example, the omega sequence located at the 5' end of the TMV genome increases the translation of the TMV 126 kD replicase gene, and will also enhance the translation of heterologous mRNAs in both plant and animal systems (Gallie et al., 1987). Further characterization of these sequences and their interaction with the host translational machinery may lead to new strategies for control of gene translation.

Undoubtedly new plant viral regulatory elements will be described as our basic knowledge of these agents increases. One needs only to look at the unexpected and complex collection of regulatory elements recently discovered with the AIDS virus (for review see Haseltine and Wong-Stall, 1988) to realize that a wealth of elements have yet to be characterized in plant viral systems. The potential to exploit these elements should lead to the development of novel mechanisms of gene control in plants and their viruses.

SUBGENOMIC PROMOTERS OF PLANT RNA VIRUSES

     An intriguing approach to virus specific expression of a gene involves the use of subgenomic RNA promoters. Many plant RNA viruses express one or more of their gene products via subgenomic RNAs. The generation of subgenomic RNAs from polycistronic genomes of (+) stranded RNA viruses represents a mechanism for expression and amplification of internal cistrons, providing an effective system for production of proteins needed in large quantities. The most thoroughly characterized plant viral subgenomic RNA promoter is that of brome mosaic virus (BMV). The coat protein gene of BMV is translated from a monocistronic mRNA which is produced by internal initiation by the BMV replicase at a subgenomic promoter site on the (-) strand of the dicistronic RNA 3 (Miller et al., 1985). Large quantities of the coat protein mRNA are produced via transcription from this promoter. Mutational analysis has defined specific sequences required for operation and regulation of this promoter (Marsh et al., 1988). Its structure appears to be quite different from the promoters responsible for replication of BMV genomic RNA.

     The subgenomic promoter regions from a diverse source of RNA viruses have been compared and found to have defined sequence motifs (Marsh et al., 1988; Strauss and Strauss, 1988). Analysis of the BMV subgenomic promoter region suggests that it comprises four functional domains of approximately 62 nucleotides (Marsh et al., 1988). These domains are core sequences (UCUA, GUCCUAA, and GCGUA), a poly (A) tract, an upstream (U-A) block, and a downstream (A-U)tract. The generality of this model is tested as one compares new viruses. We have found that the untranslated sequence upstream of the translational start site of the coat protein gene of barley yellow dwarf virus (BYDV)

contains several, but not all of these sequence motifs.
This region is associated with the production of a
subgenomic RNA in BYDV infected plants (Gerlach et
al.,1987). This strongly suggests that the sequences
are also functional as a subgenomic promoter region for
expression of the coat protein mRNA for BYDV.

We are attempting to take advantage of the BYDV
subgenomic promoter to express a foreign protein in
cereal cells only in the presence of BYDV. The
rationale for this experiment is that transcription
directed by the BYDV subgenomic promoter is strictly
dependent on the presence of active viral replicase.
Therefore, a constitutive, low level, expression of an
(-) sense RNA molecule possessing the sub genomic
promoter sequence linked to a reporter gene might be
transcribed and amplified into a (+) sense mRNA only in
the presence of BYDV replicase. To test this model of
subgenomic promoter directed transcription of a mRNA,
the subgenomic promoter fragment of BYDV has been cloned
upstream of the GUS reporter gene. This construct was
further placed under the control of the constitutive
CaMV 35S promoter in such a way that transcription from
the 35S promoter results in a (-) sense BYDV subgenomic
promoter-GUS RNA (Fig. 1). Triticum monococum
protoplasts will be coinoculated with this construct and
assayed for the expression of GUS in the presence or
absence of BYDV. One would predict that only when
protoplasts were coinoculated with BYDV would GUS
activity detected. This would strongly suggest that the
detected GUS activity results from transcription by the
BYDV replicase from the subgenomic promoter to yield the
(+) sense GUS mRNA.

Expression of resistance genes from viral
subgenomic promoters analogous to the system described
above may provide a novel approach to conferring
protection against an invading plant virus. As one
example, production of a mutant viral replicase which
competes with wild type replicase, analogous to
experiments performed with bacterial phage QB (Inokuchi
and Hirashima, 1987), could provide a self regulating
system for resistance. The mutant replicase could be
expressed as a (-) strand transcript, so that
production, amplification, and translation of the (+)
strand mRNA would only occur from the subgenomic

promoter in the presence of wild type replicase. Once
the mutant replicase outcompeted the wild type replicase
its own expression would cease. Other sequences which
interfere with plant virus expression could also be
expressed as a (-) strand transcript which could be
amplified upon infection by the virus. Self regulating
expression of such resistance factors would have the
advantage of avoiding possible side effects of
constitutive expression of resistance genes on the host
and the reduced possibility of RNA recombination of
resistance factors with other genes.

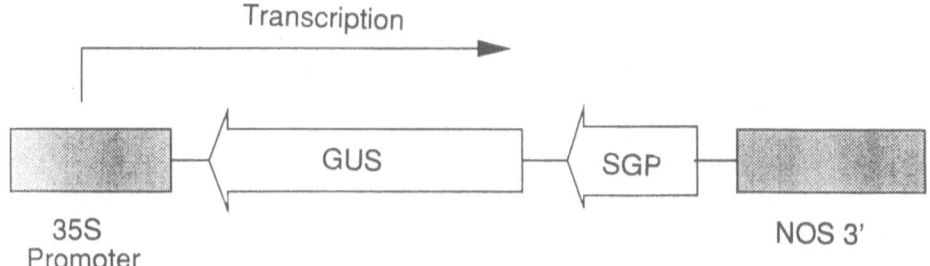

Fig. 1.   Schematic representation of BYDY subgenomic
          promoter (SGP) directed expression of GUS mRNA.
          Transcription directed by the CaMV 35S promoter
          results in the expression of (-) sense RNA
          containing GUS and SGP sequences. The SGP
          sequences encoded on the (-) sense RNA should
          provide signals for (+) sense GUS mRNA
          expression only in the presence of the BYDV
          replicase.

SATELLITE RNAs AND RELATED SEQUENCES

     Satellite RNAs of plant viruses are small single
stranded RNAs which rely on their helper virus for
propagation through replication and encapsidation (For
review; Francki, 1985). They replicate in virus
infected cells but are not necessary for replication of
the helper virus. They do not encode for protein
products and lack significant sequence homology with
their supporting virus. Biologically, the presence of
satellite RNAs can greatly modulate viral symptom

expression. Some satellites reduce symptom severity
while others enhance symptom expression. The mechanism
by which satellites effect this alteration in symptom
expression is unknown. They do replicate at the expense
of the helper virus, but the protection is not simply a
function of altering virus levels in the infected cells.

We have taken advantage of the biological
properties of satellite RNAs to confer viral disease
resistance to plants. The presence of the satellite RNA
of tobacco ringspot virus (STobRV) greatly reduces
symptom expression caused the tobacco ringspot virus
(TobRV) alone (Schneider, 1977). We have used
Agrobacterim tumifaciens to introduced multimeric cDNA
copies of STobRV into both tobacco (*Nicotiana tabacum*
var. Samson) (Gerlach et al., 1987) and lettuce
(*Lactuca sativa*)(M. Young, unpublished results).
Expression of STobRV RNA in these transgenic plants is
under control of the constitutive 35S CaMV promoter.
When transgenic plants are inoculated with the tobacco
ringspot virus (TobRV) there is dramatic increase in the
levels of STobRV, presumably resulting from replication
by the virus, and attenuation of viral symptom
expression. Others have used a similar strategy using
the satellite of cucumber mosaic virus to confer
protection against cucumber mosaic virus and the closely
related virus tomato aspermy virus (Harrison et al.,
1987). Two important conclusions can be drawn from
these experiments. First, the use of non-protein coding
satellite RNA sequences in transgenic plants can result
in biological protection against viral infection.
Secondly, satellite RNAs posses sequences which are
capable of directing high level of replication by its
supporting virus. Defining and characterizing the
sequences responsible for these properties is currently
underway and should provide the rational basis for
exploitation of such sequences for high level expression
of host or viral gene sequences.

While satellite RNAs show little sequence
similarity with their helper virus, a second class of
small RNA molecules which also act as parasites do have
significant sequence homology with their helper virus.
These are the so called defective interfering particles
(DIs). DIs were first described for certain animal

viruses and have now also been found associated with plant viruses. These include DIs associated with tomato bushy stunt (Hillman et al., 1987), sonchus yellow net virus (Ismail and Milner, 1988), and possibly for wound tumor virus (Anzola et al., 1987). The tomato bushy stunt DI has been cloned and sequenced and found to consist of complex series of internal deletions of the virus genome. Of biological significance is the observation that the presence of DI RNA in infected plants results in symptom amelioration and that coinoculation of DI RNA with virus results in phenotypic protection to the plants from the virus. Thus, it would be predicted that transgenic plants expressing DI RNAs would be protected against symptom expression. The mechanism by which DIs interfere with symptom development is unclear. However, DIs appear to reduce viral gene expression by effectively competing with the supporting virus for replication and encapsidation functions. Further characterizations of DIs could lead to synthetic DI sequences which reduce viral gene expression or even possibly host gene expression when expressed in plants.

RIBOZYMES (CATALYTIC ANTISENSE RNAs)

A property of many satellite and some viroid RNAs is the ability of these RNAs to undergo self catalysed cleavage reactions. This cleavage reaction apparently plays an important role in the replication cycle of these RNAs. Comparison of these autocatalytic RNAs suggests that the sequences surrounding the (+) cleavage site can base pair to form a characteristic secondary structure (Forster and Symons, 1987). The cleavage reactions of STobRV are among the most thoroughly characterized and are thought to lead to the production of monomeric length STobRV from multimeric forms produced during the satellite replication cycle (Kiefer et al., 1982; Prody et al., 1986; Buzayan et al., 1986). The cleavage of multimeric units of STobRV occurs from intramolecular interactions of sequences responsible for the catalysis and sequences at the actual site of cleavage. Mutational analysis of the (+) strand cleavage domain by Haseloff and Gerlach (1988) has demonstrated that the cleavage reaction can be divided into a two component system which can operate in trans.

One component consists of a short RNA sequence which
encompasses the cleavage site and acts as a substrate
for the second component which comprises the RNA
sequences responsible for the actual catalysis. From a
comparison with a range of autocatalyic sites, the only
sequence requirement of the cleavage site is at most
GUC, with cleavage occurring after the cytosine
nucleotide. The catalytic domain responsible for the
cleavage reaction comprises 24 nucleotides, embedded within the flanking sequences which are thought to base pair
and to form the secondary structure depicted in Fig. 2.
The only external requirements for the autocatalytic
reaction are a neutral or higher pH and the presence of
a suitable divalent cation such as Mg++ (Prody et al.,
1986). One possible model for the biochemical basis of
the cleavage reaction involves the association of a
divalent metal cation, possibly in the "pocket" of the
catalytic domain which is in close proximity to the site
of cleavage. The proximity of metal cation may abstract
the proton from the 2' hydroxl of the cytosine sugar
moiety, which results in the nucleophilic attack by the
oxygen on the 3' - 5' phosphodiester bond. This attack
results in the hydrolysis of the 3'-5' phoshodiester
bond with the resulting cleavage products being a 2'-3'
cyclic phosphate and a 5' hydroxyl.

    Armed with a basic understanding of the RNA
directed cleavage reaction and the ability perform the
reaction in trans, Haseloff and Gerlach (1988) designed
new ribozymes which direct to the site, specific cleavage
of a chosen target mRNA *in vitro*. The ribozymes consis-
ted of the 24 nucleotide catalytic unit, flanked by
complementary sequences which align the catalytic unit
with the GUC cleavage site in the target RNA (Fig. 2).
In vitro cleavage of chloramphenicol acetyl transferase
(CAT) mRNA by ribozymes directed against three different
GUC target sites in the mRNA demonstrated the activity,
site specificity, and enzymatic nature of these
synthetic ribozymes.

    Over the past several years attempts have been made
to use antisense RNA transcripts to generate dominant
mutations by inhibiting the expression of specifically
targeted mRNA species (for reviews; Weintraub, 1985;
Inouye, 1988; van der Krol, 1988). The mechanism by
which antisense RNA inhibits mRNA expression in

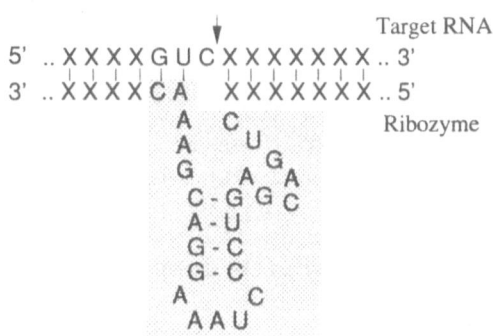

Fig. 2.   Schematic representation of the three
          components of ribozyme design. These
          components include (1) the target RNA
          containing the GUC sequence motif with cleavage
          occurring after the cytosine nucleotide. (2)
          The catalytic domain (shaded) and (3) the
          flanking base pairing required to align the
          catalytic domain with the cleavage site.

eukaryotes are not completely understood but is thought
to involve the formation of a dsRNA complex which
interferes with translation, (Melton, 1985; Harland and
Wientraub, 1985), processing (Strickland et al., 1988),
stability (Crowley et al., 1988), or translocation (Kim
and Wold, 1895) of the targeted mRNA. While there have
been some successes in attempts to use antisense RNAs to
produce mutant phenotypes, not all attempts to
inactivate targeted mRNAs have been successful. There
are examples of ribosome complexes being capable of
translating through ds RNA complexes (Ligelbach and
Dobberstein, 1988). More recently there has been the
discovery of unwindase activities which are capable of
disociating dsRNA complexes (Bass and Weintraub, 1986;
Rebagliatis and Melton, 1987; Wagner and Nishikura,
1988), and in one case, covalently modifying the

diassociated RNAs to prevent reassociation (Bass and
Weintraub, 1988). These may affect the use of antisense
RNA as a tool to inactivate gene expression.

Ribozymes, or more precisely catalytic antisense
RNAs, may circumvent some of these limitations of
conventional antisense approaches to inactivate gene
function. Cleavage of the targeted RNA should
inactivate the targeted RNA and the presence of an
unwindase activity could potentially increase the
turnover of the enzyme-substrate complex.

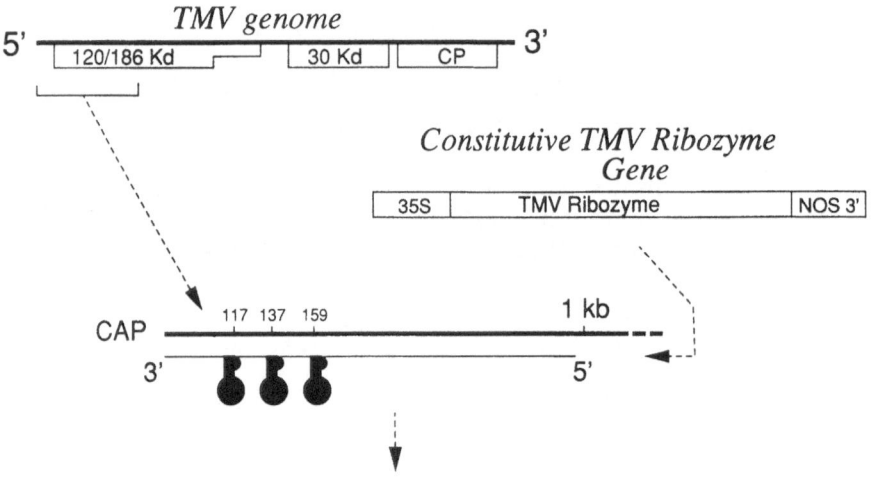

Fig. 3.   The experimental approach to assay *in vivo*
          ribozymes directed against the RNA genome of
          tobacco mosaic virus (TMV). A construct
          encoding for a 1000 nucleotide catalytic
          antisense containing three ribozymes directed
          against three naturally occurring GUC sequences
          located within the first TMV gene was placed
          under the control of the CaMV 35S promoter.
          This construct has been introduced into
          Nicotiana tabacum var Samsun using
          <u>Agrobacterium</u> <u>tumefaciens</u> T-DNA transformation.
          The ability of transgenic plants expressing the
          catalytic antisense to prevent TMV infection
          will be determined.

We are currently testing the ability of ribozymes
to inactivate RNAs *in vivo*. Three different RNAs are

being targeted in plant cells; an mRNA for CAT in
transgenic tobacco (*Nicotiana tabacum* var. Samsun),, a
viroid RNA in tomato (*Lycopersicon esculentum*), and
tobacco mosaic virus (TMV) RNA in infected plants. The
strategy we are using to test ribozyme activity directed
against TMV is illustrated in Fig. 3. A cDNA
encompassing approximately 1000 nucleotides at the 5'
end of TMV has been used as a template for introduction
of three ribozyme catalytic domains by site directed
mutagenesis. The three domains are aligned to different
GUC cleavage recognition sites, located in the first TMV
cistron, which most likely codes for the viral
replicase. *In vitro* cleavage of the TMV genome by a
ribozyme RNA prepared from this construction is active
against TMV RNA *in vitro*, producing the expected
cleavage products. This TMV ribozyme, under the control
of the CaMV 35S promoter, has been introduced into the
tobacco genome using *A. tumifaciens* mediated
transformation. Transformed plants will soon be
analysed for their ability to resist infection by TMV.
Similar strategies are being used to test the ability of
ribozymes to inhibit citrus exocortis viroid infection
and introduced CAT gene expression in plants.

CONCLUSION

     This paper provides a brief overview of the use and
potential use of non-protein coding sequences derived
from plant viral and related nucleic acids to alter gene
expression. Some of these sequences (promoters,
subgenomic promoters, enhancers, translational
enhancers) may be useful for increasing the expression
of viral or host genes. Other sequences, such as
ribozymes, DIs, and satellite RNAs may be manipulated to
reduce gene expression. As our understanding of the
function of these non-protein coding elements increases,
so too will our ability to manipulate them to control
gene expression. This should result in new plant
phenotypes, including the development of novel
mechanisms of resistance to plant pathogens.

Acknowledgements: Experimental results from the work of
a number of our colleagues are used in this chapter. In
particular, we thank J. Haseloff, P. Keese, P.

Waterhouse, P. Larkin, R. Perriman, L. Graf, R. Holliday, and L. Kelly who have all made contributions.

REFERENCES

Abel, P. P., Nelson, R. S., De, B., Hoffmann, N., Rogers, S. G., Fraley, R. T., and Beachy, R. N., 1986, Delay of disease development in transgenic plants that express the tobacco mosaic virus coat protein gene, Science, 232:738-743.

Anzola, J. V., Xu, Z., Asamizu, T., and Nuff, D. L., 1987, Segment-specific inverted repeats found adjacent to conserved terminal sequences in wound tumor virus genome and defective interfering RNAs, Proc. Natl. Acad. Sci. USA, 84:8301-8305.

Bass, B. L., andWeintraub, H., 1987, A developmentally regulated activity that unwinds RNA duplexes, Cell, 48:607-613.

Bass, B. L., and Weintraub, H., 1988, An unwinding activity that covalently modifies its double-stranded RNA substrate, Cell, 55:1089-1098.

Bujarski, J. J., Ahlquist, P., Hall, T. C., Dreher, T. W., and Kaesberg, P., 1986, Modulation of replication, aminoacylation and adenylation in vitro and infectivity in vivo of BMV RNAs containing deletions within the multifunctional 3' end, EMBO J., 5:1769-1774.

Bujarski, J. J., Dreher, T. W., and Hall, T. C., 1985, Deletions in the 3'-terminal tRNA-like structure of brome mosaic virus RNA differentially affect aminoacylation and replication in vitro, Proc. Natl. Acad. Sci. USA, 82:5636-5640.

Buzayan, J. M., Gerlach, W. L., and Bruening, G., 1986, Non-enzymatic cleavage and ligation of RNAs complementary to a plant virus satellite RNA, Nature (Lond.), 323:349-353.

Buzayan, J. M., Gerlach, W. L., Bruening, G., Keese, P., and Gould, A. R., 1986, Nucleotide sequence of satellite of tobacco ringspot virus RNA and its relationship to multimeric forms, Virology, 151:186-199.

Crowley, T. E., Nellen, W., Gomer, R. H., and Firtel, R., 1985, Phenocopy of discoidin I-minus mutants by antisense transformation in Dictyoselium, Cell, 43:633-641.

Fang, R., Nagy, F., Sivasubramaniam, S., and Chua, N., 1989, Multiple cis regulatory elements for maximal expression of the cauliflower mosaic virus 35S promoter in transgenic plants, Plant Cell, 1:141-150.

Forster, A. C., and Symons, R. H., 1987, Self cleavage
    of plus and minus RNAs of a virusoid and a struc-
    tural model for t-e active sites, Cell, 49:211-220.
Francki, R. I. B., 1985, Plant virus satellites, Ann.
    Rev. Microbiol., 39:151-174.
French, R., Janda, M., and Ahlquist, P., 1986, Bacterial
    genes inserted in an engineered RNA virus:
    efficient expression in monocotyledonous plant
    cells, Science, 231:1294-1297.
Gallie, D. R., Sleat, D. E., Watts, J. W., Turner, P. C.,
    and Wilson, M. A., 1987, The 5'-leader sequence of
    tobacco mosaic virus RNA enhances the expression
    of foreign gene transcripts in vitro and in vivo,
    Nucl. Acids Res., 15:3257-3273.
Gerlach, W. L., Llewellyn, D., and Haseloff, J., 1987,
    Construction of a plant disease resistance gene
    from the satellite RNA of tobacco ringspot virus,
    Nature (Lond.), 328:802-805.
Gerlach, W. L., Miller, W. A., and Waterhouse, P. M.,
    1987, Molecular genetics of barley yellow drawf
    virus, Barley Yellow Dwarf Newsletter, 1:17-19.
Harland, R., and Weintraub, H., 1985, Translation of mRNA
    infected into Xenopus oocytes is specifically
    inhibited by antisense RNA, J. Cell Biol.,
    101:1094-1099.
Harrision, B. D., Mayo, M. A., and Baulcome, D. C., 1987,
    Virus resistance in transgenic plants that
    express cucumber mosaic virus satellite RNA,
    Nature (Lond.), 328:799-802.
Haseloff, J., and Gerlach, W. L., 1988, Simple RNA
    enzymes with new and highly specific endoribo-
    nuclease activities, Nature (Lond.), 334:585-591.
Haseltine, W. A., and Wong-Staal, F., 1988, The molecular
    biology of the AIDS virus, Scientific American,
    34-42.
Hillman, B. I., Carrington, J. C., and Morris, T. J.,
    1987, A defective interfering RNA that contains a
    mosaic of a plant virus genome, Cell, 51:427-433.
Inokuchi, Y., and Hirashima, A., 1987, Interference
    with viral infection by defective RNA replicase,
    J. Virol., 61:3946-3949.
Inouye, M., 1988, Antisense RNA: its functions and appli-
    cations in gene regulation - a review, Gene,
    72:25-34.
Ismail, I. D., and Milner, J. J., 1988, Isolation of
    defective interfering particles of sonchus yellow
    net virus from cronically infected plants. J. Gen.
    Virol., 69:999-1006.
Kiefer, M. C., Daubert, S. D., Schneider, I. R., and
    Bruening, G., 1982, Multimeric forms of satellite

of tobacco ringspot virus RNA, Virology, 121:262-273.

Kim, S. K., and Wold, B. J., 1985, Stable reduction of thymidine kinase activity in cells expressing high levels of anti-sense RNA, Cell, 42:129-138.

Kuhlemeier, C., Green, P., and Chua, N.-H., 1987, Regulation of gene expression in higher plants, Annu. Rev. Plant Physiol., 38:221-257.

Lingelbach, K., and Dobberstein, B., 1988, An extended RNA/RNA duplex structure within the coding region of mRNA does not block translational elongation, Nucl. Acids Res., 16:3405-3414.

Marsh, L. E., Dreher, T. W., and Hall, T. C., 1988, Mutational analysis of the core and modulator sequences of the BMV RNA 3 subgenomic promoter, Nucl. Acids Res., 16:981-995.

Melton, D. A., 1985, Injected antisense RNAs specifically block messenger RNA translation in vivo, Proc. Natl. Acad. Sci. USA, 82:144-148.

Miller, W. A., Dreher, T. W., and Hall, T. C., 1985, Synthesis of brome mosaic virus subgenomic RNA in vitro by internal initiation on (-)-sense genomic RNA, Nature (Lond.), 313:68.

Odell, J. T., Knowlton, S., Lin, W., and Mauvais, C. J., 1988, Properties of an isolated transcription stimulating sequence derived from the cauliflower mosaic virus 35S promoter, Plant Mol. Biol., 10:263-273.

Odell, J. T., Nagy, F., and Chua, N.-H., 1985, Identification of DNA sequences required for activity of the cauliflower mosaic virus 35S promoter, Nature (Lond.), 313:810-812.

Prody, G. A., Bakos, J. T., Buzayan, J. M., Schneider, I. R., and Bruening, G., 1986, Autolytic processing of dimeric plant virus satellite RNA, Science, 231:1577-1580.

Rebagliati, M. R., and Melton, D. A., 1987, Antisense DNA injections in fertilized frog eggs revals an RNA duplex unwinding activity, Cell, 48:607-613.

Schneider, I. R., 1977, Defective plant viruses, in: "Bletsville Symposia on Agricultural Research-Virology in Agriculture", J. R. Romberger, ed., Allenheld Osmun, New Jersey (1977), pp. 201-219.

Strauss, J. H., and Strauss, E. J., 1988, Evolution of RNA Viruses, Ann. Rev. Microbiol., 42:657-683.

Strickland, S., Huarte, J., Belin, D., Vassalli, A., Rickles, J. R., and Vassalli, J., 1988, Antisense RNA directed against the 3' noncoding region prevents dormant mRNA activation in mouse Oocytes, Science, 241:680-684.

Tumer, N. E., O'Connell, K. M., Nelson, R. S.,
    Sanders, P. R., Beachy, R. N., Fraley, R. T.,
    and Shah, D. M., 1987, Expression of alfalfa
    mosaic virus coat protein gene confers cross-
    protection in transgenic tobacco and tomato
    plants, EMBO J., 6:1181-1187.
van der Krol, A. R., Mol, J. M. N., and Stuije, A. R.,
    1988, Antisense genes in plants: an overview,
    Gene, 72:45-50
van Dun, C. M., Bol, J. F., and Van Volten-Doting, L.,
    1987, Expression of alfalfa mosaic virus and
    tobacco rattle virus coat protein genes in
    transgenic tobacco plants, Virology, 159:299-305.
Wagner, R. W., and Nishikura, K., 1988, Cell cycle
    expression of RNA duplex unwindase activity in
    mammalian cells, Mol. Cell. Biol., 8:770-777.
Weintraub, H., Izant, J. G., and Harland, R. M., 1985,
    Antisense RNA as a molecular tool for genetic
    analysis, Trends Genet., 1:23-25.

MAPPING IN MAIZE USING RFLPs

D.A. Hoisington[1] and E.H. Coe, Jr.[1,2]

[1]Department of Agronomy and [2]USDA-ARS
University of Missouri
Columbia, Missouri 65211

INTRODUCTION

The development of genetic maps for an organism can
be traced back to the first genetic experiments of Stur-
tevant. The discovery that loci can be placed into ordered arrangements, based on the observation of recombination events, has led to the development of genetic
linkage maps for a number of plant and animal species.
The only requirements are the availability of distin-
guishable alleles at a locus and the abilities to perform genetic crosses and to analyze the resulting progeny. Among plant species, maize (_Zea_ _mays_ L.) has one
of the best developed genetic linkage maps available.
The species has a long history of genetic investigations, due in part to the ease with which it can be manipulated genetically and to its important role as a
major agronomic crop species. The current genetic database contains nearly 600 unit factors, of which nearly
400 have sufficient linkage information available to
place the locus to a specific genetic position on the
linkage map. These factors include loci defined by morphological variations (conventional loci) and by biochemical variants (isozymes and protein polymorphisms).
In addition, there are the nearly 900 loci defined by
DNA polymorphisms (Restriction Fragment Length Polymorphisms, RFLPs).

In addition to the conventional, biochemical and molecular database for maize, there exist numerous cytogenetic stocks (i.e. knobs, heterochromatic regions,
chromosomal rearrangements and deletions) (Carlson,

Gene Manipulation in Plant Improvement II
Edited by J. P. Gustafson
Plenum Press, New York, 1990

1988). In particular, a wealth of reciprocal translocations exists, both between members of the A chromosome
set (Longley, 1961) and between the supernumerary B
chromosomes and the A chromosomes (Roman, 1947; Beckett,
1987). These latter translocations provide a rather
efficient means for rapidly identifying genetic factors
to a particular chromosome segment and have aided in
further refinement of linkage maps for the maize genome
(Roman and Ullstrup, 1951). The A-A reciprocal translocations have been useful in generating dosage series for
particular chromosome segments (Birchler, 1980) and for
locating both unit factors as well as complex traits to
chromosome segment (Burnham, 1982). Perhaps the most
important feature of these translocations is the fact
that stocks for nearly all of them have been maintained
by and are available from the Maize Genetics Cooperation
Stock Center (Department of Agronomy, University of Illinois, Urbana, IL.).

It has been relatively easy to develop linkage maps
based on RFLPs in maize, due to its diploid nature; the
presence of numerous homozygous inbred lines; the high
degree of polymorphism present between these lines; the
large amount of DNA present in any one individual; the
availability of cytogenetic stocks for enhanced localization of molecular loci; and the extreme interest in
use of such molecular maps for the mapping of complex
traits. Several laboratories, both public and private,
have isolated RFLP clone sets and developed corresponding linkage maps for maize (Helentjaris, 1987; Burr et
al., 1988; Hoisington, 1989; M. Murray and D. Grant,
personal communication). Efforts in the public sector
were initiated to develop a publicly derived core set of
RFLP markers for use in correlating the various private
maps and for deriving an integrated map of the maize
genome.

The existence of these various genetic stocks has
raised the problem of integrating the markers into common maps. Over the years, this has been performed on an
ad hoc basis by assembling genetic information supplied
by numerous researchers using diverse stocks and approaches. The first genetic maps developed by Emerson,
Beadle and Fraser (1935) represent the most adequately
documented maps available. More recent compilations by
Rhoades (1955), Coe and Neuffer (1977), Coe, Neuffer and
Hoisington (1988) and the current working maps of Hoisington (1989) have relied on the compilations of information supplied on an informal basis. The recent
availability of molecular loci and their relative ease

of use in maize opens the possibility of highly integrated maps for the maize genome. This paper will present efficient methods for the integration of the numerous markers, both genetic and cytogenetic, into a core
RFLP map. Efforts are currently underway at the University of Missouri to utilize these approaches in the development of a comprehensive and integrated genetic map
for maize.

DEVELOPMENT OF A CORE RFLP LINKAGE MAP

    A core RFLP map was developed by screening a genomic
library constructed from fully developed maize leaves.
Total genomic DNA was digested with PstI. Fragments
from 1000-2000 bp were isolated by sucrose gradient cen-
trifugation and ligated into pUC19, and the resulting
plasmids transformed into a suitable host. Transformed
colonies were originally screened for repeated sequence
inserts by colony hybridizations; however, this procedure was eliminated when it was determined that fewer
than 10% of the clones were repetitive. Isolated inserts from all of the resulting clones were subsequently
screened for use as RFLP probes by hybridizing to Southern blots of Tx303 and CO159 inbred DNA digested with
EcoRI, HindIII and EcoRV. Nearly 50% of the almost 400
clones screened proved to be low-copy and polymorphic
between the two lines with at least one enzyme. Each of
the three enzymes tested proved to be equally effective
in depicting polymorphisms, and all three were chosen
for use in map development. An F2 linkage map was developed using 46 F2 individuals from a self of an F1
cross of the above two inbred lines. It should be noted
that 46 individuals were chosen simply based on gel morphology (on a 50 lane gel, 2 lanes for parents, 1 lane
for F1, and 1 lane for MW markers, leaving 46 lanes for
F2 samples). Given the co-dominant nature of most RFLP
loci, 46 F2 individuals provides a 95% probability of
detecting a recombination between loci 3 cM apart.

    The current RFLP linkage map presented in Figure 1
consists of 262 RFLP loci, 208 of which are defined by
clones developed within the public sector, either in our
lab or by Burr et al. (1988). In addition to these
loci, an original set of loci defined by clones developed by Helentjaris et al. (1987) was mapped into the F2
to aid in the rapid location of new loci to chromosome
linkage group. Also on the map are a number of "known
sequence" loci for which the function of the clone has
been defined, either in maize or in another plant spe-

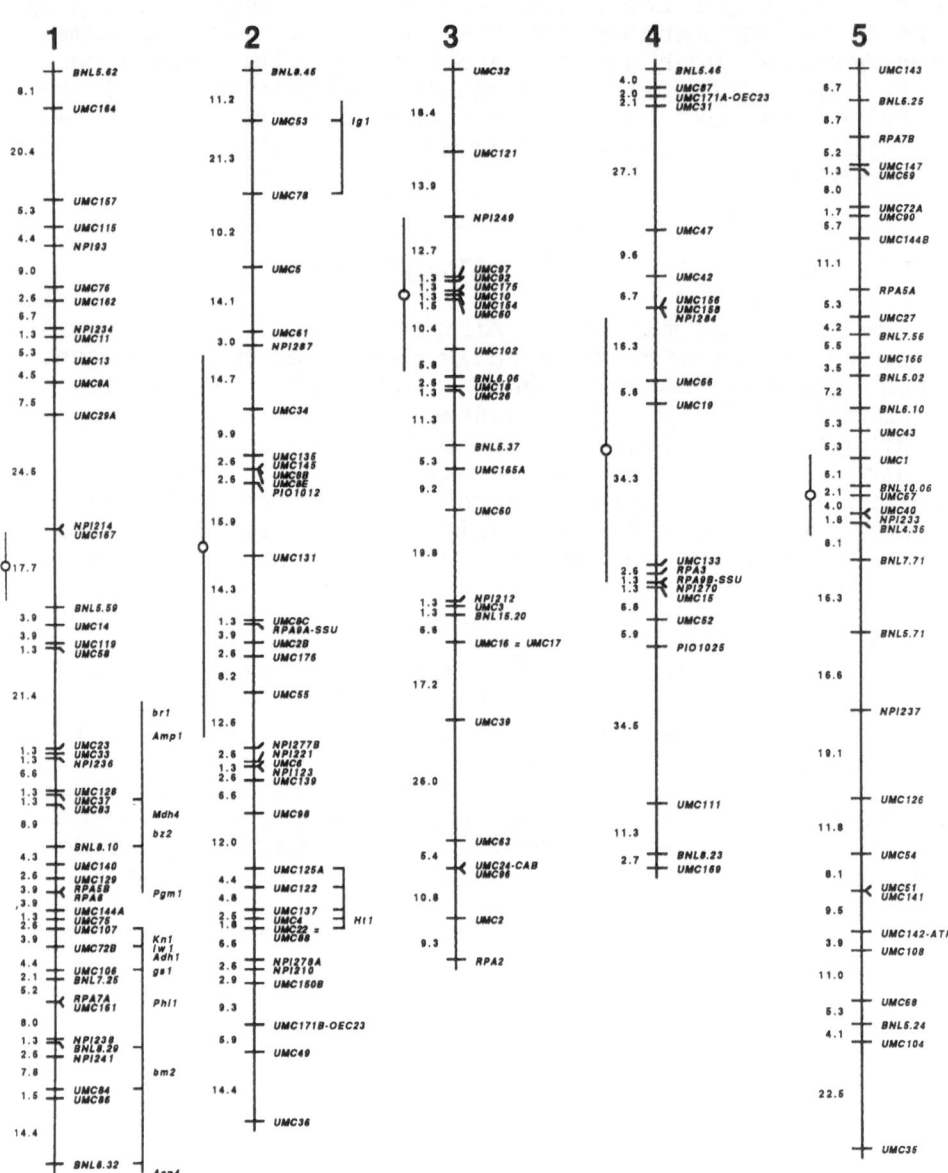

Fig. 1. University of Missouri Maize RFLP Linkage Map –
April, 1989. All map orders and distances were
determined using MAPMAKER (Lander et al., 1987)
and are expressed as cM after applying Kosambi's
mapping function to the % recombination.

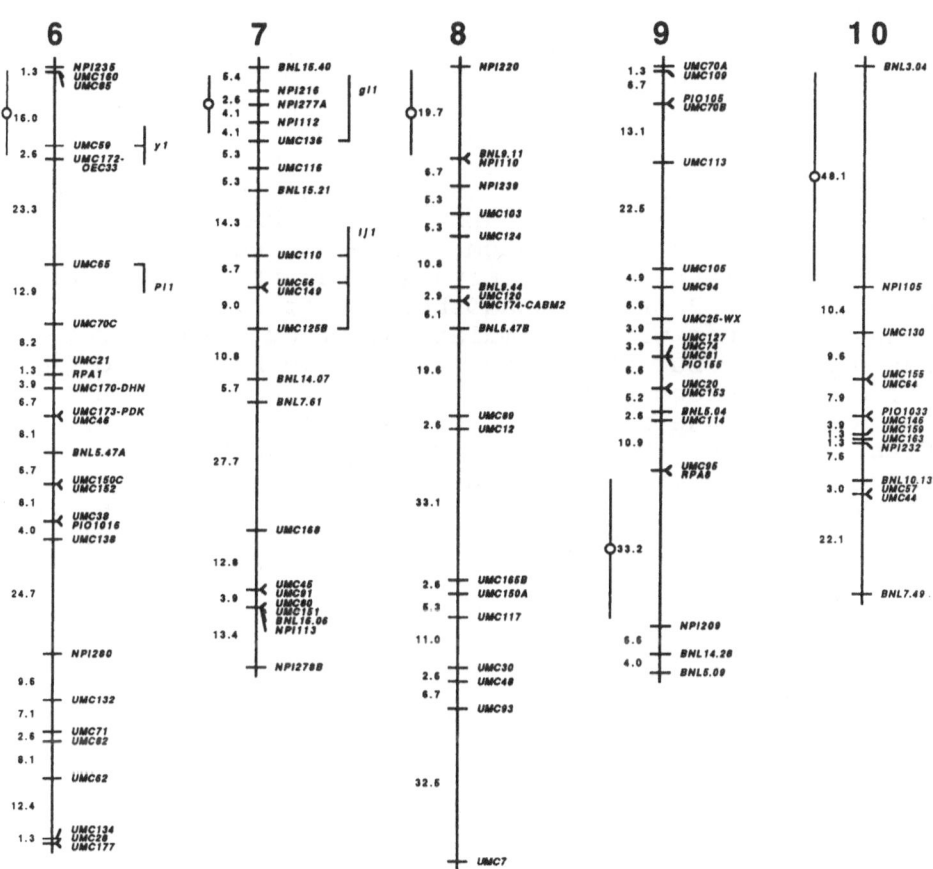

Fig. 1. UMC RFLP Linkage Map, continued.

cies.   All gene orders and distances were determined
using the multi-point maximum likelihood method (Lander
and Green, 1987) performed through the computer package
MAPMAKER (Lander et al., 1988).

The major goal of the initial phase of the project
was to develop a 10 cM map based on simple, polymorphic
clones.   As Figure 2 demonstrates, the majority of the
interlocus distances are less than 10-15 cM, with an
average distance of 7.6 cM (see Table 1 for total, average, and maximum distances for each chromosome).   Only a
few interlocus distances above 15 cM remain, although
there is at least one interlocus distance greater than
20 cM on each chromosome and two chromosomes (1S and
10S) contain distances larger than 30 cM.   The decrease
in total number of interlocus distances for each range
in Figure 2 agrees with an exponential decay curve (r^2 =
0.91), indicating the random nature of the locus spac-
ing.   Thus, it can be anticipated that the development
of another RFLP map from another set of clones would
result in a map with a similar number of large interlocus distances but that these gaps would be located in
different regions of the map.   Initial correlation of
the RFLP linkage maps developed by Helentjaris, Burr and
Grant indicates this to be true, although the exact coverage possible is unclear until further integration of
the various maps is complete.   Given the duplicated nature of the maize genome (Helentjaris et al., 1988), it

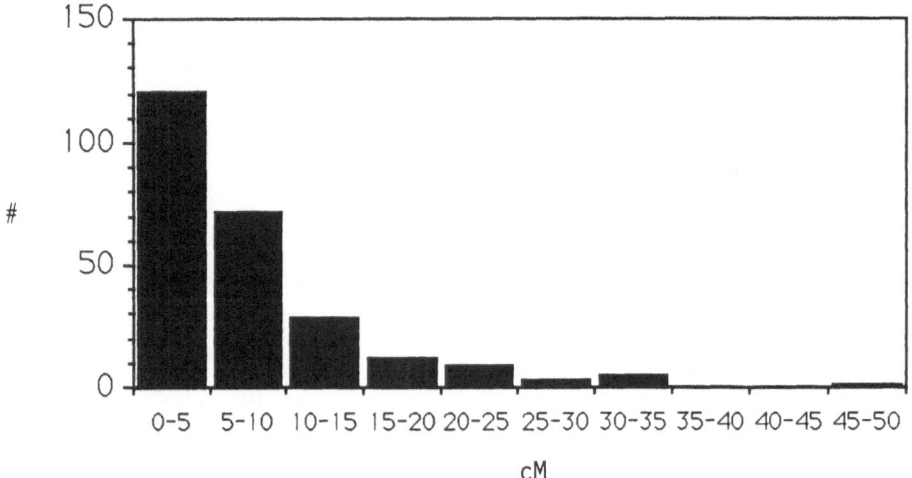

Fig. 2. Interlocus recombination distances for the cur-
        rent RFLP linkage map.

is possible that specific genomic regions may be under-
represented since sequences in these regions are repetitive and selection of single copy probes lowers or
eliminates the possibility of detecting loci in this region.

If the RFLP map is to be used to correlate the conventional and physical features of the maize genome, it
is important that the map be as comprehensive as possible. While exact correlations to the conventional map
are only beginning, some idea as to the extent of
genomic coverage can be gained from comparing the length
of the RFLP map of each chromosome arm to that of the
conventional linkage map. Figure 3 presents a comparative histogram of the total linkage distances for each
of the twenty chromosome arms. For 17 of the 20 arms,
the RFLP linkage map equals or exceeds that of the conventional map. Only for the short arms of chromosome 3
and 7 and the long arm of chromosome 9 is the conventional linkage map of greater length than the corresponding RFLP map. While the exact coverage of the conventional map will be determined following extensive
interval mapping (see discussion later), the RFLP map
appears to be initially adequate for comprehensive mapping in maize. It also should be pointed out that the
current RFLP map was developed over the course of two
technician years, as compared to the many years and many
scientist hours which have gone into the development of
the current conventional map.

Table 1. Total, average, and maximum recombination
         distance (in % recombination) for each
         chromosome linkage group.

Chromosome	Total	Mean	Maximum
1	296	6.9	32.4
2	233	7.3	20.9
3	216	8.3	24.9
4	173	8.6	28.4
5	243	7.4	20.9
6	177	7.7	22.2
7	131	6.6	26.0
8	154	8.6	26.0
9	136	6.5	29.5
10	108	8.3	33.2
ALL	1849	7.6	33.2

A second correlation possible is to compare the
length of the RFLP map to the physical length of each
chromosome arm. In Figure 4, the physical length of
each chromosome arm based on measured pachytene spreads
(see Sheridan, 1982) was plotted against the recombination length of both the conventional map and the RFLP
map. While the correlation coefficients (see Figure)
are not extremely large, there is a general trend for
the linkage distance of both maps to increase as the
physical length of the arm increases. One of the major
properties to identify for any linkage map is the most
distal locus located on each chromosome arm. The presence and position of this locus would insure that the
end of the linkage group has been reached (at least using low copy loci) and would accurately define the size
of the genome. Currently, it is not possible to perform
in situ hybridizations in maize using the types of

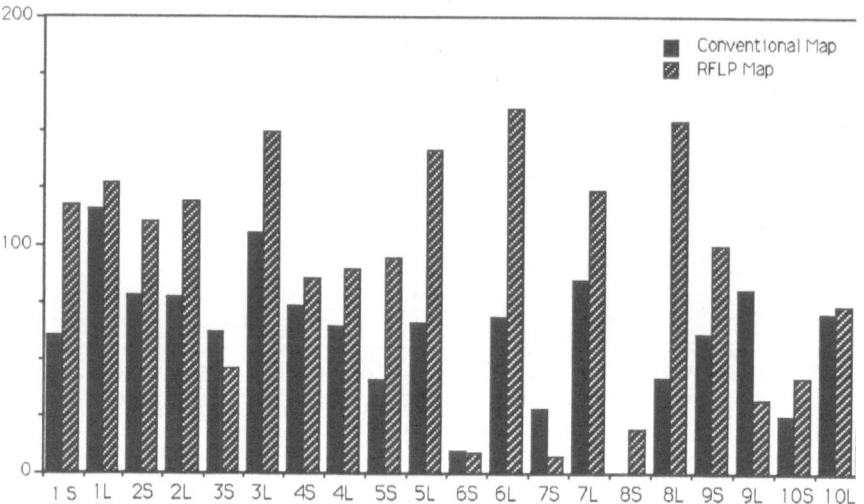

CHROMOSOME

Fig. 3. Comparison of recombination distances between
        the conventional and RFLP linkage maps. Total
        recombination distance (as % recombination) was
        charted for each chromosome arm. Centromeres
        were assumed to be located midway between loci
        uncovered by the most proximal B-A transloca-
        tion.

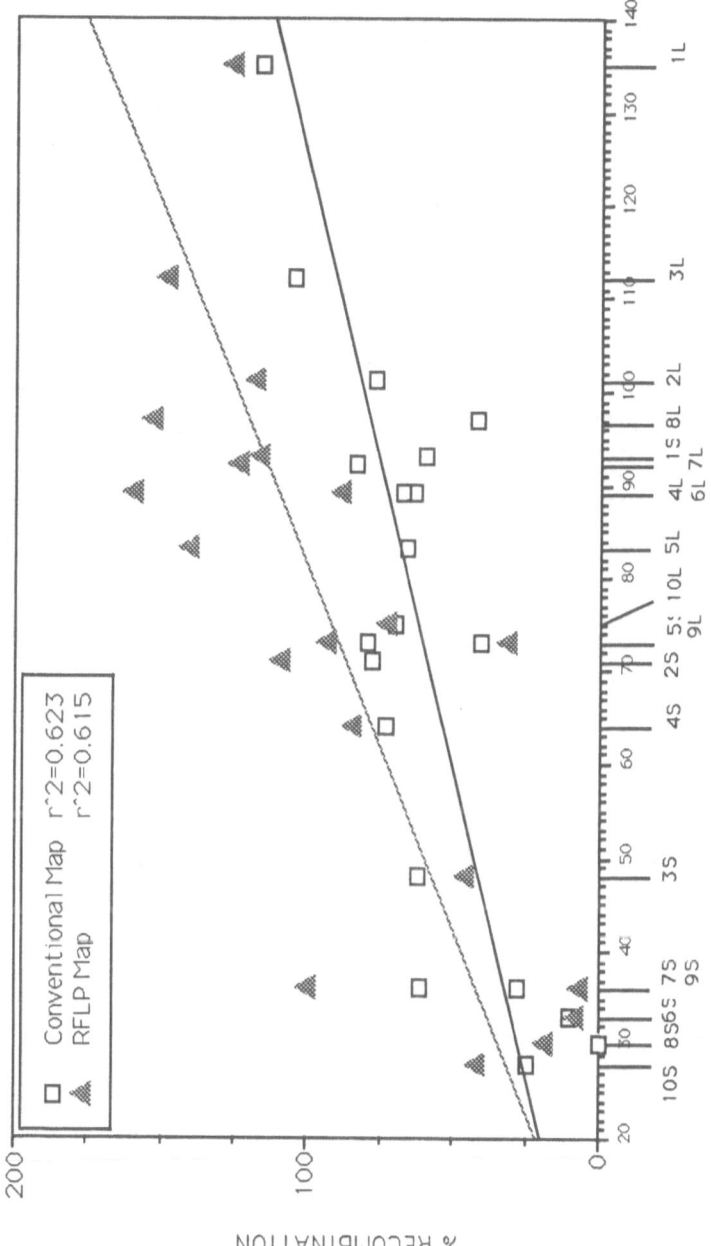

PHYSICAL CHROMOSOME LENGTH

Fig. 4. Comparison of the physical versus recombination distance of the conventional and RFLP linkage maps. Chromosome arm distances were based on measured pachytene values.

clones used to detect most RFLP loci (i.e. single-copy
and relatively small insert sizes). Wu and co-workers
(Shen et al., 1987; Wu, 1989) have been successful in
statistically detecting single-copy probes of larger
lengths in pachytene spreads of maize. However, as the
insert length decreases, the correlations become much
weaker and probes less than 2000 bp are generally not
mappable by in situ hybridizations.

B-A translocation-generated hypoploids have been
used by Burr et al. (1988), Weber and Helentjaris
(1989), and ourselves to identify RFLP loci to a chromosome arm; however, the number of translocation stocks is
limited and most of the translocations have breakpoints
close to the A chromosome's centromere. Therefore,
while they are extremely useful in determining chromosome arm locations, they are less useful in determining
the ends of the chromosomes. The A-A translocation-
based deletion mapping strategy outlined later in this
article should provide the physical correlations necessary for more accurately determining the extent of
physical coverage.

INTEGRATION OF CONVENTIONAL MARKERS WITH RFLP LOCI

While the development of a linkage map based on molecular markers will, in itself, prove extremely valuable for correlation of very complex traits such as
quantitative trait loci (QTLs), information as to the
relative position of the more conventional "naked-eye
polymorphism" loci will prove to be necessary for discrete genomic manipulations. The possibility that QTLs
represent natural variation of "wild-type" alleles at
qualitative loci (Robertson, 1985) further strengthens
the need for localization of conventional loci within
the RFLP framework. As stated earlier, maize has a particularly large number of single or duplicated conventional genetic loci already identified and classified.
An additional 2000-3000 single gene loci have been isolated following various mutagenic treatments of maize
pollen (Bird and Neuffer, 1987) and are being maintained
and investigated by Neuffer and co-workers. Any strategy for integrating such a diverse and extensive collection of markers must be extremely efficient and univer-
sal in application. RFLP loci present an alternative
approach to mapping not feasible with conventional recessive and dominant loci. Because of the high level of
polymorphism present in maize, most RFLP loci will be
segregating in any particular cross. Therefore, the

need to develop a particular linkage tester for mapping
is eliminated. In addition, the multi-site marking
available with RFLPs provides an added advantage for
mapping new loci into an existing map.

Homozygosity mapping has been proposed as the method
of choice for mapping traits when the number of affected
individuals is low (Lander and Botstein, 1986, 1987).
The same strategy is adaptable to mapping conventional
loci in maize. Because of the large number of loci to
be mapped, it is important that the number of individuals necessary for mapping each locus be kept to a minimum. Interval mapping is based on the co-segregation of
pairs of adjacent RFLP loci, rather than single RFLPs,
with the trait locus. In essence, the method uses the
additional power of three-point crosses over two-point
crosses to extract strong evidence of order. Specifically, in meioses in which no crossovers occurred between adjacent RFLPs, the interval between them must

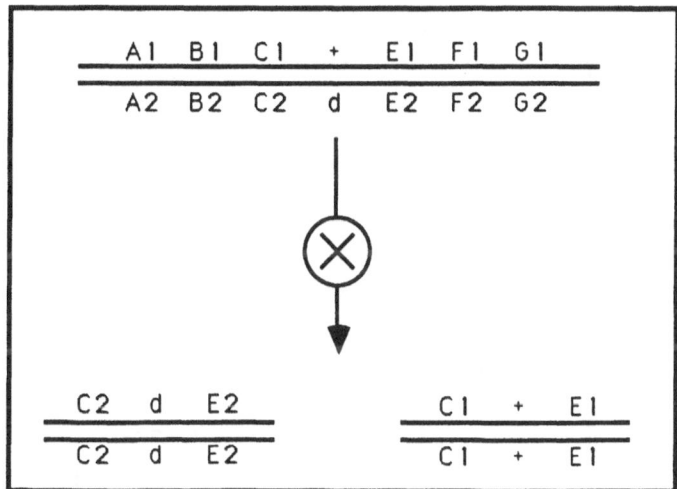

Fig. 5. Interval mapping strategy. Given a locus to
        map (**d**) within a multiply marked set of co-
        dominant markers (**A**, **B**, **C**, **E**, **F** and **G**), the
        selection of individuals of known homozygous
        genotype (either **d/d** or **+/+**) from the self of
        a heterozygote will decrease the number of
        progeny required to map the locus. Essen-
        tially a three-point cross is performed such
        that the two flanking markers should have a
        high probability of being co-inherited.

have been inherited as a single block (apart from very rare double recombinants). Thus, there will be a nearly perfect correlation between the inheritance of the trait and of the interval containing the trait in such meioses.

In an F2 population, a further decrease in the number of individuals to genotype for flanking RFLPs can be achieved by pre-selecting homozygous individuals for the trait to be mapped. Given a single recessive locus, selection for homozygous recessive (and, thus, expressing) or homozygous "wild-type" (non-segregating) individuals would predict co-inheritance of flanking RFLP markers for the interval in which the recessive locus resides (see Figure 5). The only difference between the two homozygous classes is the RFLP allele inherited.

Figure 6 presents a graph developed in collaboration with Eric Lander similar to the one presented in Lander and Botstein (1987) demonstrating the number of affected individuals necessary to select from a segregating F2 in

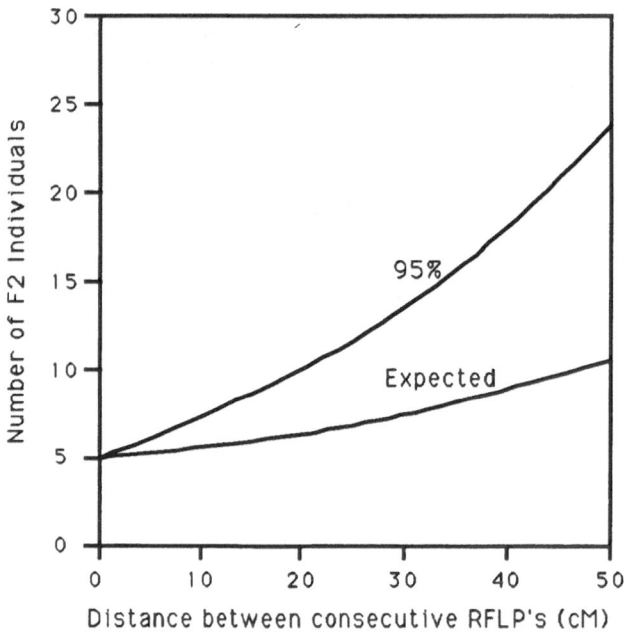

Fig. 6. Number of selected F2 individuals required to map a trait locus to an RFLP interval.

Table 2. F2 individuals to select for interval mapping.

Genetic Nature of Trait	Individuals to Select
Viable Recessive	m/m (Expressing)
Non-viable Recessive	+/+ (Non-mutant)
Dominant	+/+ (Non-mutant)
Reciprocal Translocation	+/+ or T/T (Non-semisterile)

order to map a locus within an RFLP interval of defined
distance. As can be seen, only 12 affected individuals
are necessary for 20 cM mapping resolution of any trait
with a 95% level of probability. If the locus is not
located at all in the genome, then approximately 25 affected individuals would be adequate to locate a locus
to a 50 cM segment of the map. For low-penetrant
traits, or those which are difficult to locate using the
existing B-A translocation method, interval mapping with
RFLPs may prove to be the method of choice.

This method has an additional advantage beyond the
small number of individuals required in that it is uni-
versally applicable. As Table 2 presents, any type of
trait, whether recessive or dominant, viable or lethal,
can be mapped with the same level of efficiency. Mapping of several recessive lethals to the same 10 cM region of the genome by more conventional means would require rather complex analyses. With interval mapping,
each locus would be mapped independently. Essentially,
any F2 population segregating for a locus of interest is
all that is needed to map that locus (a very common
population type developed in all genetic programs).

We have begun an extensive interval mapping effort
involving nearly 300 loci. A few loci have already been
mapped using this technique and are presented to the
right of the RFLP linkage map. Since interval mapping
depends on the detection of a crossover event on each
side of the locus to be mapped, selection of those recombinant individuals a priori would further decrease
the absolute number of individuals needed and increase
the ultimate resolution of the map. If a number of
heterozygous recessive loci have been linked in coupling
(as depicted in Figure 7), and the individual selfed,
recombinant individuals from the self can be identified
as expressing the desired trait (b/b in the Figure)

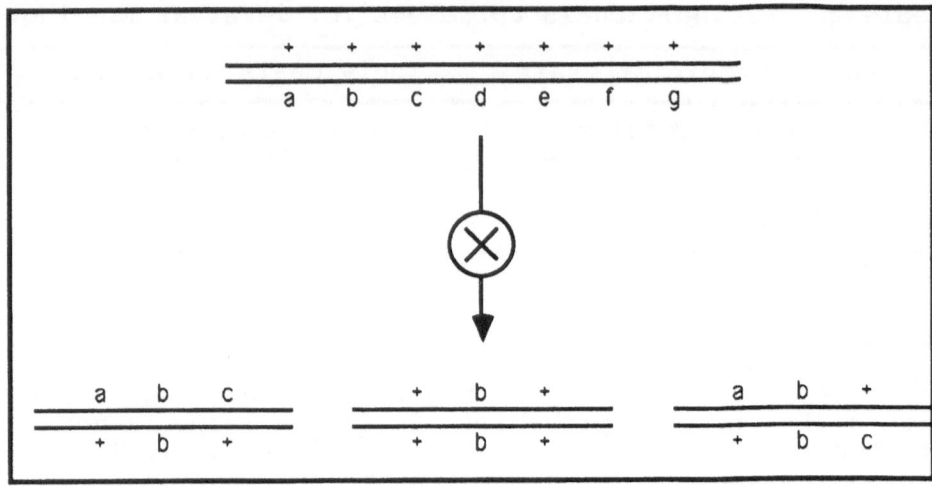

Fig. 7. Selection of recombinant individuals for interval mapping. From the self of a multiply marked
linkage tester, selection of individuals expressing the trait to be mapped (**b**) and not expressing those of the flanking loci (**a** and **c**)
will ensure at least a single recombination
event has occurred flanking the locus to be
mapped.

while not expressing those of the flanking loci (a and c
in the Figure). These individuals would have had at
least one crossover on either side of the locus of
interest.

Single recombinant individuals (those lacking expression of only the locus either distal or proximal)
would contain a recombinational event on only one side
of the locus. The resolution of such individuals would
be at least as great as the distance between the flanking markers (often as small as 1 cM). Thus a single affected individual could provide 1 cM resolution.

CORRELATION TO PHYSICAL MARKERS

As was mentioned earlier, knowledge as to the physical location of RFLP loci is desired and is perhaps as
important as the correlation to conventional genetic
loci. Since in situ hybridization is difficult in maize
and would be very tedious for the large number of loci
currently available, another approach to physical inte-

gration is necessary. Maize has a particularly extensive set of reciprocal translocations involving the A
chromosome set (Longley, 1961). The current set of
translocations numbers over 800 and provides a defined
breakpoint almost every 1% of the length of each chromosome arm. Figure 8 identifies the breakpoints of available translocations involving all ten chromosomes and
shows the extensive coverage available. Often there are
several translocations that have had their breakpoint
identified at the same position. These translocations
will provide a high degree of saturation for that particular region of the chromosome when employed in the
deletion mapping strategy outlined below.

Interval mapping, as described previously, can also
be applied to mapping of the translocation breakpoints,
as shown in Table 2. One potential problem does exist
in using interval mapping as a strategy for locating
translocation breakpoints, since translocations, in
maize, inhibit crossing over in a large portion of the
chromosome surrounding the breakpoint (Carlson, 1988).
This would reduce the resolution possible for the physical correlations since very few recombinants would occur
near the breakpoint. At least three other methods are
possible for locating physical breakpoints using RFLPs.

The combination of multiple translocation stocks involving the same chromosome arm has been successfully
used by Birchler (1980) to locate accurately the Adh1
locus on 1L to a physical segment. Crosses between
pairs of translocations that have both breakpoints in
the same arms generate meiotic segregational products
(depending on the relative order of the breakpoints)
that may be duplicate for the segment. RFLP loci can be
used to examine progeny from such crosses and individuals duplicate for the segment identified (rather than
deficient, which is usually lethal to the gametophyte).
All loci duplicate in the same individual would define
the RFLP segment between the respective breakpoints of
the translocations involved.

As was used in the development of the core RFLP
linkage map, translocations between the supernumerary B
chromosome and an A chromosome can also be used to detect the physical location of RFLP loci. The ability of
the B-A chromosome to undergo non-disjunction in the second mitotic division of the pollen grain produces embryos in the resulting seeds which lack the parental
allele at all loci along the translocated chromosome.
Unfortunately, only a limited (although comprehensive)

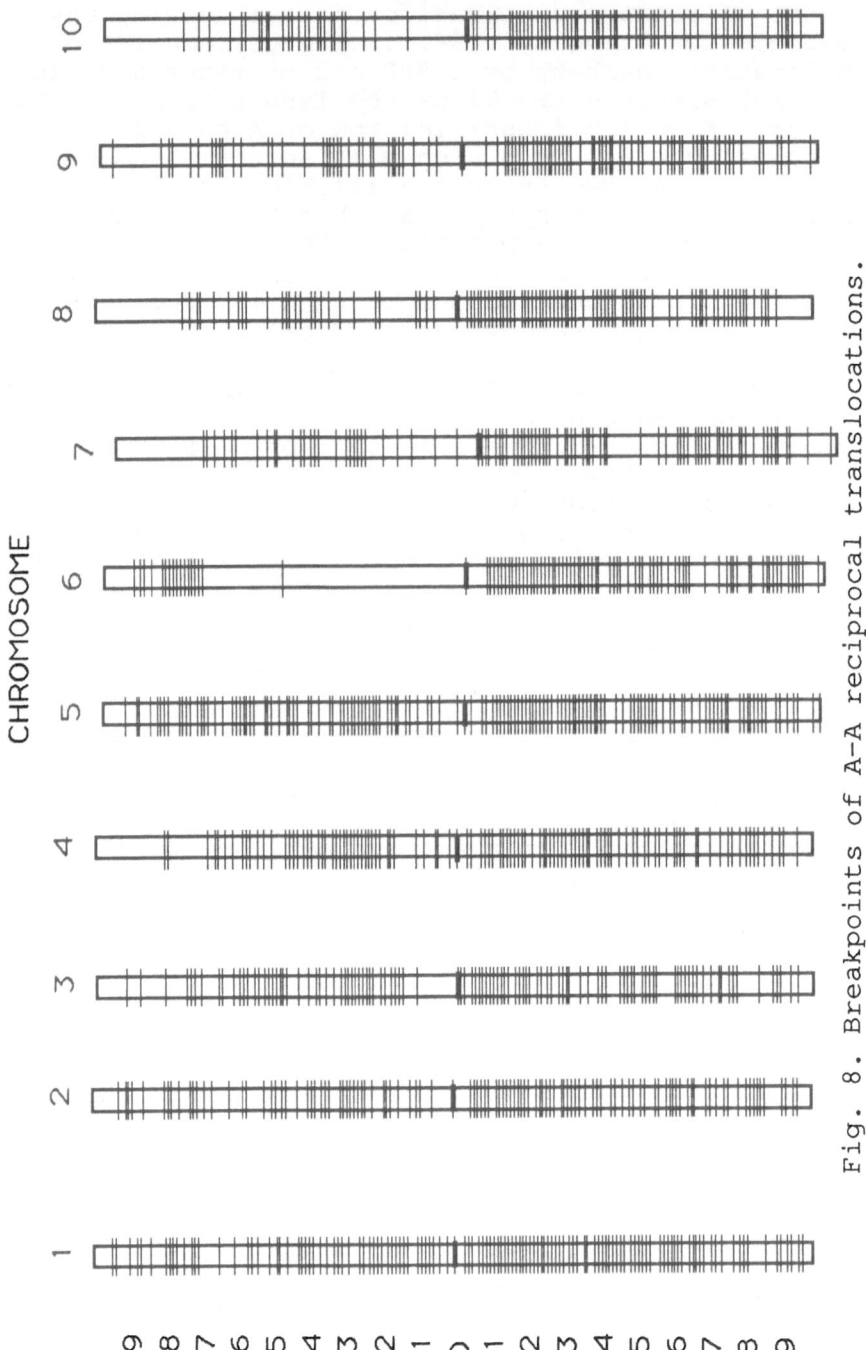

Fig. 8. Breakpoints of A-A reciprocal translocations.

set of B-A translocations is available and the production of new stocks is difficult and involves several
generations to produce.

A third method is proposed which would allow every
translocation breakpoint to be used for mapping and involves a minimum of generations to produce. This method
involves the combination of a monosomic generating system in maize, the r-x1 system, and reciprocal A-A
translocations to produce double hemizygous individuals
(see Figure 9). The r-x1 system has been extensively
studied by Weber and co-workers (1986) and shown to be
useful in locating RFLP loci to chromosome, particularly
when the clone detects multiple loci (Helentjaris, et
al., 1986). The r-x1 system is characterized by generating non-disjunction of random chromosomes during the
mitotic division of the egg (Lin and Coe, 1986; Simcox
et al., 1987). If such an egg is fertilized, the resulting individual will be monosomic for the particular
chromosome involved. All loci on the affected chromosome will then be missing the maternal allele. This is
in contrast to B-A translocations in which the paternal
allele is lost due to the non-disjunctive event occurring in the male.

By generating a stock in which the non-disjunction
could involve a translocated chromosome, individuals
monosomic for the translocated segment can be generated.
To generate such individuals (see Figure 9), stocks containing a single reciprocal A-A translocation are
crossed to a stock heterozygous for the r-x1 allele r-
x1 is not transmitted through the male gamete). The F1
individual (identified as having semi-sterile pollen and
ears due to the reciprocal translocation) is then
crossed by an appropriate tester stock. The optimal
tester stock would be a line containing the two most
distal seedling recessive markers on one of the chromosomes involved in the translocation. Progeny from the
tester cross would then be planted and the resulting
seedlings screened for expression of the recessive loci.
An individual expressing both of the tester loci would
most probably be a monosomic individual for the entire
chromosome on which the loci are located. The individual of choice would be one which expresses only one of
the two loci on the chromosome. This individual would
either be a partial monosomic for that chromosome or
would be monosomic for the entire translocated chromosome. This latter possibility would mean that all loci
distal to the translocation breakpoint on each chromosome would be missing the maternal allele. All loci

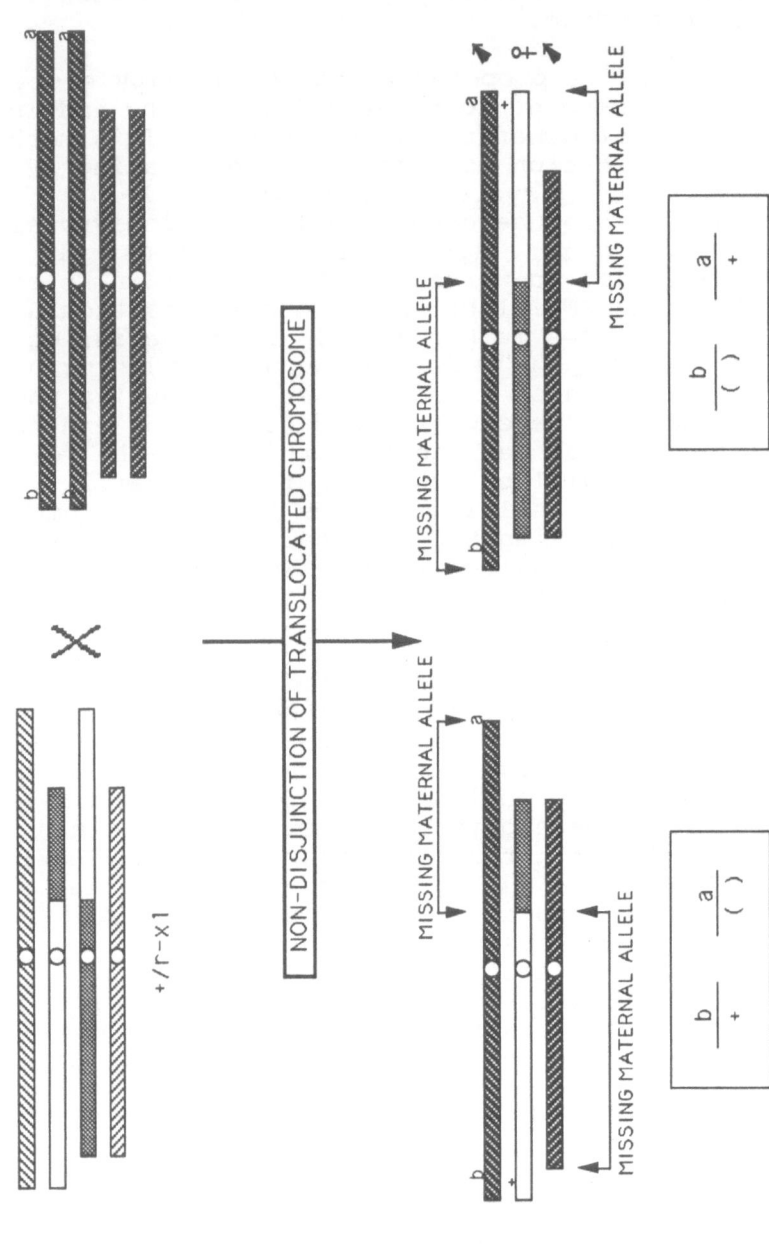

Fig. 9. Generation of double hemizygotes for deletion mapping of RFLP loci.

proximal to the breakpoint would contain both the maternal and paternal alleles. Thus the position of the breakpoint could be easily identified as residing between two adjacent RFLP loci by just the presence of the maternal allele at one locus and the absence at the next.

In essence, this type of mapping is similar to any type of deletion mapping. Neither the order of the translocation breakpoints nor the order of the RFLP loci need be known, since presence/absence information will determine unambiguously the order of both. In this manner, not only will correlations to the physical map be possible but also a definitively ordered RFLP map will be developed - a map whose order is not based on the observation of recombinational events. Because of the numerous translocations which exist in maize, it is anticipated that the physical resolution currently exceeds the density of the RFLP map. Thus, the majority of the RFLP loci should be ordered using this technique. Currently, all 879 translocation stocks are being crossed to a standard r-x1 containing line and the appropriate tester stocks derived to provide the progeny for screening for double hemizygotes during the summer of 1990.

CLOSING

While a great amount of work remains to integrate both the conventional and the physical maps of the maize genome with the rapidly developing molecular map, the actual development of an RFLP map has led to the development of more efficient and broad-based mapping strategies. While just a few years ago such an integration was not feasible, through the use of interval and deletion mapping using RFLP based markers, a comprehensive and integrative map of the maize genome should be possible within the next few years. This map will then provide the basis for future integration of more complex traits which provide the foundation of most plant breeding efforts (i.e. resistance factors and QTLs). In addition, knowledge as to the location of flanking molecular markers and the physical size of intervening chromosome segments will aid in developing more efficient cloning techniques for specific genetic loci.

ACKNOWLEDGEMENTS

    The excellent technical support of Jack Gardiner,
Randall Grogan and Susan Melia-Hancock is greatly appreciated. This research was supported in part by grant
No. DIR 8721921 from the National Science Foundation to
DAH and EHC, and by gifts to DAH from Pioneer Hi-Bred,
Int., Ciba-Geigy/Funks Seeds and ICI/Garst Seeds. The
authors would also like to thank all of their fellow
maize geneticists, without whose past contributions to
maize genetics, and without whose current support, the
development of this project would not have been possible. It is to them that this research is dedicated.

    Contribution from the Missouri Agricultural Experi-
ment Station, Journal Series Number 10,903.

REFERENCES

Beckett, J.B., 1978, B-A translocations in maize,J.
    Hered., 69:27-36.
Birchler, J.A., 1980, The cytogenetic localization of
    the alcohol dehydrogenase-1 locus in maize,Genet-
    ics, 94:687-700.
Bird, R. McK., and Neuffer, M.G., 1987, Induced muta-
    tions in maize, in: "Plant Breeding Reviews, Vol.
    5," J. Janick, ed., Van Nostrand Reinhold, New York
    (1987), pp. 139-180.
Burnham, C.R., 1982, The locating of genes to chromosome
    by the use of chromosomal interchanges,in: "Maize
    for Biological Research," W.F. Sheridan, ed., Plant
    Mol. Biol. Assoc., Charlottesville, Virginia (1982),
    pp. 65-70.
Burr, B., Burr, F.A., Thompson, K.H., Albertsen, M.C.,
    and Stuber, C.W., 1988, Gene mapping with recombi-
    nant inbreds of maize, Genetics, 188:519.
Carlson, W.R., 1988, The cytogenetics of corn, in:
    "Corn and Corn Improvement, 3rd Edition," G.F. Spra-
    gue and J.W. Dudley, eds., American Society of
    Agronomy, Madison, Wisconsin (1988), pp. 259-343.
Coe, E.H., Jr. and, Neuffer, M.G., 1977, The genetics of
    corn, in: "Corn and Corn Improvement, 2nd Edition,"
    G.F. Sprague, ed., American Society of Agronomy,
    Madison, Wisconsin (1977), pp.111-223.

Coe, E.H., Jr., Hoisington, D.A., and Neuffer, M.G., 1987, Linkage map of corn (maize) (Zea mays L.), in: "Genetic Maps, 1987: A Compilation of Linkage and Restriction Maps of Genetically Studied Organisms, Vol. 4," S.J. O'Brien, ed., Cold Spring Harbor Laboratory, Cold Spring Harbor, New York (1987), pp. 685-707.

Coe, E.H., Jr., Neuffer, M.G., and Hoisington, D.A., 1988, The genetics of corn, in: "Corn and Corn Improvement, 3rd Edition," G.F. Sprague and J.W. Dudley, eds., American Society of Agronomy, Madison, Wisconsin (1988), pp. 81-257.

Emerson, R.A., Beadle, G.W., and Fraser, A.C., 1935, A summary of linkage studies in maize, Cornell Univ. Agr. Expt. Stn. Memoir, 180.

Helentjaris, T., 1987, A genetic linkage map for maize based on RFLPs, Trends Genet., 3:217.

Helentjaris, T., Weber, D., and Wright, S., 1988, Identification of the genomic locations of duplicate nucleotide sequences in maize by analysis of restriction fragment length polymorphisms, Genetics 118:353-363.

Hoisington, D.A., and Coe, E.H., Jr., 1989, Methods for correlating RFLP maps with conventional genetic and physical maps in maize, in: "Development and Application of Molecular Markers to Problems in Plant Genetics," T. Helentjaris and B. Burr, eds., Cold Spring Harbor Laboratory, New York (1989), pp. 19-24.

Hoisington, D.A., 1989, Working linkage maps, Maize Genet. Coop. Newsletter, 63:141-151.

Lander, E.S., and Botstein, E.H., 1986, Strategies for studying heterogeneous traits in humans by using a linkage map of restriction fragment polymorphisms, Proc. Nat. Acad. Sci. U.S.A, 83:7353-7357.

Lander, E.S., and Botstein, D., 1987, Homozygosity mapping: a way to map human recessive traits with the DNA of inbred children, Science, 236:1567-1570.

Lander, E.S., and Green, P., 1987, Construction of multilocus genetic linkage maps in humans, Proc. Nat. Acad. Sci. U.S.A, 84:2363-2367.

Lander, E.S., Green, P., Abrahamson, J., Barlow, A., Daly, M., Lincoln, S., and Newburg, L., 1987, MAPMAKER: an interactive computer program for constructing genetic linkage maps of experimental and natural populations, Genomics, 1:174-181.

Lin, B.Y., and Coe, E.H., Jr., 1986, Monosomy and trisomy induced by the r-x1 deletion in maize, and associated effects on endosperm development, Can. J. Genet. Cytol., 28:831-834.

Longley, A.E., 1961, Breakage points for four transloca-
    tion series and other corn chromosome aberrations
    maintained at the California Institute of
    Technology, U.S. Dept. Agr.-Agr. Res. Serv, 34-16.
Rhoades, M.M., 1955, The cytogenetics of maize, in:
    "Corn and Corn Improvement," G.F. Sprague, ed., Aca-
    demic Press, New York (1955), pp. 123-219.
Robertson, D.S., 1985, A possible technique for isolat-
    ing genic DNA for quantitative traits in plants, J.
    Theor. Biology, 117:1-10.
Roman, H., 1947, Mitotic nondisjunction in the case of
    interchanges involving the B-type chromosome in
    maize, Genetics, 32:391-409.
Roman, H., and Ullstrup, A.J., 1951, The use of B-A
    translocations to locate genes in maize, Agron. J.,
    43:450-454.
Shen, D., Wang, Z., and Wu, M., 1987, Gene mapping on
    maize pachytene chromosomes by in situ
    hybridization, Chromosoma, 95:311-314.
Sheridan, W.F., 1982, Maps, markers and stocks, in:
    "Maize for Biological Research," W.F. Sheridan, ed.,
    Plant Mol. Biol. Assoc., Charlottesville, Virginia
    (1982), pp. 37-52.
Simcox, K.D., Shadley, J.D., and Weber, D.F., 1987, De-
    tection of the time of occurrence of nondisjunction
    induced by the r-X1 deficiency in Zea mays L.,
    Genome, 29:782-785.
Weber, D.F., 1986, The production and utilization of
    monosomic Zea mays in cytogenetic studies, in: "Gene
    Structure and Function in Higher Plants," G. Reddy
    and E. Coe, eds., Oxford and IBH Publishing Co., New
    Delhi (1986), pp. 191-204.
Weber, D., and Helentjaris, T., 1989, Mapping of RFLP
    loci in maize using B-A translocations, Genetics,
    121:583-590.
Wu, M., 1989, High-resolution gene mapping by in situ
    hybridization on maize pachytene chromosomes, in:
    "Development and Application of Molecular Markers to
    Problems in Plant Genetics," T. Helentjaris and B.
    Burr, eds., Cold Spring Harbor Laboratory, New York
    (1989), pp. 153-157.

RFLP MAPPING IN WHEAT - PROGRESS AND PROBLEMS

Michael D. Gale, Shiaoman Chao and Peter J. Sharp

Institute of Plant Science Research, Cambridge Laboratory
Maris Lane, Trumpington, Cambridge CB2 2JB, UK

INTRODUCTION

The genetic linkage map of bread wheat (*Triticum aestivum* L. em Thell., 2n = 6x = 42) has always been less well developed than those of diploid cereals such as maize (*Zea mays* L.) and barley (*Hordeum vulgare* L.). Progress in wheat has been hindered by several problems inherent in it being an inbreeding hexaploid species of recent origin. Most importantly, recessive mutations such as morphological and pigment variants, male steriles and lethals, which comprise the backbone of genetic maps in diploid plants have not often been available to wheat geneticists. Although a few have been recognised, and presumably many have been generated in mutation studies, their phenotypic expression is usually masked by effective alleles at the homoeoloci in the other two genomes. In addition, linkages between the few recessive mutations obtained are rare, as wheat has a relatively large number of linkage groups, having 21 pairs of chromosomes. Moreover, naturally occurring variation between varieties as measured at conventional biochemical marker loci appears to be less common in wheat than in many other crops. This is as expected in an inbreeding species that may have arisen from a single entirely homozygous spontaneously doubled tri-haploid only some 10,000 years ago.

Restriction fragment length polymorphism (RFLP) mapping appears to offer the strongest possibility to overcome some of these difficulties. However, other problems were expected when the initiative to construct an extensive, informative and exploitable RFLP map of wheat was commenced. The patterns of hybridisation were expected to be complex, because of wheats polyploid status. The size of the wheat

Gene Manipulation in Plant Improvement II
Edited by J. P. Gustafson
Plenum Press, New York, 1990

genome at 35 pg relative to, for example, 5.6 pg in maize or 1.2 pg in rice (*Oryza sativa* L.) was expected to pose technical difficulties. Interference from the high proportion of repeated sequences present was expected. The degree of variability was expected to be low, but it was hoped that it would be greater than observed with other techniques. Mapping 21 pairs of chromosomes at the same time was expected to give problems, but it was anticipated that these could be overcome with a large scale operation.

As work has progressed, it has become clear that some early fears were unfounded. However difficulties have been experienced. These include the degree of variation exposed by cDNA probes, problems arising from the presence of duplicated sequences, and complications due to the many chromosomal interchanges that distinguish wheat varieties.

In this report we summarise some of the results obtained and describe the changes in our strategy to overcome the difficulties presently perceived.

PROGRESS

Mapping has been undertaken with some 80 low, or single, copy anonymous cDNA clones and 30 known function probes including some genes isolated from other plant species. The strategy being employed has been outlined by Sharp et al. (1989) and Gale et al. (1989) and the most comprehensively mapped chromosomes, those of homoeologous group 7, have been described by Chao et al. (1989).

Wheat-alien homology

In the *Triticeae*, it has been known for a long time (Huskins, 1931) that the chromosomes of various species are closely related, so that the chromosomal location of a marker gene in one genome is good evidence that it will be located on the related chromosomes in other genomes, including those of wheat (Miller and Reader, 1987). Alien chromosome addition lines have been used to predict the homoeologous chromosome group location of RFLP loci, using three sets derived from *H. vulgare* cv Betzes (Islam et al., 1981), *Secale cereale* cv. Imperial (Driscoll and Sears, 1971), and *Aegilops umbellulata* (Kimber, 1967), identifying chromosomes in the H, R and U genomes, respectively.

The similarities between genomes as judged by whole chromosome locations are extremely close, with the single chromosome addition lines predicting the chromosomal group locations of RFLP loci in wheat with almost total accuracy, once intragenomic translocations are accounted for (see also Sharp et al. 1989). Of the three genomes assayed in this way, the H genome, as present in the Betzes addition lines, has given the most consistent predictions, indicating that translocations between the H genome relative to the three genomes of hexaploid wheat are rare.

Of the low copy wheat cDNA clones analysed to date only two appear not to have homoeologous DNA fragments in Betzes at the level of stringency used. A few other cases of non-correspondence in chromosomal location have been noted, particularly with duplicated loci. These include the nitrate reductase loci identified by the barley Nar1 probe. Only one hybridising fragment was found in barley on 6H, two sets of fragments were found in wheat on homoeologous groups 6 and 7, and the three fragments found in rye, located on 4R, are probably located in the region of that chromosome translocated from 7R (Kleinhofs et al., 1988).

Other indications of significant independent evolutionary changes are emerging from comparisons of RFLP maps between related species. The evidence available indicates that 7H (barley chromosome 1) has many of the same genes arranged in the same order with similar recombinational intervals (Kilian et al., 1989), as compared to the wheat group of chromosomes. Similar comparisons of 6R (Masojc, pers comm 1989), 6H and the wheat chromosome 6B (Kleinhofs et al., 1988) indicates that substantial intrachromosomal rearrangements may have occurred (Gale et al., 1989).

The value of mapping chromosomes from alien species with genomes in common to wheat has been demonstrated by the analysis of single chromosome recombinant lines of wheat 7D with the 7D present in the breeders line VPM1. VPM1 was derived by crossing *T.persicum* L. (AABB) with *Ae.ventricosa* L. (DDM'M') and backcrossing to bread wheat (Maia, 1967), resulting in a line carrying a gene(s) for resistance to the cycspot pathogen (*Pseudocercosporella herpotrichoides*), transferred from the alien 7D chromosome. The analysis by Chao et al. (1989) showed that almost the entire 7D chromosome in VPM1 is derived from *Ae. ventricosa* Acc No 11 and that the gene order and recombination frequencies are as expected by comparison to those found for 7A and 7B. However, RFLPs were much easier to find. Analysis of 18 cDNAs exposed 14 RFLPs in the comparison of the VPM1 7D and the 7D of Hobbit 'S', while the same probes exposed only two 7D RFLPs in four segregating intervarietal mapping populations.

Hexaploidy

The fact that wheat has three genomes, which each carry DNA fragments hybridising to wheat cDNA probes, has proved to be an advantage rather than a problem, although an extra analytical step, relative to the similar work in diploid species, is needed to define the chromosomal location of any fragment(s). This involves the use of the nullisomic-tetrasomic (NT) aneuploids (Sears, 1954) for each restriction enzyme used in mapping. However, once this has been accomplished, each probe will identify at least three loci in different, but known, homoeologous linkage groups. The identification of polymorphic bands in other varieties has, so far, been possible by stepwise comparisons from Chinese Spring and other standard varieties. The further use of ditelosomic lines enables the positive location of the centromeres on RFLP maps.

The analyses of 7A, 7B and 7D indicated that, within wheat genomes, gene order is extremely well conserved. The map in Fig 1 is a consensus of the data presented by Chao et al. (1989) which draws on gene orders obtained in barley for the distal regions of the short arms.

Recombination

The group 7 map, as incomplete as it is, allows some observations concerning genetic distances and the distribution of recombination events among the chromosomes.

First, the overall genetic length of the group 7 chromosomes, presently at 230 cM, is not compatible with published data for chiasma frequencies for these chromosomes. The relevant mean fiugre for these chromosomes obtained by Sallee and Kimber (1979) in 3.25 chiasma/PMC, for both arms of each group 7 chromosomes, including 7BS (see below). This is equivalent to only 163 cM, and it is likely that further significant genetic length can be expected between the most distal loci and the telomeres, then extruding the observed map even further.

Second, the distribution of markers along the chromosomes indicate that recombination is not random. The consensus map (Fig. 1) clearly shows that wheat RFLP loci are to some extent clustered in the region of the centromere. This does not appear to be a feature of other maps constructed with cDNAs, e.g. tomato, *Lycopersicon esculentum* L. (Helentjaris et al., 1986) or with genomic probes, e.g. rice, *Oryza sativa* L., (McCouch et al., 1988), although the precise location of the centromeres in these species is difficult to ascertain, as telocentric stocks are not readily available.

The present evidence in wheat, based on a few cytological and genetic markers on chromosomes 6B (Dvorak and Chen, 1984; Kleinhofs et al., 1988) and 1B (Flavell et al., 1987), demonstrated that recombination tends to occur in the more distal regions of the

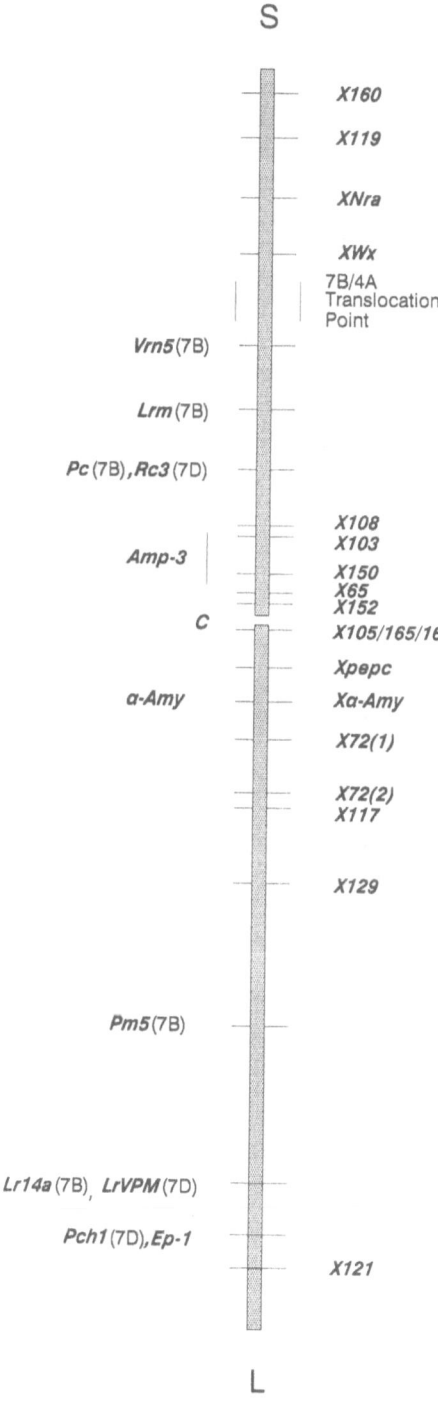

Fig. 1. A concenus map of wheat chromosomes 7A, 7B and 7D.

The map combines data obtained for the three chromosomes individually, detailed in Chao et al. (1989).

The linkage data was obtained from seven recombinant populations, including F_2s, random inbreds, doubled haploids and single chromosome recombinants. Genetic data for loci other than RFLP loci, shown on the left of the map, were already available for the CS x CS(Hope 7B) recombinants (Law, 1971, Gale, et al. 1983, Worland, unpublished data) and Hobbit'S' x Hobbit 'S'(VPM1 7D) recombinants (Koebner et al., 1988). The distances are approximate but span about 100 cM on the short arms and 130 cM on the long arms.

Amp-3, *α-Amy-2* and *Ep-1* are isozyme loci, *Pm5*, *Lr*, *Lrm* and *Pch1* are disease resistance loci, *Vrn5* is a vernalisation locus and *Rc3* and *Pc* are anthocyanin loci.

The gene order over all three chromosomes was consistent, except for some ambiguity in the *Xpsr 152, 65, 150* region. The gene order in the distal region of the short arms is ambiguous with the data available in wheat. However the order from *Xpsr108* to *160* has been inferred from, and is consistent with, data obtained in barley (A. Kilian, unpublished).

The 4AS.4AL-7BS translocation is primitive and probably occurs in all hexaploid wheats. Thus *XWx* to *Xpsr160* are to be found on 4AL, rather than 7BS.

chromosomes. Snape et al. (1985) suggested that the large blocks of centromeric heterochromatin of these chromosomes contribute to this distortion. If recombination is more frequent in the distal regions of the group 7 chromosomes, then the clustering observed indicates that the cDNA probe loci are more randomly spread along the chromosomes than chiasmata. In relation to this, it is notable that chromosomes 7A and 7D do not have large blocks of heterochromatin around the centromere, so that this is plainly not the only factor reducing crossing over in the centromeric regions.

Further data are needed in order to clarify the situation regarding non random distribution of recombination events, but the answer is of more than academic interest. The mean figure for molecular DNA length per recombination unit is 3.8×10^6 bp/cM in wheat (assuming a mean length of 200 cM/chromosome). With differences in recombination frequency between the distal and centromeric regions, this figure will be an underestimate for centromeric regions, and conversely, an overestimate for distal segments. Knowledge of the precise distance will be crucial when RFLPs become entry points for chromosome DNA "walking" and "jumping" techniques to identify and clone genes of agronomic interest.

Primitive translocations

Naranjo et al. (1987) postulated, based on meiotic pairing data, that three translocations, 4AS.4AL-7BS*, 5AS.5AL-4AL* and 7BL.7BS-5AL had occurred during the evolution of hexaploid wheat.

Two of these can now be considered confirmed. The portion of 7BS transferred to 4AL and the location of the breakpoint can be seen in Fig. 1. In addition this region also carries the isozyme locus *Per-B4* (Kobrehel and Feillet, 1975). The segment of 4AL transferred to 5AL carries another isozyme locus *β-Amy-A1* (Ainsworth et al., 1983) and at least one RFLP marker (F.J. Nicoll, pers. comm). No confirmation of the third translocation (5AL-7BS) has yet been obtained from the location of any RFLP or isozyme marker, but there is no evidence contrary to its existence.

The evidence from both isozyme and RFLP loci is compatible with the translocations having occurred before the emergence of the hexaploid species, since no varieties have yet been found to differ from CS

---

*The chromosome nomenclature used here is that agreed by the 7th International Wheat Genetics Symposium where the designations of 4A and 4B were reversed.

with respect to these translocations. The fact that they do not involve the
D genome suggests that they may have been present in the tetraploid
AABB which hybridised with *Ae. squarrosa* (DD) during the origin of
hexaploid wheat, as suggested by Naranjo et al. (1987).

PROBLEMS

RFLP variability

In view of the recent origin of hexaploid wheat, with some
evidence indicating that the event occured only 9,000 years ago (Miller
1987), and its likely derivation through spontaneous chromosome doubling
of an inter-specific hybrid, one might expect the wheat genome to exhibit
relatively little polymorphism.

This indeed turns out to be the case. As judged by comparisons
of 18 cDNA probes (54 loci) on a few varieties, and thus the figures need
to be treated with caution, wheat is the least variable of the three species
being investigated in this laboratory. In rye, an ancient outbreeding
species, an average of 48% of pairwise comparisons between inbred lines
with each probe x enzyme combination revealed polymorphism (P. Masojc,
pers. comm.). In barley, an ancient inbreeder, this occurred at a
frequency of 28%. In wheat the comparable figure is only 8%.
In lettuce, (*Lactuca sativa* L.) 24% of such combinations are polymorphic
(Landry et al. 1987) and in maize the figure is over 75% (Burr et al.,
1983). Moreover there are large differences between the three wheat
genomes, with the B genome loci exhibiting 17% variability and those in
the A and D genomes showing only 5% (Chao et al., 1989). Although no
clear explanation is available yet for the higher mutation rate apparently
occurring in the B genome, the same trend is evident after consideration
of the level of polymorphism displayed in the three genomes at protein
marker loci which are similarly observed as sets across the three genomes
(see list in McIntosh, 1988).

The prospective difficulty of mapping such a conserved genome, is
offset by the considerable differences that have been found between
probes. For example, one probe, PSR160 which identifies the *Xpsr160* loci
(Fig.1) gives polymorphism values of 21%, 64% and 49% for the A, B and
D genomes, while a number of other probes have not yet revealed
intervarietal polymorphism. The conclusion is that highly informative
probes should be identified and selected for mapping so that it will be
possible to produce maps of greatest value to geneticists and breeders.

It is probable that the differences in variability observed between
probes relates directly to the variability tolerated by the plant in the gene

products. A study by Sharp et al. (1988) showed clearly that conservation and variability of proteins at three β-amylase loci in wheat were closely associated with the degree of RFLP found at the same loci.

Duplicated loci

Several of the cDNAs used appear to detect 'duplicated' loci, in that they give a minimum number of six or more DNA fragments after hybridisation with a range of different restriction digests. For mapping purposes, providing that duplicated loci are on different chromosome groups, or even on different arms of the same chromosome, this poses no problem since the identification of loci can be determined by aneuploid lines. Where loci are duplicated within chromosome arms we have, as is the convention for protein loci in wheat, proceeded on the assumption that, until segregational evidence to the contrary is observed, they should be treated as members of closely linked multigene families. This assumption can lead to ambiguous results.

In one mapping population significant recombination, 9%, has been observed between two sets of polymorphic DNA fragments detected by probe PSR72, so that two loci $Xpsr72$-$7B(1)$ and $Xpsr72$-$7B(2)$ are therefore defined. In other families and using different restriction enzymes, RFLPs have been detected with PSR72, but we have no means of distinguishing the two $Xpsr72$ loci (Chao et. al. 1989).

This must be a common problem, not restricted to wheat, which can lead to misinterpretation or ambiguous data. In wheat, while RFLPs are relatively infrequent and the maps are not detailed enough to identify particular members of duplicated loci by their position on linkage maps, it is clear that intrachromosomally duplicated sequences should be avoided or, at least, treated with caution.

Intervarietal translocations

Wheat, again because of its hexaploid status, has the potential to tolerate and accumulate more chromosomal aberrations than most other crop species. One such type of rearrangment, which is becoming well documented, but certainly not described for all varieties, is interchromosomal translocation. Based on meiotic configurations in intervarietal hybrids, Schlegel and Schlegel (1989) have compiled a compendium of evidence for translocations present in wheat. An estimate of the frequency of translocation differences between varieties can be made using data available from hybrids involving Chinese Spring as one parent. Of 171 crosses, 21% showed no evidence of translocation, 65% showed one quadrivalent, 2% a hexavalent, 10% two quadrivalents, and

1% three quadrivalents. More complicated relationships are found in other hybrids, with octavalents, evidence of four different chromosomes being involved in linked interchanges not being unusual. Within the small number of varieties in which the identity of the chromosomes involved in translocations have been determined, 19 of the 21 wheat chromosomes have so far been found to have been translocated. Such translocations will plainly effect the mapping process and will be significant for the interpretation and exploitation of the map data in the various genetic and breeding situations where it may be employed. In mapping with an F_2 from a translocation heterozygote, changes in the order of loci on the linkage maps involved are to be expected. In addition, the genetic distances between loci not involved in the translocation, but located near the translocation sites, may well be affected. The nature of these changes in wheat has yet to be defined. In other plant species both increased and decreased recombination and interchromosomal interactions have been observed. In maize, recombination is reduced to the extent of that translocation breakpoints can be used to show linkage between loci which would normally segregate independently (Hoisington pers. comm.). On the other hand, experimental data from *Hypochoeris radicata* (Parker et. al., 1982; Parker, 1987) indicate that an increased frequency of chiasmata may be expected in the segments on either side of the translocation breakpoints.

A conclusion from this is that construction of a detailed genetic map in wheat should be accompanied by cytological screening for the presence of translocations in the segregating lines being used. When *in situ* hybridisation of single-copy genomic fragments becomes possible, RFLP probes which map on either side of known translocation breakpoints could be used as diagnostic tools to screen varieties for particular translocations. Clearly any work with European varieties should consider the 3B/3D, 7B/2D, 7A/7D and 5B/7B translocations which appear to be common.

CONCLUSIONS

RFLP mapping in wheat will probably proceed more slowly than in some other plant species. The use of related species such as rye, barley and *Ae. squarrosa* will be beneficial as bridges to the wheat map. However the low levels of RFLPs, the extreme variability in the amount of information gained from different probes and the possible structural rearrangements within chromosomes between species indicate that the work must also be carried out in the object species.

ACKNOWLEDGEMENTS

Much of the RFLP work to date has been funded by the AFRC New Initiative scheme. We are also grateful to many others involved in the project, particularly   A. Kilian, P. Masojc, A.J. Worland, J.W. Snape, C.N. Law, F.J. Nicoll, R.L. Harcourt, A. Kleinhofs, R.M.D. Koebner and T.E. Miller.

References

Ainsworth, C.C., Gale, M.D. and Baird, S. (1983). The genetics of β-amylase isozymes in wheat. I Allelic variation among hexaploid varieties and intrachromosomal gene locations. Theor. Appl. Genet. 66, 39-49.

Burr, B., Evola, S.V., Burr, F., and Beckmann, J.S., 1983, The application of restriction fragment length polymorphisms to plant breeding, Genet. Engineering, 5: 45-59.

Chao, S., Sharp, P.J., Worland, A.J., Warham, E.J., Koebner, R.M.D., and Gale, M.D., 1989, RFLP based genetic maps of wheat homoeologous group 7 chromosomes, Theor. Appl. Genet., in press

Driscoll, C.J., and Sears, E.R., 1971. Individual addition of the chromosomes of "Imperial" rye to wheat, Agron. Abstr. p 6.

Dvorak, J., and Chen, K-C., 1984, Distribution of non-structural variation between wheat cultivars along chromosome arm 6Bp : evidence from the linkage map and physical map of the arm, Genetics, 106: 325-333.

Flavell, R.B., Bennett, M.D., Seal, A.G., and Hutchinson, J., 1987. Chromosome structure and organisation, in : "Wheat breeding - its scientific basis", F.G.H. Lupton, ed., Chapman and Hall, London, pp 211-268.

Gale, M.D., Law, C.N., Chojecki, A.J. and Kempton, R.A. (1983). Genetic control of α-amylase production in wheat. Theor. Appl. Genet. 64, 309-316.

Gale, M.D., Sharp, P.J., Chao, S., and Law, C.N., 1989, Applications of genetic markers in cytogenetic manipulations of the wheat genomes, Genome, 31: in press.

Helentjaris, T., Slocum, M., Wright, S., Schaefer, A., and Nienhuis, J., 1986, Construction of genetic linkage maps in maize and tomato using restriction fragment length polymorphisms, Theor. Appl. Genet., 72: 761-769.

Huskins, C.L., 1931, A cytological study of Vilmorins' unfixable dwarf wheat, J. Genet., 25: 113-124.

Islam, A.K.M.R., Shepherd, K.W., and Sparrow, D.H.B., 1981, Isolation and characterization of euplasmic wheat-barley addition lines, Heredity 46: 161-174.

Kilian, A., Chao, S., Sharp, P.J., Kleinhofs, A., and Gale, M.D., 1989. An RFLP map of barley chromosome 1 (7H), (in preparation).

Kimber, G., 1967, The addition of the chromosomes of Aegilops umbellulata into Triticum aestivum (var. Chinese Spring), Genet. Res., 9: 111-114.

Kleinhofs, A., Chao, S., and Sharp, P.J., 1988, Mapping of nitrate reductase genes in barley and wheat, in : "Proc. 7th Int. Wheat Genet. Symp." T.E. Miller, and R.M.D. Koebner, eds., pp 541-546, IPSR, Cambridge.

Kobrehel, K., and Feillet, P., 1975, Identification of genomes and chromosomes involved in peroxidase synthesis of wheat seeds, Can. J. Bot., 53: 2336-2344.

Koebner, R.M.D., Miller, T.E., Snape, J.W. and Law, C.N. (1988). Wheat endopeptidase: genetic control, polymorphism, intrachromosomal location and alien variation. Genome 30, 186-192.

Landry, B.S., Kesseli, R., Leung, H., and Michelmore, R.W., 1987, Comparisons of restriction endonucleases and sources of probes for their efficiency in detecting restriction fragment length polymorphisms in lettuce (Lactuca sativa L.), Theor. Appl. Genet., 74: 646-653.

Law, C.N. (1971). 7B linkage group. Plant Breeding Inst. Ann. Rep. 1970, pp 89.

Maia, N., 1967, Obtention de bles tendres resistants au pietin-verse (Cercosporella herpotrichoides) par croisments interspecifiques, C.R. Acad. Agric. France, 53: 149-154.

McIntosh, R.A., 1988, Catalogue of gene symbols for wheat, in : "Proc. 7th Int. Wheat Genet. Symp.", T.E. Miller and R.M.D. Koebner, eds., pp 1225-1323, IPSR, Cambridge.

McCouch, S.R., Kochert, G., Yu, Z.H., Wang, Z.Y., Khush, G.S. Coffman, W.R., and Tanksley, S.D., 1988, Molecular mapping of rice chromosomes, Theor. Appl. Genet. 76: 815-829.

Miller, T.E., 1987, Systematics and evolution, in: "Wheat breeding - its scientific basis", F.G.H. Lupton, ed., pp 1-30, Chapman and Hall, London.

Miller, T.E., and Reader, S.M., 1987, A guide to the homoeology of chromosomes within the Triticeae, Theor. Appl. Genet., 74: 214-217.

Naranjo, T., Roco, A., Goicoechea, P.G. and Giraldez, R., 1987, Arm homoeology of wheat and rye chromosomes, Genome, 29: 873-882.

Parker, J.S., 1987, Increased chiasma frequency as a result of chromosome rearrangement, Heredity, 58: 87-94.

Parker, J.S., Palmer, R.W., Whiteborn, M.A.F., and Edger, L.A., 1982, Chiasma frequency effects of structural chromosome change, Chromosoma, 85: 673-686.

Sallee, P.J., and Kimber, G., 1979, An analysis of the pairing of wheat telocentric chromosomes, in "Proc. 5th Int. Wheat Genet. Symp.", S. Ramanujam, ed., pp 408-419, Indian. Soc. Genetics and Plant Breeding, New Delhi.

Schlegel, G., and Schlegel, R., 1989, A compendium of reciprocal intervarietal translocations in hexaploid wheat. Die Kulturpflanze 37: (in press).

Sears, E.R., 1954, The aneuploids of common wheat, Missouri Agric. Exp. Sta. Res. Bull., 572: 1-58.

Sharp, P.J., Chao, S., Desai, S. and Gale, M.D., 1989, The isolation characterisation and application in the Triticeae of a set of 14 wheat RFLP probes identifying each homoeologous chromosome arm, Theor. Appl. Genet., (in press).

Sharp, P.J., Desai, S. and Gale, M.D. (1988). Isozyme and DNA polymorphism at the β-amylase loci in wheat. Theor. Appl. Genet. 76: 691-699.

Snape, J.W., Flavell, R.B., O'Dell, M., Hughes, W.G., Payne, P.I., 1985, Intrachromosomal mapping of the nucleolus organiser region relative to three marker loci on chromosome 1B of wheat (Triticum aestivum), Theor. Appl. Genet., 69: 263-270.

enhanced routes for gene transfer among sexually incompatible
plant species, this role may be of minor importance in the long
run.   Instead, the most important promise of molecular biology
as it pertains to agriculture may be the possibility of making
a transition from empirical to mechanistic approaches, with the
corresponding transition from ignorance to an understanding of
the biology of crops and their environment. It is this
transition that will be the most revolutionary and exciting,
and it is the failure to make the intellectual and technical
leaps that are necessary to implement these approaches that is
currently so frustrating. It is one of the unfortunate
characteristics of modern molecular biology that a desire for
mechanistic understanding is frequently and mistakenly equated
with reductionist attitudes. This confusion underlies the
difficulties of approaching agriculture with molecular biology.

Molecular biology, and the associated technologies that
are loosely grouped under the term recombinant DNA methods,
have evolved in concert with and parallel to the model systems
in which they were developed. Escherichia coli and
bacteriophage, the basic source of these methods - and still
the pied a terre for most approaches - are extremely simple
compared to the real world of agriculture. They are made much
more so by their selection and development as model systems -
and thereby being removed from their own complex ecologies.
For example, the bacterium E. coli has played a key role in
the development of molecular biology, partly because of the
ease with which it can be moved from its normal environment -
the rich ecosystem of the intestine - into a defined and simple
(if unrealistic) environment - the laboratory. This has been
both its strength and its weakness. Because it has been a
model under these conditions it has facilitated the development
and fine-tuning of the enormously powerful techniques of
genetics, biochemistry and their ultimate fusion, molecular
biology. But because of the necessary removal of a model
system from its existing ecology, we are unable to easily make
a methodological or conceptual transition that allows the
system to be assessed and studied in situ.

The approaches that have been and are being developed
are clearly suited for addressing some of the complexities
presented by simple model systems - where, by definition,
complexities are reduced. While the difficulty of
understanding the biology of eukaryotes can be greatly reduced
by the use of suitable model systems - yeast, Drosophila,
Caenorhabditis, and Arabidopsis afford excellent examples -
molecular biological techniques that have evolved from the
spatially and temporally simple systems such as bacteria, are
becoming stretched to their limits to address sophisticated
questions of gene action in the four dimensional context of
these complex eukaryotes, even when in their model
environments.

Agriculture presents a remarkable quantum leap in complexity that cannot be reduced to model systems. By definition, agriculture takes place in the real world, and exists not as plant biology, but as the enormously complex interaction between plants, insects, bacteria, viruses, fungi, protozoa, nematodes, lower and higher animals - especially humans - and the physical environment. The whole is what matters in agriculture - success with a part is not enough. However, reductionism and the study of parts has been the strength of molecular biology, and the driving force behind the development of current methodology. How then can we reconcile this tendency of molecular biology to reduce analysis to simple, manageable systems, and the need for agriculture to be assessed, as a whole?

One possible resolution is to think of current molecular biological tools and approaches as only the 22 rudimentary but necessary building blocks for a new generation of tools, that are specifically designed to pose questions in the real world - or at least a more real world than the world of model systems. But in developing such tools, the beauty and power of molecular biology to experimentally manipulate the system and thus to define precise answers, and the humility of genetics and natural history to accept the answers that arise, must be maintained.

Finally, it must be kept firmly in mind that agriculture is ultimately a human endeavor, and that agricultural research should also be kept human - and accessible to all who wish to or need to carry it out. This puts a firm focus on the need to develop powerful new methodology - and the final caveat that it will only be successful if it is intellectually sophisticated, but technologically simple enough to be used by all agricultural scientists in the world scientific community.

In this paper, I will describe a number of experiments carried out at the Institute of Plant Science Research, Cambridge, England over the last few years.   These experiments and ideas hinge around the use of gene fusions in agricultural molecular biology, and, in particular, the development and application of the B-glucuronidose (GUS) gene fusion system in transgenic plants and its testing in the laboratory and the field. I will review a few early steps in our approaches that help to illustrate the strengths and weaknesses of existing methods, and give some ideas for improvement and replacement of these methods that may prove useful in leading us toward direction of a more balanced analysis and understanding of complex systems.

GENE FUSIONS

The use of gene fusions is becoming an increasingly

common and powerful strategy in the study of gene activity.
The overall purpose of using gene fusions is to both
facilitate and expand the ways in which gene activity may be
studied. Gene fusions are DNA constructions in which DNA
sequences from two (or more) genes are combined such that the
coding sequences of one gene (the responder) are transcribed
and/or translated under the direction of another gene(s) (the
controller).   In one general class of experiments, the
purpose of incorporating the responder gene is to facilitate
and enhance detection of gene activity; in such cases the
responder gene is acting as a reporter of gene activity, as
shown in Fig.1. Reporter genes usually encode enzymes; the
measurement or detection of an enzyme reaction product is used
to make deductions about the activity of the gene fusion and
hence about the behaviour of the controller gene sequences.
In other experiments the responder gene is included with the
aim of manipulating a desired change in cell or organismal
phenotype and is thus acting as an effector. The use of
responder genes as effectors in a manipulative sense can be
exemplified by chimeric genes encoding antibiotic resistance.

GENE FUSION

(SIMPLE)

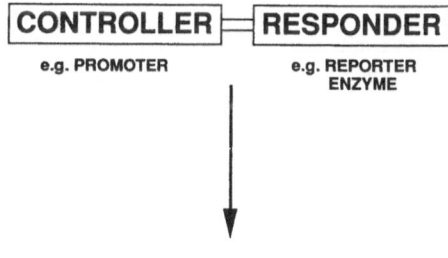

CONTROLLER	RESPONDER
e.g. PROMOTER	e.g. REPORTER ENZYME

ANALYTICAL TECHNIQUE

Fig. 1

    It is often extremely difficult and/or laborious to
devise and implement methods for the detection of a particular
gene product, usually entailing measurement of mRNA levels or

preparation of specific antibodies. By using gene fusions,
one set of methods that have been fully developed in other
systems is generally applicable to the analysis of expression
of almost any gene of interest. Moreover, a powerful reporter
gene, with simple and quantitative assays, can make the
logistics of analyzing large numbers of transgenic plants much
more reasonable, a feature which is going to become of
increasing importance as the extent of natural variation in
gene expression begins to emerge (see below).

       A powerful gene fusion system does not just make
standard forms of gene analysis available, but can actually
enable forms of analysis that are difficult or impossible by
other means. One example derives from the extreme sensitivity
of an enzyme (reporter gene) assay, which, by the inherent
amplification of signal by accumulation of an enzyme reaction
product, allows gene activity to be analyzed even in cases
where the gene product itself is present in such low amounts
that it would be almost impossible to detect by other means.
Another example is the analysis of genes that are members of
multi-gene families whose products are very similar but may be
regulated differently. By using gene fusions to individual
members of such families and introducing these fusions into the
genome, one can study the expression of individual genes
separate and distinct from the background of the other members
of the gene family. An even more exciting application of gene
fusion technology, which will become available in the near
future, is the possibility of monitoring and manipulating gene
activity in vivo and in situ in living plants in the field.

Transcriptional and Translational Fusions

       Gene fusions can be of two general types, with many
variations within types. Transcriptional fusions are defined
as fusions in which all protein coding sequences are derived
from the responder, with none from the controller. Thus,
although the mRNA produced may consist of sequences from both
controller and responder, the protein synthesized will be
encoded only by the responder. The fusions described below
between the patatin promoter and GUS, that have been analyzed
in the laboratory and the field, are transcriptional fusions.
When properly constructed, transcriptional fusions allow
discrimination between control at the level of transcriptional
initiation and control at subsequent steps of the biosynthetic
process.   Translational fusions, sometimes called protein
fusions, are defined as those in which the polypeptide
produced is the result of coding information provide by both
controller and responder, for instance the fusion of a signal
peptide encoding sequence from one gene with a reporter
enzyme, as described later. Each of these types of fusions
can be used to ask different questions, or achieve different
ends.

NEW APPROACHES FOR AGRICULTURAL MOLECULAR BIOLOGY:

FROM SINGLE CELLS TO FIELD ANALYSIS

Richard A. Jefferson

Institute Plant Science Research, Cambridge Laboratory,
Cambridge, England and Joint Division of the Food and
Agriculture Organization of the United Nations and the
International Atomic Energy Agency, Vienna, Austria

INTRODUCTION

Although the advent of molecular biology and recombinant
DNA technologies has caused tremendous excitement and
enthusiasm among many scientists and policy-makers whose goal
is the improvement of agriculture through research, the
realities of the time scale over which these technologies must
be applied, and the extent to which they are currently
inadequate is now being appreciated. The main reason for these
inadequacies is simply our overwhelming ignorance of the
biology that underlies agriculture. Traditional crop
manipulation, as typified by plant breeding, is an empirical
craft that is not generally based on a firm understanding of
the processes that govern the characters being manipulated.
Although the methodology used in an advanced plant breeding
program can be quite sophisticated, it is fundamentally limited
by the need to visualize and select for traits from existing
variation. The complexity of agriculture and the urgency of its
success or failure has generally required this empirical
approach to its manipulation, while the power of modern
molecular biology lies in the acquisition of an understanding
of the processes by experimental manipulation.

While it is possible to envision molecular biology and
transformation techniques principally as a means of providing

Gene Manipulation in Plant Improvement II
Edited by J. P. Gustafson
Plenum Press, New York, 1990

PROPERTIES OF AN IDEAL GENE FUSION SYSTEM

Gene fusion technology is useful in direct proportion to the extent to which it facilitates and/or extends the possibilities for the analysis of gene function. This is determined almost entirely by the properties of the protein encoded by the chosen responder gene. The minimal requirements for a good reporter gene are that there should be no intrinsic background activity in the organism being studied and that the enzyme can be assayed routinely and reliably in the laboratory. This means that the enzyme should be stable and active under widely varying conditions of pH, ionic environment and temperature - _in vitro_ and _in vivo_ - and should be tolerant of general mishandling. Moreover, the enzyme should not interfere with normal physiological functioning of the organism, nor affect the biochemistry adversely. It should not be adversely affected by any post-translational modifications nor subject to any other form of regulation to ensure that the assays are quantitative i.e. that the rate of enzyme end product accumulation is in direct proportion to the level of gene expression. This also means that there should be no endogenous substrates that could compete with the substrate being supplied, or any endogenous enzymes that could further metabolize the reaction product(s), as either would affect the quantitative nature of the assay.

Additional features that enhance the power of a gene fusion system are that the enzyme should tolerate amino terminal fusions, so that it can be used in translational as well as transcriptional fusions. As outlined below, this greatly facilitates the study of organellar targeting and trans-membrane movement of proteins. The enzyme should have numerous, simple assays, some of which are sensitive enough to measure gene expression of moderate to low abundance in single cells, and others which allow spatial discrimination of enzyme activity within the complex cellular patterns of tissues and organs. There should also be the possibility of using the system as a true responder, providing both reporting and effecting functions - thereby allowing genetic selections to be applied. This use, outlined in Fig. 2, is expanded upon later.

GUS

In plant molecular biology the use of gene fusions has expanded greatly over the past few years, particularly with the adoption in many laboratories of the Ã-glucuronidase (GUS) gene fusion system (eg. McCabe et al., 1988; Hinchee et al.,

1988; Klein et al., 1988; Masson and Federoff, 1989). The key
features which have led to the widespread adoption of GUS are
the almost complete absence of endogenous GUS activity in
higher plants and the variety of simple and sensitive assays
available for GUS. In particular, GUS is the only gene fusion
system presently available which allows routine spatial
localization of gene activity.

The E. coli gene encoding Ã-glucuronidase (GUS) has been
developed for use as a reporter gene in transgenic plants,
nematodes, bacteria and fungi (Jefferson, 1985; Jefferson et
al., 1986; Jefferson et al., 1987a; Jefferson et al., 1987b;
Schmitz et al., 1989; Roberts et al., 1989). GUS catalyzes
the hydrolysis of a very wide variety of Ã-glucuronides, and
with much lower efficiency some Ã-galacturonides. In fact,
almost any aglycone conjugated in a hemiacetal linkage to the
C1 hydroxyl of a free D-glucuronic acid in the Ã configuration
serves as a GUS substrate. GUS is an exo-hydrolase; it will
not cleave glucuronides in internal positions within polymers
and is inactive against Ã-glucosides, Ã-galactosides,
Ã-mannosides, or any glycosides in the Ã configuration (Levvy
and Conchie, 1966). Thus the many uronic acid residues -
generally Ã-galacturonic acid - in plant cell walls are not
substrates for GUS, being both internally situated within
polymers and in the Á configuration. The Ã-glucuronidase of
E. coli was first described in any detail by F. Stoeber (1957,
1961) who demonstrated key features of its regulation and
biochemistry. The enzyme has been purified to homogeneity,
and the gene has been cloned and sequenced from E. coli K12
(Jefferson, 1985; Jefferson et al., 1986)

GUS has many properties that make it nearly ideal for
gene fusion experiments. These properties are outlined in
Table 1. Intrinsic Ã-glucuronidase is absent from most, if
not all higher plants, fungi, and most bacteria. Reports of
intrinsic activity are few, and are usually either very
specific to a subset of substrates or highly localized and
detected under unusual conditions, but must be considered (eg.
Schultz and Weissenbock, 1987; Plegt and Bino, 1988) The
enzyme can be assayed using available substrates in numerous
spectrophotometric, fluorometric and histochemical assays.

GUS is very stable, and will tolerate many detergents,
widely varying ionic conditions, and general abuse. It has no
cofactors, nor any particular ionic requirements and can be
assayed at any physiological pH.

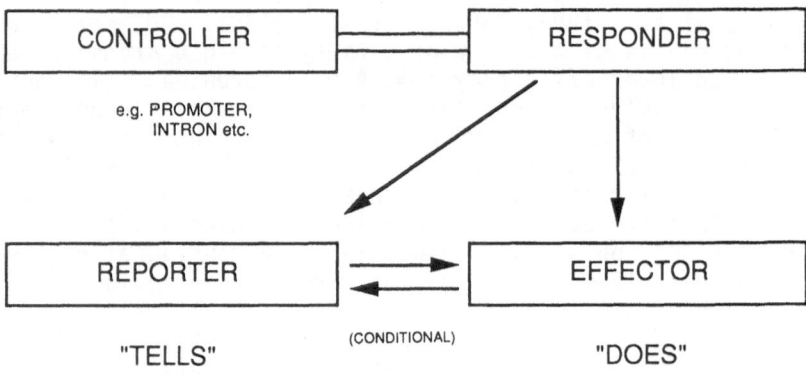

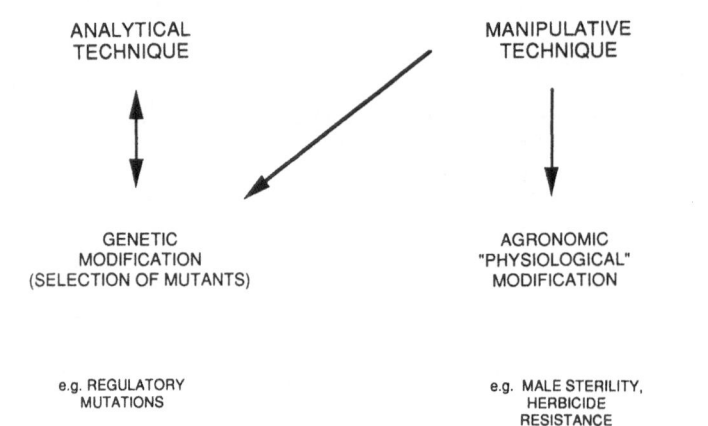

Fig. 2

Table 1. GUS *ß-Glucouronidase* (EC 3.2.1.31)

---

* Encoded by *E. coli gus*A (formerly *uid*A)

* 68 kDa Monomer Molecular Weight

* Broad pH Optimum

* Resistant to Many Proteases, Salt and Detergents

* Stable and Tolerates Large Amino-Terminal Fusions

* Translocated into Mitochondria, Chloroplasts, and
  Endoplasmic Reticulum

* No Activity in Most Higher Plants, Fungi and Bacteria

* No Disruption of Physiology when Expressed in
  Heterologous Systems

* Numerous Substrates Commercially Available

* Unlimited Possibilities for New Substrate Development

THE USE OF GUS FUSIONS IN TRANSGENIC PLANTS: REGULATION OF
CHIMERIC PATATIN GENES IN TRANSGENIC POTATO PLANTS

Patatin and Potatoes

While the potato (Solanum tuberosum L.) is perhaps the
most important non-cereal staple crop in the world, at least
in terms of acreage under cultivation and tonnage, it is one
of the most poorly understood. The potato is a relatively new
crop, with only a modest history of genetic improvement when
compared to cereals such as wheat (Triticum aestivum L.em
Thell.) and maize (Zea mays L.) Potatoes are propagated
vegetatively, and most commercial cultivars are poorly
characterized tetraploids of limited fertility, hence
conventional breeding is not simple or straightforward. These
factors strengthen the argument that potatoes will be one of
the first crops to benefit from improvement by non-traditional
means, and in particular by genetic engineering. As a
solanaceous plant, potato is highly susceptible to infection
by Agrobacterium tumefaciens, and genetic transformation by
this means has become routine. However, the ability to
introduce new genes into a crop only makes more urgent the
need to understand how these genes function, so that sensible
strategies for crop improvement by these powerful new tools
can be developed.

The goal of the experiments described here was to develop an understanding of the behaviour and regulation of a gene encoding patatin - a major protein of the potato tuber - in laboratory, glasshouse and field grown potato plants. Towards this end, transgenic approaches using GUS fusions were used to increase the resolution of the analysis and to facilitate the experiments.

Patatin is a 40 kDa glycoprotein that accumulates in tubers to up to 40% of the total soluble protein (reviewed by Park, 1986). The function of patatin is still completely obscure, but it is clearly associated with storage tissue. Induction of patatin mRNA transcription can also be observed in vitro when single node cuttings are subjected to tuberization conditions, even when tuberization itself does not occur (Paiva et al., 1983). It is encoded by a gene family with more than a dozen members, meaning that in the normally tetraploid cultivated potato there are upwards of fifty possible genes bearing patatin coding sequences in many allelic and non-allelic loci (Mignery et al., 1988; Twell and Ooms, 1988). Each of these genes may behave differently, some regulated to express at high levels, others very low or even inactive - and perhaps with different spatial and temporal control regimes. How then can one distinguish between the contribution of one gene from that of another, or even its allele? The easiest way, and the way that allows experimental manipulation of the DNA, is to mark one isolated gene in vitro such that it can be readily studied distinct from others, and re-introduce it to the host genome - that is to construct a transgenic plant using a chimeric gene, or gene fusion as described above.

To investigate the transcriptional regulation of the patatin gene in transgenic potato, I started by generating simple gene fusions between a promoter from a patatin gene and GUS. The patatin gene that I used was cloned and sequenced by Mike Bevan and his colleagues at the Institute for Plant Science Research (Bevan et al., 1985) using a cDNA probe obtained from Bill Park at Texas A&M. The initial DNA constructions were simple transcriptional fusions in which varying lengths of the 5′ upstream regions of the patatin gene were fused to the coding sequence of GUS contained on a plasmid called pRAJ260 (Jefferson et al., 1986). Some of these are outlined in Fig. 3. These constructions were subcloned within a binary Agrobacterium vector, pBIN19 (Bevan, 1984) and transformed into a commercial potato cultivar by a tuber-disc method (Sheerman and Bevan, 1987), selecting for resistance to the antibiotic kanamycin. Many independent transformants were obtained for each of the promoter lengths, and subjected to further analysis.

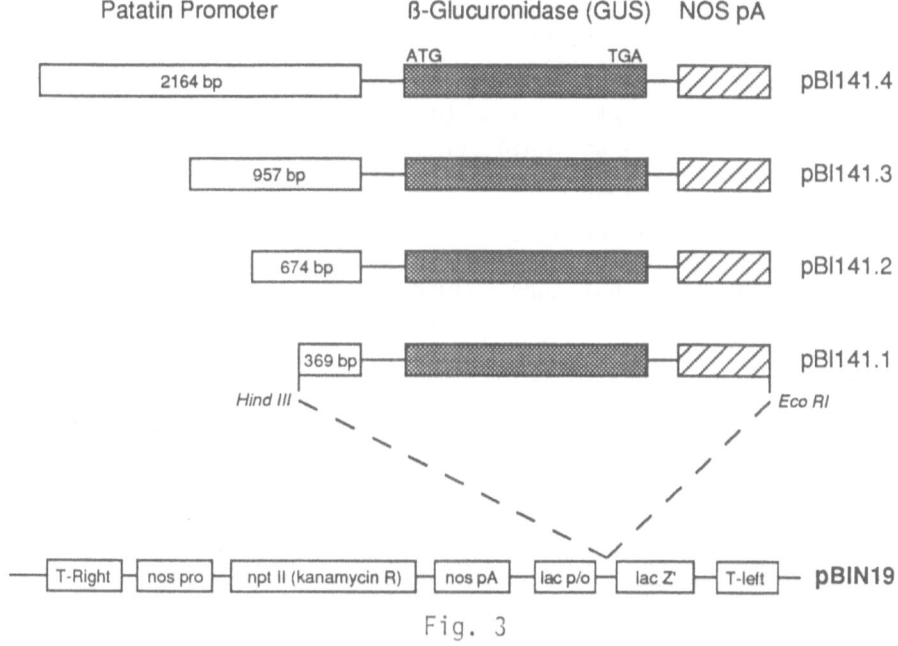

Fig. 3

IN VITRO INDUCTION EXPERIMENTS

The first experiments carried out on the transformants were in vitro inductions of patatin-GUS activity. It had previously been shown by Paiva et al. (1983) that accumulation of patatin protein could be induced in cuttings. To determine whether our patatin-GUS fusions were transcriptionally regulated in the same way, small single-node cuttings of the primary transformants (approximately 100) were placed on media containing either 3% sucrose without cytokinin or 7% sucrose with 2 mg/l benzyladenine, under either normal lighting regime or in the dark. Under these conditions, about 30 - 60% of the cuttings in the dark gave small microtubers within 2 weeks, while in the light the induced cuttings accumulated high levels of starch but showed no signs of tuberization. After two weeks incubation on either of the two media, the cuttings were homogenized using mortar and pestle, and GUS levels were determined with a fluorogenic assay. Most transformants from each of the promoter classes showed strong induction of GUS activity under conditions of high sucrose and cytokinin. Interestingly, GUS levels were highly elevated in the induced cuttings irrespective of whether they were grown in the light, or in the dark with concomitant tuberization. This observation is consistent with the experiments of Paiva et al. (1983) and Bourque et al. (1987) who showed that patatin synthesis could be uncoupled from tuberization. These

experiments demonstrated clearly that our patatin promoter
would respond to induction phenomena that were in some sense
distinct from those for tuberization - representing a subset
of the tuberization induction conditions. An alternative way
to interpret this is that patatin-GUS fusions are induced by
all the same conditions as those for tuberization, but do not
respond to a "normal" light-inhibition of tuberization.

To better define the nature of the induction signal,
different sugars and osmotica were used, including fructose,
glucose, and mannitol, but none showed a significant induction
- sucrose alone gave high GUS activity. In subsequent
experiments we observed no reproducible stimulation of
patatin-GUS induction by cytokinin so it was omitted.

An example of the induction data obtained is shown in
Table 2. This table shows the GUS levels accumulated in a
cutting grown in vitro for two weeks under non-induced
conditions, or under conditions that induced patatin with or
without concommitant tuberization (dark and light induced,
respectively). The data shown are for a single patatin-GUS
deletion derivative consisting of a 2164 bp promoter fragment
directing GUS expression.

While it is clear that this promoter is fully capable of
inducing GUS under these conditions, it is also clear that
there is enormous variation between individual transformants.
Some of this variation may be due to physiological differences
in the cuttings and some must be due to "position-effect"
influences on the gene fusions. "Position-effect" is a very
important but poorly defined issue in transgenic science that
refers to variation in gene expression that can be ascribed to
the influence of differing sites of integration of the foreign
DNA (eg. Dean et al., 1988).   All currently available methods
for transformation of plants result in apparently random
integration of the foreign DNA into the genome of the host
plant. This inability to target DNA to its homologous site or
at the very least a reproducible site, and the resulting
uncertainties caused by neighboring sequences, local structure
constraints or even three dimensional positioning in the
genome can result in a confusing and frustrating variation in
gene expression. The nature and mechanism (s) of this
influence is not at all clear, nor is the extent of the
influence.

Table 2   Analysis of Patatin-GUS Induction of Cuttings In Vitro for 17 Independent Transformants With pBI141.4, Consisting of a 2164 bp Promoter Fusion to Gus.

Transformant	Induction Conditions			Fold Induction	
	Uninduced[c]	Dark[d]	Light[e]	Dark[d]	Light[e]
pBI141.4 - 1	15	40	160	3	10
#2	6	170	650	28	108
3	12	85	135	7	11
4	10	25	250	3	25
5	1	75	60	75	60
6	7	35	250	5	35
7	6	9	300	2	50
8	1	10	70	10	70
9	60	12	250	0.2	4
10	7	70	150	10	21
11	40	900	1700	22	42
#12	15	150	3000	10	200
13	4	60	700	15	175
14	9	6	700	0.7	78
15	10	350	2300	35	230
16	40	320	1900	8	47
#17	20	6700	7300	335	365
	Uninduced	Dark	Light	Dark	Light
Average n=17	15	530	1169	33	90
Minimum	1	6	50	0.2	4
Maximum	60	6700	7300	335	365

Cuttings were all approximately 1 cm long, with a single leaf node with no obvious axillary bud growth at time zero. Cuttings were placed in petri plates containing MS media plus 3% sucrose (c.), 7% sucrose & 2 mg/l BAP (d. and e.) in the light (c. and e.) or dark (d.) for 14 days. The cuttings were removed and ground in a mortar and pestle and assayed for GUS fluorometrically. Values are expressed as nM MU/hr at 37 C. All values are per cutting, and not normalized. # denotes those plants chosen for statistical study and field trial.

To investigate the extent of variation that was not due to "position effect", but rather to other factors, 72 clonal single node cuttings were taken from a single set of identically maintained and aged in vitro grown plantlets that were in turn all derived from a single transformant. This transformant, 141.4-17 had been propagated for at least two years by nodal cuttings prior to analysis, and so was certainly not chimeric. The cuttings were divided into two sets of 36 cuttings, and placed on media containing either 3% sucrose or 8% sucrose in the light, noting from which region of the parental plant they were derived. After two weeks, they were homogenized and assayed for GUS activity by fluorometric measurements. The results clearly indicated that the variation from one cutting to another was as high as one thousand-fold, although the means of the uninduced and induced cuttings were different by a factor of ten. Moreover, the distribution of GUS activity among the cuttings was so broad that a third of the highest expressing uninduced cuttings overlapped a third of the lowest expressing induced cuttings. In addition, there was absolutely no correlation between the position of the cutting on the plant and the response of the cutting to induction. The scatter was such that it would easily be possible, by inadvertantly selecting a small number of plants, to deduce that the patatin promoter was repressed under conditions that clearly give rise, on the whole, to a high induction of transcription. When a particular transformant was analyzed at several different times - as close as a month apart or as long as two years apart - the reproducibility of induction was very poor. While a highly active transformant (4-17 was our highest expressing transformant) was usually very high, the actual extent of induction varied tremendously.

These observations suggested the possibility that the influence of the physiological state of the plant or cutting was contributing perhaps as much or more to the variation in patatin-GUS expression as the genetic component of "position-effect". It is possible that the variation among clonally propagated plants is either a characteristic of patatin gene regulation, a general feature of gene regulation in sporophytic development, or indeed even a feature of tissue culture conditions. It is easy to invoke models in which random, but stable methylation occurs to promoter sequences during the propagation in tissue culture that alters the ability of the gene fusion to express.

This calls into question the value of attempting to discriminate between the effects of mutations on the patatin promoter - in our case deletions - without a very large, statistically significant, synchronous population. Even with that caveat it is uncertain that assertions about relative "quality" of induction can reliably be made. In the early experiments, based on single sets of measurements of nearly one hundred independent transformants, I proposed a model for the patatin promoter that invoked both "enhancer"-like and "silencer"-like components. While this initial model may still be true, the data on which it was based have not been reproducible. An argument can be readily raised that this variation is only an artifact of inductions in tissue culture, and do not reflect a real variation in planta.

Patatin-GUS Expression in Planta

        The next set of experiments was designed to address this issue, and further, to investigate the temporal patterns and spatial localization of patatin transcription - as measured by GUS fusion activity - during the growth and development of the plant. A representative set of transformants - three from each promotor class, (360 bp of 5' sequence to 3500 bp) - were moved to glasshouse conditions in soil. The plants matured indistinguishably from untransformed control plants. Tubers, roots and aerial tissues were assayed for GUS activity quantitatively with fluorogenic assays and histochemically using the indigogenic substrate X-Gluc (5-bromo-4-chloro-3-indolyl-Ã-D-glucuronide).

        As with the cuttings induced in vitro, there was a very wide range of variation in GUS activity observed in planta, when assayed quantitatively. In spite of this, there was a very strong trend for GUS levels to be consistently much higher in tubers than in other organs. However, the extent of the variation again precluded any simple assertions to being made regarding the effects of inclusion of differing lengths of promoter in the constructions. One of the goals of the field trial described below was then to discriminate between the different influences - genetic and non-genetic - and quantitate their relative contributions to gene expression.

        Qualitative histochemical assays were performed to visualize GUS activity within the three dimensional context of the plant. These assays clearly demonstrated one of the most powerful uses of the GUS system, the ability to routinely and reliably distinguish individual cells that express GUS. The histochemical analyses were successful at illuminating a number of interesting properties of the transcription of the patatin-GUS fusion. Sections of plant materials were prepared, fixed, and stained for GUS activity using X-Gluc. This chromogenic dye deposits an indigo precipitate at the site of enzyme activity. GUS was observed at very high levels in individual cells of the ground tissue of the tuber, but never observed in the cells of the phelloderm or epidermis. Additionally, GUS was observed in discrete locations in developing stolons and in aerial tissues. For instance, GUS activity was reproducibly found in cells immediately adjacent

to the phloem bundles in aerial stem sections - even when the
sections were prepared from stem segments near the apex of the
plant. These observations, coupled with histochemical
analysis of uninduced and induced cuttings, led us to
speculate that high local concentrations of sucrose - or a
metabolite of sucrose - were responsible for induction of
patatin transcription, irrespective of the location or
developmental origin of the affected cell. The cells in the
vicinity of the phloem will presumably be effecting sucrose
loading and unloading from the phloem, and hence would be
expected to experience very high levels of sucrose. The idea
of "tissue-specificity" may therefore be inappropriate in this
case, and physiological and environmental regulation of gene
action can be invoked, relating to source - sink relationships
within the plant. This hypothesis is strengthened by the
quantitative assessments of GUS in potato plants grown in the
field.

ANALYSIS OF PATATIN-GUS EXPRESSION IN THE FIELD

The variation in expression that we observed in cuttings
induced to express patatin in vitro, both between individual
transformants and within populations of genetically identical
individuals, indicated that it would be necessary to collect a
great deal of data to assess with statistical significance the
factors that influence gene action and variation. We were
interested in the physiological and developmental control of
the expression of the patatin gene - which is certainly greatly
affected by environmental conditions - and reasoned that this
gene would therefore serve as an excellent model to study under
field conditions, where large scale replication and
accumulation of biometric data should be possible. In
addition, it is clearly important to understand the process of
tuberization in order to eventually manipulate the process for
agricultural purposes. Finally, because of the wide range of
expression that we had observed in the transgenic plants, and
the many subtle influences on that expression, we anticipated
that development of methods for large-scale field analysis
using the patatin gene as a model would present us with a good
spectrum of the difficulties that we will eventually need to
overcome if our methods are to be truly general.

Some of the questions that can be addressed with such
large trials include: How much variability in expression of a
single gene occurs within a population? What is the genetic
component of this variation? What effect does the site of
integration have on the expression of foreign DNA? What is
the contribution of "position-effect" to variation that can be
observed? What degree of the variation is due to stochastic
processes, or is otherwise uncontrollable and what degree can
be ascribed to physiological or environmental variation that
can be controlled or quantitated?

The answers to all of these questions will, of course, vary from system to system, but will be important for providing a solid understanding of the behaviour of genes in individuals and in populations. It is also important to acquire this information for the more pragmatic purposes of designing agronomically useful programs that involve gene transfer.

DESIGN OF THE FIELD TRIAL

The Patatin-GUS field trial (Fig. 4) can be considered as three separate trials, designed to address different aspects of the questions outlined above. The first trial, called GUS I , is designed to measure the degree of variation in gene expression observed between independent transformants - that is, the "genetic" component of variation in chimeric gene expression. GUS II is designed to ask how much variation is not genetic, but is due to other factors, such as environmental, physiological or stochastic influences. GUS III is designed to ask how patatin-GUS expression - both quantitatively and qualitatively - changes during the course of the growing season.

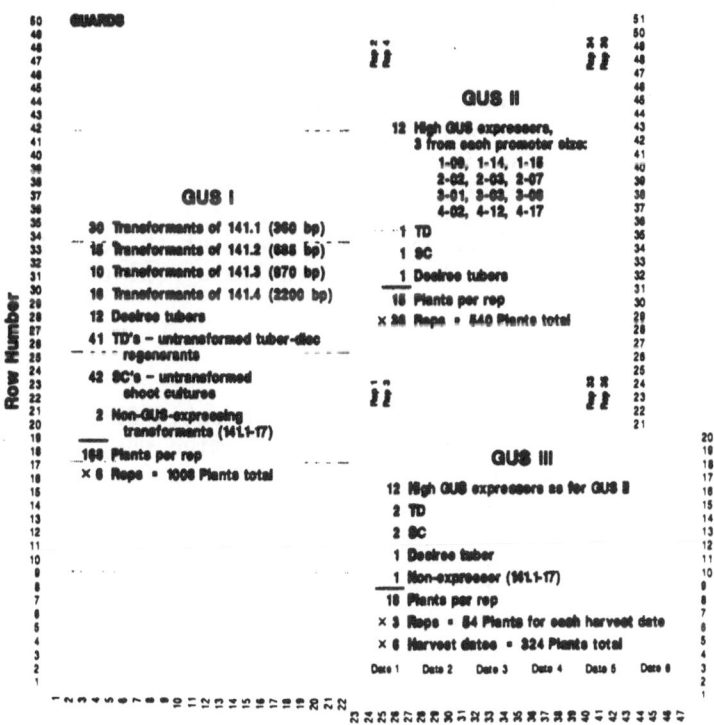

Fig. 4 Patatin-Gus Field Trials

GUS I

     To investigate the extent of variation in patatin gene
expression that could be ascribed to "position-effect" or
otherwise characteristic genetic variation, a large collection
of independent transformants was used. These transformants
contained one of four different patatin-GUS gene fusions as
outlined in Fig. 3, and were not pre-selected other than for
the expression of detectable levels of GUS. Each transformant
was propagated vegetatively in vitro by nodal cuttings to
generate 6 clonal replicates. A total of 71 independent
transformants was used, giving a total of 426 transgenic
plants. Control plants that were not transgenic, but that had
been regenerated from shoot cultures or tuber disc, were also
used. The trial was replicated in six blocks, each
self-contained but internally randomized.

GUS II

     To determine the distribution of patatin-GUS gene
expression and measure the extent of variation within a
particular transformant,ie. that which is not caused by genetic
differences between individuals, but rather by environmental,
physiological or stochastic differences, we selected 12
independent transformed plants, three from each of the four
promoter deletion classes. The plants were pre-selected based
on easily measureable GUS activity and strong induction of
patatin-GUS under tuberization conditions in vitro. These
twelve transformants were propagated and multiplied
vegetatively to give 36 clonal replicas of each plant. The
trial was replicated in 36 internally randomized blocks and
included shoot culture and tuber-disc regenerant controls.
Since six additional replicates of each of these plants was
represented in GUS I, the total available replication was up
to 42.

GUS III

     To investigate the changes in patatin-GUS expression
during the growing season, and to determine whether
differences in organ-specificity occured during growth, the
same twelve pre-selected transformants as in GUS II were
propagated in vitro to give eighteen replicate plants for
each. These eighteen replicates of twelve transformants plus
controls were distributed in six blocks, each containing three
replicates randomized internally. Every two weeks during the
growing season, one of these blocks was randomly selected and
harvested for analysis of GUS activity.

CONTAINMENT CONSIDERATIONS

     In the design of the field trial, and during its

execution, attention was given to minimizing the possibility
of movement of the transgenic plant material off-site
according to the recommendations of the Advisory Council on
Genetic Manipulation, and by the Plant Breeding Institute
Safety Committee. This entailed placing the test plot well
away from any other potato plot, planting guard rows of
untransformed plants, containing material during harvest and
processing, removal of flower buds to prevent pollen formation
and potential spread, and destruction of transgenic plant
material after completion of the field trial. In addition,
the trial plot was sterilized and allowed to lie fallow for an
extended time after harvest. The trial was performed under
MAFF license no. PHF 48A/114(57).

PLANTING, GROWTH AND HARVEST PROCEDURE

Young potato plantlets derived from tissue culture were
grown in peat pots in a glasshouse to a height of
approximately 5 - 10 cm, and planted in the test plot on June
1, 1987. The planting date was chosen to minimize the
possibility of a late frost, which would have deleterious
consequences on young plantlets. The usual agricultural
practice of planting seed tubers was not employed due to
logistical constraints of producing the seed material, and
time constraints. The validity of the outcome was not
compromised by this planting regime, as the purpose of the
trial was not to compare yield or agronomic performance with
potatoes grown under normal practice, but to investigate
variation within and between similarly propagated populations.

During growth of the potato plants, manual weeding and
spraying with commercial fungicides and aphidicides was
undertaken at regular intervals, and the plants were regularly
irrigated. Immediately before harvesting, vigour was scored
for all plants.

Harvesting of GUS III plants was performed every two
weeks during the growing season beginning with an initial
harvest on July 28. GUS I & II were harvested over the period
from October 14 - 23. Harvesting was carried out manually.
For GUS III, the entire plant, including aerial structures,
root and tubers was harvested at the same time to ensure that
the relative GUS levels in each tissue were directly
comparable. For GUS I & II, aerial structures were harvested
first, followed about one week later with the tuber and root
samples. This protocol was followed to minimize the
possibility that the tuber population would be contaminated by
fungal spores or other disease agents that would decrease
their suitability for storage and replanting in a subsequent
trial. In addition, this method most closely approximates
agricultural practice in which the aerial structures are
removed well before harvest.

SAMPLING AND ASSAY OF GUS ACTIVITY

The logistics of sampling and analysis were greatly
enhanced by the use of microtitre plates, and apparatus
designed to use the characteristic 8 X 12 array. These
included 1 ml polypropylene tubes in racks with the microtiter
array (Micronic), microtiter plate carriers for the
centrifuge, multi-well pipettors and microtiter plate
absorption and fluorescence spectrometers (Titertek Multiskan
Plus and Fluoroskan II). Software for direct reading of the
microtiter plates, correction of mis-labelled or incorrectly
specified wells, data analysis and manipulation was developed
expressly for this study ("Plates" Micro-Manipulator) by David
Wolfe.

RESULTS FROM THE FIELD ANALYSIS

Variation in Patatin GUS Expression Among Clonal Propagants

Fig. 2 of the most striking results from the field trial
are summarized in Figs. 5 and 6. Figs. 5 and 6 show the organ
distribution of GUS activity (expressed as specific activity)
within a large population of clonally propagated 4-17
transformants. This data, abstracted from GUS II, shows
clearly that even when grown in the field under conditions
that approximate agricultural practice, the net expression of
the patatin promoter is highly variable. It must be
remembered that the GUS protein, being reasonably stable,
reflects an integral of gene expression over the lifetime of
the reporter enzyme, not the instantaneous levels of
transcription at the moment of harvest.

This property is quite useful to minimize the variation
that must occur daily if not hourly. The trend for expression
to be by far the highest in the tubers, relative to other
organs is still maintained, but the extreme quantitative
variation (more than one hundred fold) can be mirrored by
qualitative variation. The plants that are highest in the
tuber are rarely those that are highest in aerial tissues.
Hence, not only is absolute expression level modulated among
clones, but relative expression in different organs as well. A
possible model to explain a component of this variation emerges
in considering some of the results from GUS III together with
the laboratory analyses.

VARIATION IN "ORGAN SPECIFICITY" DURING GROWTH

Fig. 7 shows the mean GUS specific activity in five organ
samples (tuber, root, node, stem and leaf) from up to nine
plants carrying the same gene fusion. These samples were taken
from lifts of three drills at two week intervals from GUS III.
The results clearly show that "organ-specificity" is

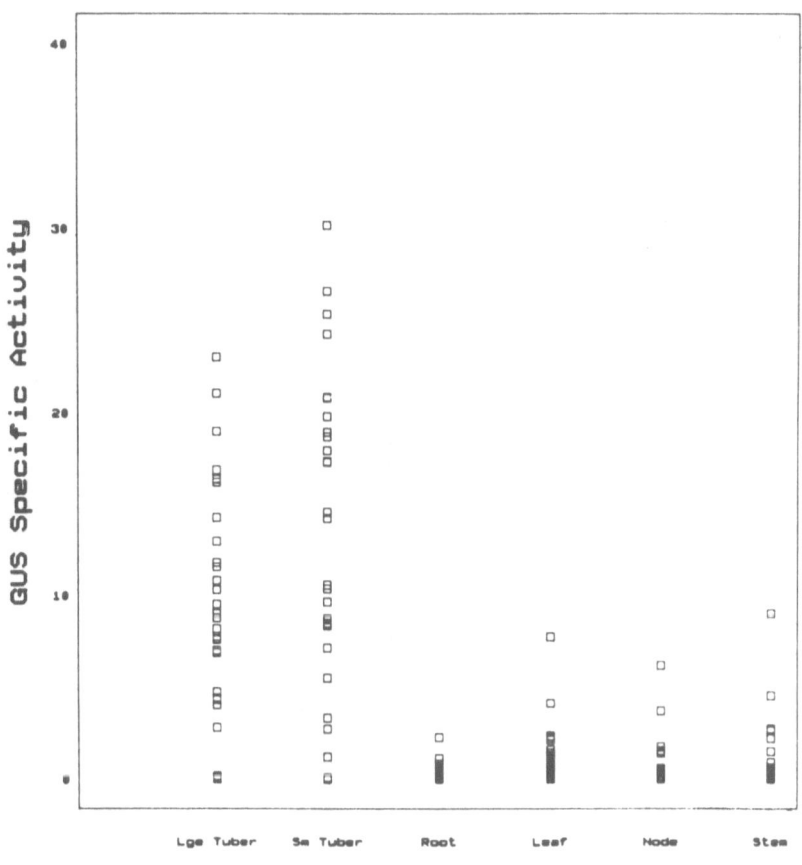

Fig. 5.   Tissue Distribution for Gus Activity PBI 141.1-14.

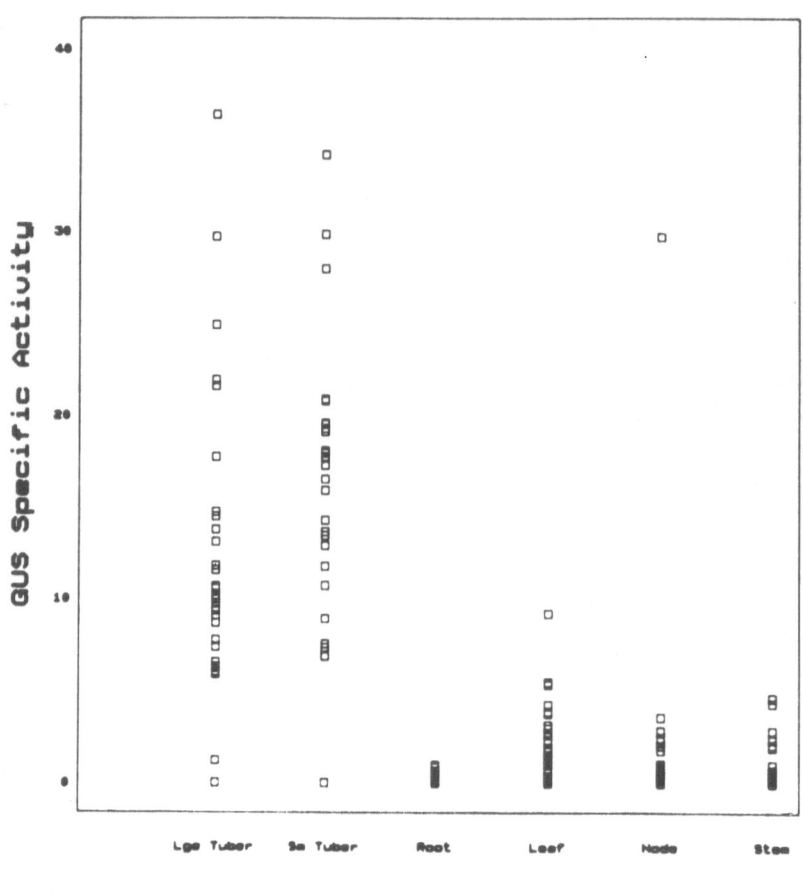

Fig. 6.   Tissue Distribution for Gus Activity For PBI 141.12.

dependent on when the plants are harvested! Early harvests
show a high degree of tuber specificity, while lifts in the
middle of the growing season show an increasing degeneracy in
expression. The third, fourth and fifth harvest dates show
very high level expression in all aerial tissues - in a large
number of plants the stem and node levels are higher than
tuber levels - while at the final harvest the tuber levels are
predominant. It is interesting to note that the specific
activity of GUS in the tubers remains roughly constant during
the growing season. This implies that the accumulation of GUS
protein closely follows the net accumulation of total protein
in the tuber - thereby maintaining a constant ration. This
indicates that the stability of GUS during growth is very
similar to the overall stability of patatin, since patatin
accounts for the highest proportion of soluble proteins in the
tuber.

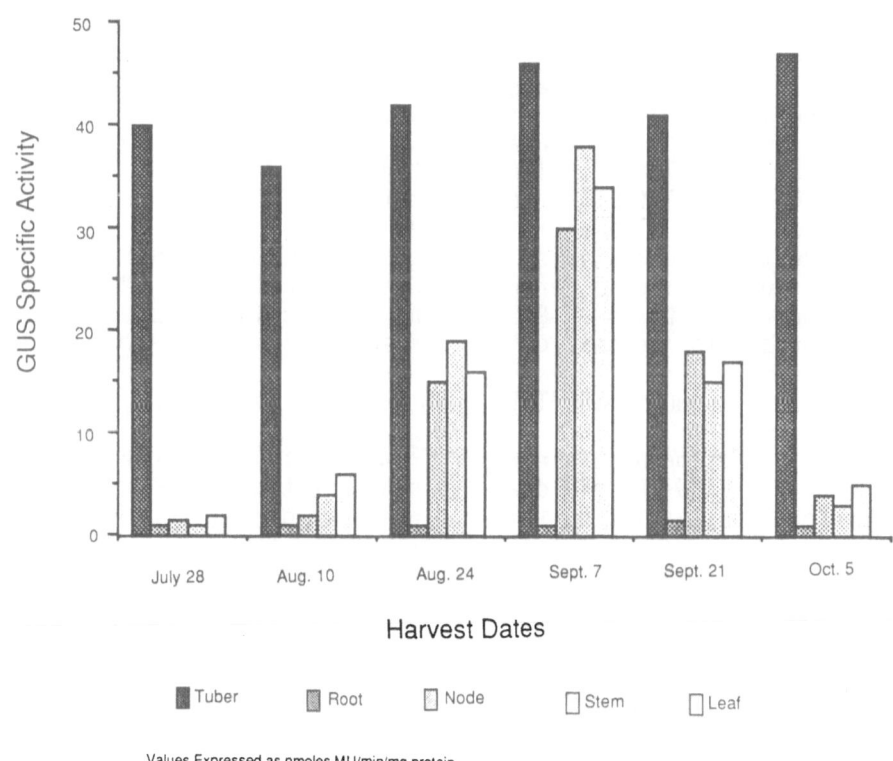

Values Expressed as nmoles MU/min/mg protein

Fig. 7.   Patatin-Gus Expression Changes In Organ-Specificity.
           During Growth Under Field Conditions.

The analysis of the field trial is still in progress, and it is clear there is much to be learned from the data already accumulated. But the general message to take away is that our methodologies have not yet evolved to cope with the analysis of gene action in populations - and by inference - are not even adequate to accurately answer questions of gene action in individuals. If a large field trial must be performed to accurately assess gene action for each DNA construction, then progress will clearly be minimal. What are the solutions? If the variation observed in patatin expression is typical, or even relatively common, how can we proceed? Certainly new approaches are required, and new tools must be made available to a much wider spectrum of agricultural scientists.

WHAT CAN GUS PROVIDE-WHY IS IT CALLED A SYSTEM?

The GUS gene fusion system as it stands is clearly a powerful tool for the generation and analysis of transgenic plants. We can now measure gene action quantitatively - albeit destructively - and localize gene activity to individual cells within tissues.   However, the potential of the system has not yet been realized, and future developments will expand its utility many fold. GUS is called a fusion system because it is not simply an isolated gene encoding a useful hydrolase, but rather a gene involved with an entire metabolic pathway of enormous significance to organisms and their attendant microflora, almost all aspects of which can be further manipulated to increase the power of the tool and to expand its utility into new directions. These directions include the development of methods for real time non-destructive analysis, for fusion genetics and positive and negative selections, for novel physiological and biochemical manipulations based on GUS action, and possibly even novel mechanisms for detoxification of xenobiotics in plants.

THE NEED FOR NON-DISRUPTIVE, NON-DESCTRUCTIVE IN VIVO ANALYSIS

As has been seen earlier from both the laboratory inductions and the field analysis of patatin-GUS expression, the enormous variation that can occur in gene action between plants that are genetically identical (as far as we know) makes comparison of gene action at different times a very dubious endeavour when pursued with destructive or non-quantitative methods. Thus when uprooting, grinding up and assaying one plant at one time point, followed by another plant at a further time point, it is now unlikely - based on our observations of extreme variation - that the second plant will give the same answer that the first plant would have given if allowed to grow to the second time point. Of course there are bound to be many reasons for this variation. Micro-environmental influences that differ between plants - and cannot be adequately quantitated or controlled - will

greatly affect the behaviour of many genes that are involved
with the life of the sporophyte. Additionally, stochastic
processes - random variation - may well be built into the
mechanisms for the control of gene action for many plant genes
allowing for responses to the viscitudes of the environment.
These processes, in turn, may be fixed and modified by
epigenetic mechanisms such as transient (or even long term)
methylation of the controlling elements.

With this problem in mind, one approach to minimize the
uncertainties generated by this variation is the use of
statistically large populations for each time point.   Towards
this end, having simple, easily quantitated and automated
tools such as GUS is quite useful. However, the routine use
of such numerical methods is unlikely to be chosen for both
logistical and aesthetic reasons. In addition, molecular
biologists, by and large, loathe and misunderstand statistics
- even though its importance and the biology that it can
reveal is unquestionable.

What we clearly need is a way of asking how one plant's
gene(s) is behaving at a given time without disrupting the
plant in any way and later asking the same question of the
same plant - thereby achieving an accurate picture of how gene
action in a particular plant is changing with time and
environment. This will then allow us to experimentally
manipulate both the gene and its environment, and achieve some
understanding of how they interact. However the Heisenberg
caveat must be applied - if we disrupt the system in studying
it, then continuous in vivo analysis becomes of limited value.
This is a tall order. Developing methods that will allow this
type of real-time monitoring must also deal with the further
constraints that the methods will only be generally used if
they are cheap and cheerful - that is if they do not require
delicate technology and can be done by most scientists.

HOW CAN THE GUS SYSTEM BE DEVELOPED TO ADDRESS THIS PROBLEM?

GUS expression does not apparently affect the physiology
of the host plant in any situation thus far reported (over one
hundred published papers reflecting GUS expression in dozens
of different plant species); GUS substrates are not usually
components of the biochemical makeup of plants. Because
methods for fluorescence analysis can be non-destructive and
very sensitive, it would therefore seem reasonable to ask
whether we could introduce fluorogenic substrates for GUS into
living cells and use fluorescence imaging techniques to
visualize GUS action in real-time. Again, there are certain
criteria that must be fulfilled to make this approach work.
To make the system quantitative, which it must be, we will
need to ensure that the substrates for GUS have access - in
excess of the K_m for GUS - to all sites of activity. Without

this assurance, the GUS levels observed may only reflect
substrate levels - and hence be valueless for quantitation or
even estimation of gene activity. This means that we must
ensure active transport of GUS substrates into all living
cells. Additionally, even though the substrates for GUS may
not be toxic, we must ensure that the product of GUS action is
comparably innocuous, remains associated with the site of
activity, and is not interfered with by quenching compounds in
the plant. And we must ensure that the detection methods are
versatile enough to circumvent the optical limitations
presented by plant systems, but again are not byzantine or
ridiculously expensive. These criteria may be met, or at
least approached through the long-term development of the GUS
system, and in particular the combination of novel substrate
synthesis and the manipulation of the glucuronide permease, a
newly discovered component of the gus operon of E. coli.

UPTAKE OF GLUCURONIDES INTO LIVING CELLS - THE GLUCURONIDE
PERMEASE

        E. coli has evolved to live within the rich environment
of the vertebrate large intestine, in an extraordinarily rich
mixture of diverse micro-organisms. To do so, it has evolved
a complex array of biochemical capacities which allow E. coli
to successfully compete and establish populations in the gut.
One of these capacities, which is restricted to E. coli and a
few other micro-organisms, is the ability to actively
accumulate and digest many diverse Ã-glucuronides.

GLUCURONIDATION AS A PRINCIPAL DETOXIFICATION PATHWAY IN
VETEBRATES

        Vertebrates consume vast quantities of organic compounds
that are not integrated into their biochemical makeup or
physiology. Many of these foreign compounds arise from
ingestion of plant materials (or other animals that have
consumed plant materials) and are plant secondary products.
These compounds, for instance phenolics, flavonoids,
coumarins, and assorted complex molecules, are frequently
taken up by the intestine, modified and then re-excreted by
any of a variety of mechanisms. The principal mechanism for
this modification and excretion is by the conjugation of the
foreign compound to glucuronic acid and excretion of the
conjugate in the urine or bile. This mechanism of
glucuronidation and excretion is ubiquitous among vertebrates,
and the range of compounds that are conjugated is enormous.
Not only most plant secondary compounds, but almost all
endogenous steroid hormones, thyroxine, bilirubin (the
breakdown product of heme and the source of the pigment in
faeces), antibiotics and drugs as well are conjugated as
glucuronides. The excreted material is often eliminated
through the bile, and thence into the large intestine. In the

large intestine, this impressive array of diverse glucuronides encounters a comparably impressive array of bacterial endosymbionts. Among these bacteria are almost always found a few species that have acquired the ability to feed on the glucuronide conjugates. E. coli is often the predominant aerobic bacterium in the intestine of animals (although the absolute number of aerobes is often quite small relative to the obligate anaerobes). E. coli is one of the few bacteria that can metabolize glucuronides, and the gus operon is the locus responsible for this ability.

The gus operon consists of the Ã-glucuronidase gene itself, encoding GUS (gusA - formerly uidA) a specific repressor, gusR (formerly uidR) and two newly discovered genes, gusB and gusC. gusB encodes a specific transport system - the glucuronide permease that allows E. coli to actively accumulate an extraordinarily wide spectrum of Ã-glucuronides. This gene has now been cloned and sequenced and its biochemistry and molecular biology are being studied (Jefferson, in preparation; Jefferson and Liang, in preparation). The glucuronide permease is an integral membrane protein of molecular mass about 49 kDa. The primary amino acid sequence shows 25% identity to another well characterized carbohydrate transport system of E. coli, the melibiose permease, encoded by the melB locus. The glucuronide permease is most probably a sodium/hydrogen symporter and functions by coupling the transport of a glucuronide substrate to the transport of a proton or sodium ion, thus utilizing the transmembrane proton and/or ionic gradient that exists across the plasma membrane of almost all living cells.

One of the most exciting features of the permease is the extraordinarily broad spectrum of substrates that can be taken up. While comparable permeases such as the lactose permease or the melibiose permease seem relatively restricted to transporting compounds with structures not too different from the disaccharides or glycosides that they normally encounter, the glucuronide permease will accumulate many different compounds, including aliphatic glucuronides, phenolic glucuronides, coumarin glucuronides, indole glucuronides, phenoxazine glucuronides, and even auxin glucuronides. These compounds include highly fluorogenic and colorogenic substrates such as 4-methylumbelliferyl glucuronide (MUG) and 5-bromo-4-chloro-3-indolyl glucuronide (X-Gluc), and even intensely fluorogenic, large, substituted three ring compounds such as resorufin glucuronide. The permease is, however, highly specific for the sugar residue, not recognizing galactosides, arabinosides, glucosides or other conjugates. This extraordinary combination of specificity with promiscuity, while unprecedented, is consistent with our teleological perception of the role of GUS and the glucuronide

permease in the natural history of <u>E. coli</u> - that is the
acquisition and metabolism of the wide range of novel
glucuronides that are presented to it through the detoxication
pathway that results in excretion in the bile.

Since the glucuronide permease is a single polypeptide,
and since it functions by energetic coupling to a
transmembrane gradient that will exist across almost all cell
membranes, it is therefore quite reasonable to think that the
glucuronide permease will function in heterologous systems
such as plants. We are embarking on a major research project
to engineer the glucuronide transporting capabilities into
plants. This project will progress by both conventional
molecular biological approaches and through the use of yeast
genetics. Because the information available about complex
processes such as plasma membrane targeting and membrane
gradient coupling is limited, we are unable to directly design
strategies that are certain to function with a bacterial
protein. Instead, or rather concurrently, we will take a
genetic approach, by setting up genetic selections for
permease function in eukaryotes in the fission yeast
<u>Schizosaccharomyces pombe</u>. We believe that such an approach
will yield a successfully integrated and coupled permease
together with the information necessary to ensure its
functioning in plants. Once we are able to obtain plants in
which the glucuronide permease is working, we will prepare
homozygous permeant seed to distribute to researchers who wish
to use the GUS system in vivo.

IF YOU CAN'T BRING MOHAMMED TO THE MOUNTAIN, BRING THE MOUNTAIN
TO MOHAMMED - SECRETION OF GUS FROM LIVING CELLS

An interesting complement, and a short term solution to
the problem of detecting GUS activity in live cells is the
possibility of secreting GUS from the cell. If GUS, normally
a cytoplasmic enzyme, could be transported through a plasma
membrane and localized on the exterior of the cell surface,
then substrate access would not be a serious limitation to the
detection of GUS-expressing cells. For example, if GUS were
elaborated between the cell membrane and the cell wall of a
plant cell, then substrate could simply be sprayed or coated
onto the surface of the tissue to localize GUS activity, and
the cell's viability would not necessarily be compromised.
This would obviate genetic selections based on differential
growth characteristics, such as typified by antibiotic
selection. Transformants could then be screened, obtained and
cultured directly without recourse to laborious, lengthy and
often impractical antibiotic selections.

Unlike bacterial Ã-galactosidase, which is notoriously
difficult, if not impossible to translocate through membranes,
GUS will apparently traverse membranes readily. Tony Kavanagh

clearly demonstrated (Kavanagh, et al., 1988) that the transit peptide together with some additional sequences from the chlorophyll a/b binding protein of tobacco, when fused to GUS is sufficient to allow efficient transport of the fusion protein into tobacco chloroplasts, both _in vitro_ and _in vivo_. Additionally, Udo Schmitz showed a similar phenomenon for transport of GUS across mitochondrial membranes of transgenic tobacco plants (Schmitz and Lonsdale, 1989) and yeast (Schmitz, et al., 1989).

The targeting of GUS through the plasma membrane via the endoplasmic reticulum was studied by Gabriel Itturiaga (Itturiaga, et al., 1989) by fusion of the patatin signal peptide to the GUS coding sequences. The patatin signal peptide targets patatin to presumptive vacuoles, via the endoplasmic reticulum system. His experiments clearly demonstrated that GUS could be efficiently transported into the endoplasmic reticulum, _in vitro_ and _in vivo_, and there processed into the predicted mature peptide. He also showed that GUS is stable to proteolysis within this organelle. However, an unexpected and interesting finding was made. When GUS was transported into the endomembrane system, virtually all GUS enzyme activity was lost, although by Western blot analysis the protein was still present and stable. Experiments with inhibitors of N-linked glycosylation showed that, although GUS is not normally a glycosylated enzyme in bacteria, it was being glycosylated and thereby inactivated in the ER. Interestingly, when N-linked glycosylation was blocked using the antibiotic tunicamycin, full GUS activity was regained. Gabriel's experiments clearly indicate that GUS can be translocated efficiently through the ER targeting system, and, if glycosylation can be avoided, retain full activity in the process. Inspection of the GUS amino acid sequence shows that there are two potential sites for N - linked glycosylation, as defined in other systems; Asn - X - Ser/Thr. These residues are the likely site of attachment of the glycosyl residues in the endoplamic reticulum, and so elimination of them by site-directed mutagenesis should allow retention of GUS activity after transport. This is underway, and preliminary experiments indicate that the change of an aparagine to a glutamine at these locations does not adversely affect GUS activity. Following this approach, we should be able to achieve secretion of GUS soon. However, unless we are very lucky and GUS continues to cooperate, optimization of the process of secretion may take somewhat longer, and may benefit from a genetic approach, perhaps in _S. pombe_ or another genetically facile eukaryote. When these constructions are finished, they will greatly accelerate transformation experiments in almost all plant species and by most methods - from particle acceleration to _Agrobacterium_-mediated transformation. They may additionally be extraordinarily useful in achieving efficient transformation of obligate

biotrophs, such as mycorhizzal fungi or downy mildews for which
there are no culture conditions or selections available.

FUSION GENETICS - POSITIVE AND NEGATIVE SELECTIONS FOR GENE FUSION ACTION

Understanding the networks of controlling factors that
regulate and mediate gene expression is clearly a principal
aim in modern biology. Achieving it is also a prerequisite
for manipulating these control networks. However, the
increasing complexity of the systems being studied has not
been paralleled by increasingly sophisticated methodologies
and approaches. The methods widely used to trace controlling
factors such as "trans-acting factors" tend to be linear in
their approach, focussing, for instance, on DNA binding
proteins. While there is certainly information to be gained
from these approaches, general solutions to networks of gene
control have historically been successful largely through the
use of genetics - that is the use of mutations that disrupt
control to focus subsequent molecular analysis. This success,
however, has been largely limited to organisms with very
facile genetic methods, such as bacteria, fungi, and to some
extent fruit flies and nematodes. Detailed genetic analysis of
higher organisms such as vascular plants, especially those
important to agriculture, is greatly restricted by the
limitations imposed by large genomes, long generation times,
difficult breeding systems and cumbersome logistics of
handling sufficiently large numbers. Even plants such as
Arabidopsis, tomato and maize with their excellent genetic and
cytogenetic properties are largely refractile to saturation
genetics.

Fusion genetics is a new and promising branch of
transgenic science with roots in early work in bacteria that
will find its greatest uses in difficult systems such as
plants and animals (Beckwith et al., 1967; Miller et al.,
1970; Bonner et al., 1984). The basic premise of fusion
genetics is that through creative development and use of
positive and negative selections, we can obtain rare genetic
mutants that cause inappropriate regulation of a previously
introduced and characterized gene fusion. Thus, if a gene
fusion is constructed that places a "responder" gene under the
control of gene sequences that are responsive to the network
that we wish to study, and we can conditionally select
mutants that are abnormal in their regulation, we can find
genes and gene products whose function is essential to that
regulation. By appropriate experimental design, either
"cis-acting" or "trans-acting" mutations can be obtained.
Development of general methods for conditional positive and
negative selections based on gene fusion action must be an
important goal, to allow scientists to pursue gene regulation
questions in a non-linear sophisticated manner, so enhancing

questions in a non-linear sophisticated manner, so enhancing the genetic methods presently available. However, this exciting approach will only be used when tools are available to implement it reliably and simply.

THE USE OF GUS IN FUSION GENETICS

As mentioned earlier, fusion genetics requires the development of selections tools - ways of obtaining growth, survival or fertility that is strictly dependent on the action of the responder gene. The basic structure of GUS substrates, Ã-glucuronides, and the ability to prepare thousands of them biosynthetically by simply feeding precursor to rabbits or undergraduates, and collecting the urine, may allow an interesting avenue for this goal.

A substrate for GUS does not have to produce a colored or fluorescent product upon enzyme action - it could also produce a bioactive compound, such as a toxin or growth regulator. Assuming that the conjugate - the combination of the bioactive aglycone and glucuronic acid in a hemiacetal linkage - does not show any bioactivity, this would give rise to the bioactivity only in situations with GUS present. Since normal plants do not have any GUS activity, this would restrict the bio-activity to those cells, tissues, plants or developmental stages that were genetically modified to produce GUS by the activity of a gene fusion.

This strategy is in fact the exact reverse of the one that has evolved within vertebrates, as described above. Instead of de-toxifying or de-activating a foreign compound by formation of a glucuronide conjugate, as is done by most vertebrates, we will be activating the conjugate by treatment with GUS. In this case, however, GUS activity can be provided by the expression of the GUS gene fusion. This will therefore allow the molecular biologist, in designing a spatially or temporally regulated GUS fusion, to direct the biological effects of a compound to the time and place of GUS activity. This activity can have many possible outcomes. If the aglycone is a plant growth regulator, we can envisage directing plant growth regulator activity to only those cells, tissues, plants or stages in which GUS is active. This could therefore provide a positive selection for the activity of the gene fusion. Similarly, if the aglycone is a toxin, such as a cell-lethal antibiotic or a herbicide, we should be able to direct a lethal effect to those same cells, tissues, whole plants or stages, thereby achieving a negative selection. The latter example will then allow the exposition of a simple fusion genetics scenario. If we can readily kill all GUS positive plants (for instance germinating seedlings) with a GUS-dependent herbicide-glucuronide conjugate, only those plants that no longer express GUS will survive. Hence, if we

are looking for mutations, even very rare mutations, that
eliminate GUS activity by interfering with the expression of a
previously introduced and characterized GUS fusion, they
should be the only survivors from such a treatment.
Strategies could be devised in which both cis-acting or
trans-acting mutations could be obtained. One of the many
beauties to this approach is that we are not restricted to any
one or even a few candidate compounds. Thousands of
compounds, with diverse biological and chemical properties can
be conjugated as glucuronides - there are already more than
three thousand Ã-glucuronides known in the literature, and
almost limitless possiblities for synthesis or biological
preparation of more.

GUS GENE FUSIONS TO MODIFY BIOCHEMISTRY AND PHYSIOLOGY OF THE
PLANT

        To demonstrate the utility of this approach, we had
synthesized a conjugate of tryptophol with glucuronic acid.
Tryptophol, indole-3-ethanol, is a weak auxin in many plant
systems. In certain media formulations, tryptophol as an
auxin can function to retard the senescence of tobacco leaf
discs. After several weeks, the discs are still green on auxin
containing media, but brown and necrotic on media lacking
auxin. The conjugate, tryptophyl-Ã-glucuronide, shows no
auxin activity in a tobacco leaf disk bioassay when assayed
using normal untransformed material. However, when the leaf
disks were derived from GUS - expressing tobacco plants,
previously transformed with a CaMV 35S - GUS fusion, the discs
remained green and healthy. This experiment, while
preliminary, showed clearly that the approach of using
synthetic conjugates to direct GUS-specific biological effects
was sound. However, tryptophol is a very weak auxin, and was
only chose for its ease of synthesis. Many other compounds
are now being synthesized to achieve both negative and
positive effects, for instance cycloheximide-glucuronide,
cytokinin glucuronides etc.

        It is clear that an offshoot of the goal of developing
fusion genetics is the use of these selections to modify plant
growth and development in a manner dependent on the gene
fusion action. Thus, if the gene fusion is only active in a
subset of cells, tissues or stages of plant development, it
will be possible to direct the effects of novel selective
agents, such as the plant growth regulators or herbicides to
those places and times. Selective expression of GUS, for
instance restricting its expression to anthers and pollen,
coupled with the use of bioactive substrates such as
herbicide-glucuronides, could give rise to numerous routes for
agronomic modifications such as male-sterility. This will
facilitate a potentially very productive avenue of plant
growth regulation and crop manipulation.

SUMMARY

There are clearly many directions in which the further development of the GUS gene fusion system can progress. Some of these have been outlined above, but others can be imagined. There are no reasons to limit our conceptions of the use of GUS gene fusions to analysis and manipulation of single genes. We can envision numerous marked genes - perhaps with several new fusion systems - giving valuable information about gene interaction, or population structure. The study of plant - pathogen and plant symbiont interactions can progress rapidly with simple quantitative markers for genes and individuals. We can imagine ways of using gene fusions to report on crop physiology or other complex phenotypes, thereby enhancing the accuracy and speed of screening. Introduction of the biosynthetic pathway for glucuronide detoxification by expressing genes for the UDP-glucuronyl transferases in plants may result in novel mechanisms for plants to deal with xenobiotics such as insecticides or herbicides. Synthesis of substrates, which until now has been performed chemically - resulting in expensive compounds - can be done biosynthetically. This should make the system not only the most powerful gene fusion system for agriculture, but also the most accessible.

ACKNOWLEDGEMENTS

I thank Andrzej Kilian, David Wolfe and Gerry Roper for their huge and selfless contributions to the field trial; Mike Bevan for his technical collaborations - especially in helping me into the world of plant transformation; Dick Flavell, Phil Dale, Peter Day, Alan Thompson for their assistance in getting the field trial started; Bob Negus, John "Geordie" Cowan and Dave Garlic for keeping the potatoes alive and helping kill them in the end; Helen McPartlan and Elaine Atkinson for skillful assistance with the tissue culture; Weijun Liang for his recent contributions to the glucuronide permease work; Kate Wilson for her work on the gus operon and help with editing; Finally I owe great thanks to Andrzej Kilian, Sara Melville, Steve Hughes and Kate Wilson for their -sometimes unwitting - help in the maturation of my ideas about science, agriculture and responsibility. I was supported by U.S. NIH Grant GM-10789-02. I am also grateful to the Gatsby Foundation and the Agricultural Genetic Company (AGC) for financial assistance.

REFERENCES

Beckwith, J.R., Signer, E.R., and Epstein, W., 1967, Transposition of the lac region of E. coli, Cold Spring Harbor Symp. Quant. Biol., 31:393.

Bevan, M.W., 1984, Binary Agrobacterium vectors for plant
    transformation, Nucl. Acids Res., 12: 8711-8721.

Bevan, M.W., Barker, R., Goldsbrough, A., Jarvis, M.,
    Kavanagh, T., and Iturriaga, G., 1986, The structure and
    transcription start site of a major potato tuber protein
    gene, Nucl. Acids Res., 14: 4625-4638.

Bonner, J.J., Parks, C., Parker-Thornberg, J., Mortin, M.A.,
    and Pelham, H.R.B., 1984, The use of promoter fusions in
    Drosophila genetics: isolation of mutations affecting the
    heat shock response, Cell, 37:979-991.

Bourque, J.E., Miller, J.C. and Park, W.D., 1987, Use of an
    in vitro tuberisation system to study tuber protein gene
    expression, In vitro Cell. Devel. Biol., 23: 381-386.

Dean, C., Jones, J., Favreau, M., Dunsmuir, P. and Bedbrook,
    J., 1988, Influence of flanking sequences on variability
    in expression levels of an introduced gene in transgenic
    tobacco plants, Nucl. Acids Res., 16: 9267-9283.

Dutton, G.J., ed., 1966, Glucuronic Acid, Free and Combined,
    Academic Press, New York.

Dutton, G.J., 1980, Glucuronidation of Drugs and Other
    Compounds, CRC Press, Boca Raton, Florida.

Hinchee, J.A.W, Connor-Ward, D.V., Newell, C.A., McDonnell,
    R.E., Sato. S.J., Gasser, C.S., Fischhoff, D.A., Re, D.B.,
    Fraley, R.T., and Horsch, R.B.,1988, Production of
    transgenic soybean plants using Agrobacterium-mediated DNA
    transfer, Bio/technology, 6:915-922.

Iturriaga, G., Jefferson, R.A., and Bevan, M.W., 1989,
    Endoplasmic reticulum targeting and glycosylation of
    hybrid proteins in transgenic tobacco, The Plant Cell,
    1:381-390.

Jefferson, R.A.,1985, DNA Transformation of Caenorhabditis
    elegans: Development and Application of a New Gene Fusion
    System, Ph.D. Dissertation, University of Colorado at
    Boulder.

Jefferson, R.A., Burgess, S.M., and Hirsh, D.,1986,
    Ã-Glucuronidase from E. coli as a Gene Fusion Marker,
    Proc. Natl. Acad. Sci. USA, 83: 8447-8451.
    Jefferson, R.A., Klass, M., Wolf, N., and Hirsh, D.,
    (1987), Expression of Chimeric Genes in Caenorhabditis
    elegans, J. Mol. Biol., 193: 41-46.

Jefferson, R.A.,1987, Assaying Chimeric Genes in Plants: The GUS Gene Fusion System, <u>Plant Molecular Biology Reporter</u>, 5: 387-405.

Jefferson, R.A., Kavanagh, T.A. and Bevan, M.W., 1987, GUS fusions: Ã-glucuronidase as a sensitive and versatile gene fusion marker in higher plants, <u>EMBO J.</u>, 6., 3901-3907.

Jefferson, R.A., 1988, Plant report genes: the GUS gene fusion system, <u>in</u>: "Genetic Engineering, Vol. 10," J.K. Setlow, ed., Plenum Press, New York.

Kavanagh, T.A., Jefferson, R.A., and Bevan, M.W., 1988, Targeting a Foreign Protein to Chloroplasts Using Fusions to the Transit Peptide of a Chlorophyll a/b Protein, <u>Mol. Gen. Genet.</u>, 215: 38-45.

Klein, T.M., Gradziel, T., Fromm, M.E., and Sanford, J.C., 1988, Factors influencing gene delivery into Zea mays cells by high-velocity microprojectiles, <u>Bio/Technology</u>, 6:559-564.

Levvy, G.A. and Conchie, J. 1966, Ã-Glucuronidase and the hydrolysis of glucuronides, <u>in</u>: "Glucuronic Acid, Free and Combined," G.J. Dutton, ed., Academic Press, New York, 301.

Masson, P., and Fedoroff, N.V.,1989, Mobility of the maize Suppressor-mutator element in transgenic tobacco cells, <u>Proc. Natl. Acad. Sci. USA</u>, 86:2219-2223.

McCabe, D.E., Swain, W.F., Martinell, B.J., and Christou, P. 1988, Stable transformation of soybean (Glycine max) by particle accelleration, <u>Bio/Technology</u>, 6:923-926.

Mignery, C.A., Pikaard, C.S. and Park, W.D., 1988, Molecular characterization of the patatin multigene family of potato, <u>Gene</u>, 62: 27-44.

Miller J.H., Reznikoff, W.S., Silverstone, A.E., Ippen, K., Signer, E.R., and Beckwith, J.R.,1970, Fusions of the lac and trp regions of the Escherichia coli chromosome, <u>J. Bacteriol.</u>, 104:1273.

Paiva, E., Lister, R.M.,and Park, W.D.,1983, Induction and accumulation of major tuber proteins of potato in stems and petioles., <u>Plant Physiol.</u>, 71, 616-618.

Park, W.D., 1986, in: "Potato Physiology," P. Li., ed., Academic Press.

Roberts, I.N., Oliver, R.P. Punt, P.J. and van den Hondel,
     C.A.M.J.J.,1989, Expression of the Escherichia coli
     Ã-glucuronidase gene in industrial and phytopathogenic
     filamentous fungi, <u>Curr. Genet.</u>, 15: 169-180.

Schmitz, U.K., Jefferson R.A., and Lonsdale, D.M., 1989, GUS as
     a gene fusion marker in the yeast, Saccharomyces
     cerevisiae, <u>Curr Genet.</u>, In press.

Schultz, M., and Weissenbock, G., 1987, Partial purification
     and characterization of a luteolin-triglucuronide-
     specific Ã-glucuronidase from rye primary leaves - <u>Secale
     cereale</u>, <u>Phytochemistry</u>, 26:933-938.

Sheerman, S., and Bevan, M.W., 1988, A rapid transformation
     method for Solanum tuberosum using binary Agrobacterium
     tumefaciens vecto, <u>Plant Cell Reports</u>, 7: 13-16.

Stoeber, F.,1957, Sur la biosynthese induite de la
     Ã-glucuronidase chez Escherichia coli, <u>C.R. Acad. Sci.</u>,
     244:950.

Stoeber, F., 1961, Etudes des proprietes et de la biosynthese
     de la glucuronidase et de la glucuronide-permease chez
     Escherichia coli, These de Docteur es Sciences, Paris.

Twell, D. and Ooms, G., 1988, Structural diversity of the
     patatin multigene family in potato cv. Desiree, <u>Mol.Gen.
     Genet.</u>, 212, 325-336.

REGULATION OF PLANT GENE EXPRESSION BY AUXINS

Tom J. Guilfoyle, Bruce A. McClure, Gretchen Hagen,
Christopher Brown, Melissa Gee, and Antonio Franco
Department of Biochemistry
University of Missouri
Columbia, MO 65211

Application of the plant hormone auxin to intact plants,
isolated plant parts, and cultured cells can have dramatic effects
on subsequent growth and developmental processes (e.g., cell
division, cell enlargement, cell differentiation, and
organogenesis). It is generally believed that endogenous auxins
play roles in plant growth and development similar to those
observed when exogenous auxins are applied to plants or plant
parts. Recent results from plants transformed with *Agrobacterium
tumefaciens* auxin-biosynthetic genes supports the above
contention (Klee et al., 1988). The mechanism(s) by which auxins
affect plant growth and developmental processes is largely a
mystery, but at least some of the processes are thought to involve
changes in gene expression (Guilfoyle, 1986). In recent years, a
number of cDNA clones for auxin-responsive mRNAs have been
identified and characterized (Baulcombe and Key, 1980; Czarnecka
et al., 1904; Hagen et al., 1984; McClure and Guilfoyle, 1987;
Theologis et al., 1985; van der Zaal et al., 1987; Walker and Key,
1982). The genes encoding these mRNAs have also been identified
in some cases (Ainley et al., 1988; Czarnecka et al., 1988; Hagen
et al., 1988; McClure et al., 1989). Experiments to determine the
functions of the auxin-responsive gene products and the signal
transduction pathways involved in auxin-induced gene expression
are only now being initiated. Here, we describe some the auxin-
responsive mRNAs and genes that have been identified.

Gene Manipulation in Plant Improvement II
Edited by J. P. Gustafson
Plenum Press, New York, 1990

cDNA CLONES AND GENES

Complementary DNA clones to auxin-responsive mRNAs have been isolated from soybean (*Glycine max*) (Baulcombe and Key, 1980; Hagen et al., 1984; McClure and Guilfoyle, 1987; Walker and Key, 1982), pea (*Pisum sativum*) (Theologis et al., 1985), and tobacco (*Nicotiana tabacum*) (van der Zaal et al., 1987). The clones isolated from soybean and pea are summarized in Table 1. The cDNA clone probes identify mRNAs that are increased or decreased in abundance following auxin administration. In most cases, the cDNA clones have been partially or totally sequenced, but the deduced amino acid sequences have not revealed the function of any auxin-responsive gene product. At least in one case, antibodies have been raised to a fusion protein constructed from an auxin-responsive cDNA (Wright et al., 1987), and these antibodies have been used to characterize and localize the protein in untreated and auxin-treated organs and cells.

Table 1. cDNA clone probes for mRNAs in soybean and pea that accumulate in response to auxin

cDNA Clone[1]	mRNA Size (nucleotides)	Fold Inducible with Auxin[2]	Organ Specificity[3]
pGH1	1700	2-3	EHS, BHS, P
pGH2/4	1000	2-25	EHS, BHS, P
pGH3	2400	15-60	EHS, BHS, P, R
pJCW1	1100	3-5	EHS>BHS>H
pJCW2	1050	5-8	EHS>BHS=H
SAURs	500	25-50	EHS, EES>BHS
pIAA4/5	950	50	EES>BES>H
pIAA6	850	50	EES>BES>H

[1] Data for pGH clones from Hagen et al. (1984), pJCW clones from Walker and Key (1982), SAURs from McClure and Guilfoyle (1987, 1989a,b), and pIAAs from Theologis et al. (1985).

[2] The fold induction with auxin varies with the type of auxin, time of exposure, and concentration of auxin applied to seedlings or organ sections.

[3] The organs of primary expression are listed. EHS = elongating hypocotyl section, BHS = basal hypocotyl section, P = plumule, R = root, EES = elongating epicotyl section, BES = basal epicotyl section, and H = hook.

In a few cases, the genes corresponding to the cDNAs have been isolated and sequenced (Ainley et al., 1988; Czarnecka et al., 1988; Hagen et al. 1988; McClure et al., 1989; Fig.1). Some of these genes have introns, while others do not. Analysis of putative promoter regions of these genes has revealed some sequence homology among genes in the same family, but little sequence homology in unrelated auxin-responsive genes. It appears, therefore, that detailed mutational analysis of the promoters will be required to identify auxin-responsive elements (i.e., assuming such elements exist) in different auxin-regulated genes.

AUXIN-SPECIFIC INDUCTION OF mRNAS

In general, studies have shown that mRNAs which hybridize to the cDNA clone probes listed in Table 1 are specifically induced by a variety of naturally occurring and synthetic auxins. Compounds with structures similar to auxins, but which lack auxin activity, generally show little or no induction of auxin-responsive mRNAs. The other plant hormones, cytokinins, gibberellic acids, abscisic acid, and ethylene, fail to induce the mRNAs by themselves, and when combined with auxin, they do not increase or decrease the auxin-inducibility. The pGH2/4 (Hagen and Guilfoyle, 1985; Hagen et al., 1988) and pCE54 (Edelman et al., 1988; Czarnecka et al., 1988) cDNA clones are exceptions to the auxin-specificity described above. These clones hybridize to the same mRNA, and this mRNA increases in abundance when soybean plants or excised organs are exposed to heat shock, abscisic acid, and certain heavy metals. In addition, the mRNA is induced by a number of non-auxin analogs that have structures similar to synthetic auxins. This mRNA, unlike the other identified auxin-responsive mRNAs, may represent an mRNA that is induced by a variety of stress agents, where high auxin concentration is one of these stress agents.

Growth promoting compounds, such as fusicoccin, do not induce accumulation of the auxin-responsive mRNAs (McClure and Guilfoyle, 1987; Theologis et al., 1985; Walker et al., 1985). Two cDNA clones from pea (Theologis et al., 1985) and a family of three

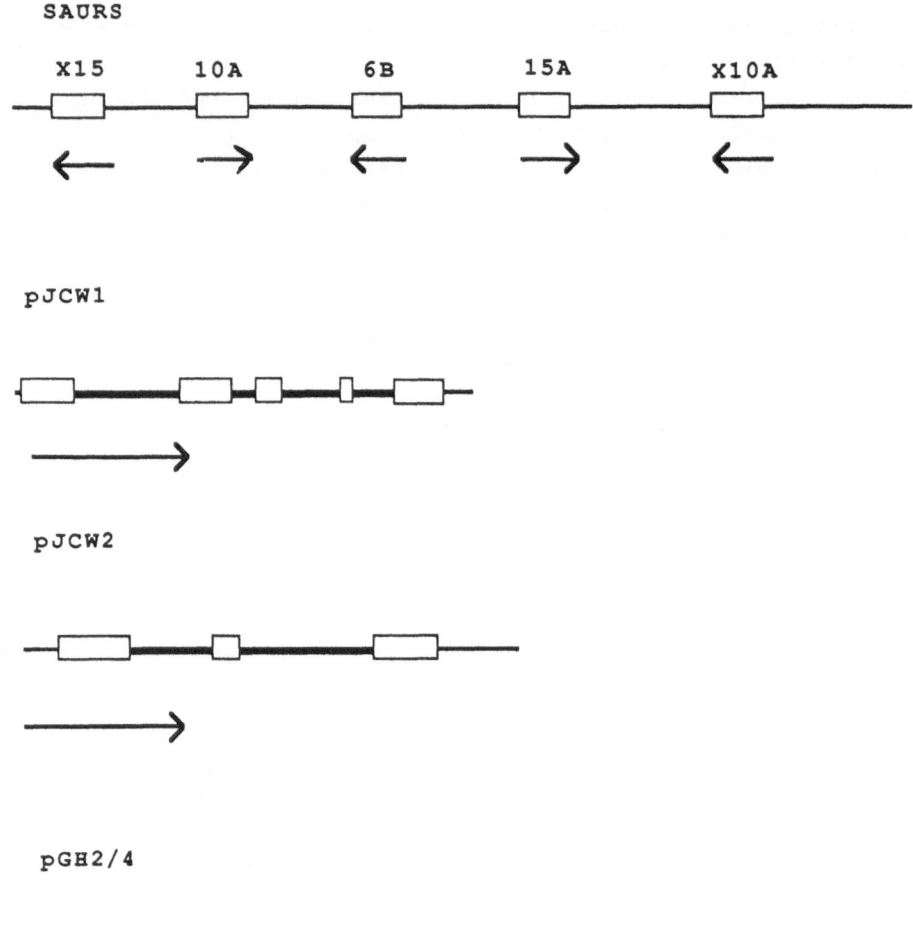

Fig. 1.   Auxin responsive genes in soybean that have been
          sequenced. Exons are shown by open boxes and introns
          are shown by thick lines. Flanking regions of genes
          are shown by thin lines. Transcripts and direction of
          transcription are shown below each gene.

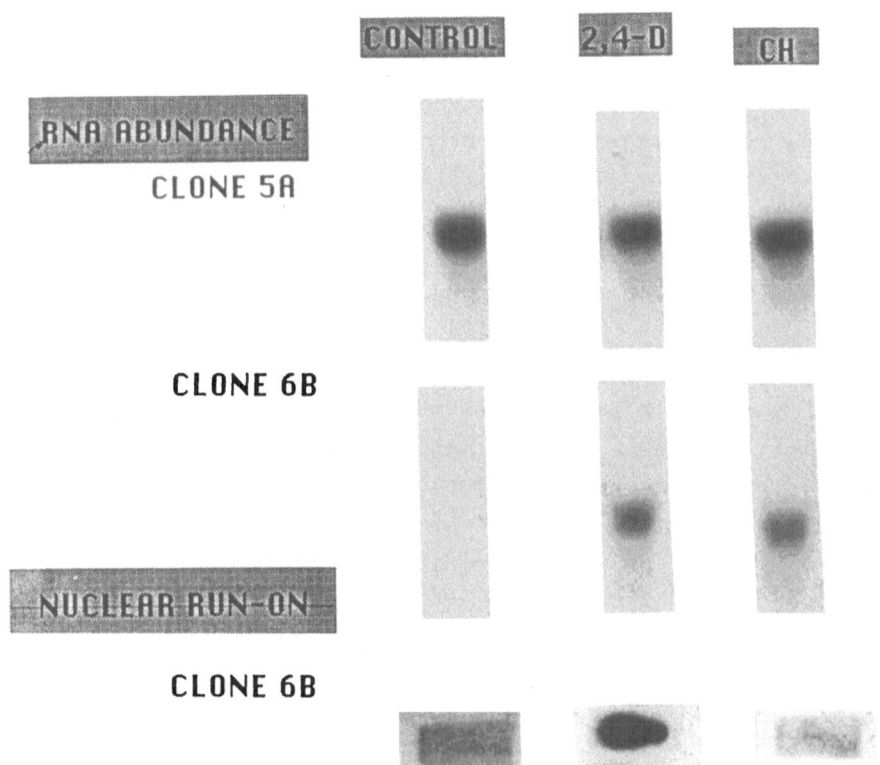

Fig. 2. Accumulation of SAUR 6B and nuclear run-on
        transcription rates in response to cycloheximide and
        2,4-D. Elongating soybean hypocotyl sections were
        preincubated for 4 hours and then treated for 1 hour
        with no addition (control), 50 uM 2,4-D, or 10 uM
        cycloheximide (CH). The 5A clone is a control probe
        that hybridizes to a mRNA of 0.5 kb which does not
        respond to auxin addition. The auxin-induced probe
        used here corresponds to the cDNA clone to SAUR gene 6B
        shown in Figure 1.

cDNA clones from soybean (McClure et al., 1989) hybridize to mRNAs that are induced by cycloheximide in the absence of added auxin (Fig. 2). In the presence of auxin, the soybean mRNAs are "super-induced" by cycloheximide treatment (A. Franco and T. Guilfoyle, unpublished results). In soybean, it has been additionally shown that while cycloheximide induces accumulation of these mRNAs, this protein synthesis inhibitor does not cause increased transcription rates on the genes encoding these mRNAs. This latter observation suggests that cycloheximide acts by stabilizing the life times of the auxin-inducible RNAs.

DOSE-RESPONSE

With most cDNA probes (see Table 1), mRNAs begin to accumulate at 10^{-8} to 10^{-6} M auxin (Hagen and Guilfoyle, 1985; McClure and Guilfoyle, 1987; Theologis et al., 1985). The dose-response profiles for induction of many of the auxin-responsive mRNAs display a nearly linear increase in mRNA abundance over several orders of log increase in auxin concentration. Many growth and developmental responses induced by exogenous hormones show similar dose-response profiles (Kende and Gardner, 1976). In a few cases, mRNA accumulation levels off or declines at high auxin concentrations (i.e., equal to or greater than 10^{-3} M), but in other cases, mRNA abundance continues to increase up to the highest auxin concentrations tested (i.e., in excess of 10^{-3} M). It should be noted that these dose-responses do not appear similar to classical hormone receptor-ligand mediated interactions (e.g., steroid receptors), where responses tend to be saturated over a couple of orders of log increase in hormone concentration. At this point, it is unclear why the dose-responses observed with the auxin-responsive mRNAs deviate from those expected for hormone-receptor mediated interactions. It is possible, however, that the complexity of tissue-specific gene expression, auxin transport systems, and auxin metabolism may all contribute to the dose-response profiles observed. With complex organs and tissues, it is difficult to estimate the effective concentration of auxin at the site of auxin action.

mRNA ACCUMULATION KINETICS

Most of the auxin-responsive mRNAs identified accumulate rapidly after auxin administration (Baulcombe and Key, 1980; Hagen et al., 1984; McClure et al., 1987; Theologis et al., 1985; van der Zaal et al., 1987; Walker and Key, 1982). However the kinetics of mRNA accumulation clearly separates the very rapidly induced mRNAs from the rapidly induced mRNAs. The mRNAs hybridizing to the bulk of the cDNA clones isolated begin to accumulate within about 15 minutes after auxin addition, but they continue to increase in abundance for one or more hours following hormone induction (Fig. 3).

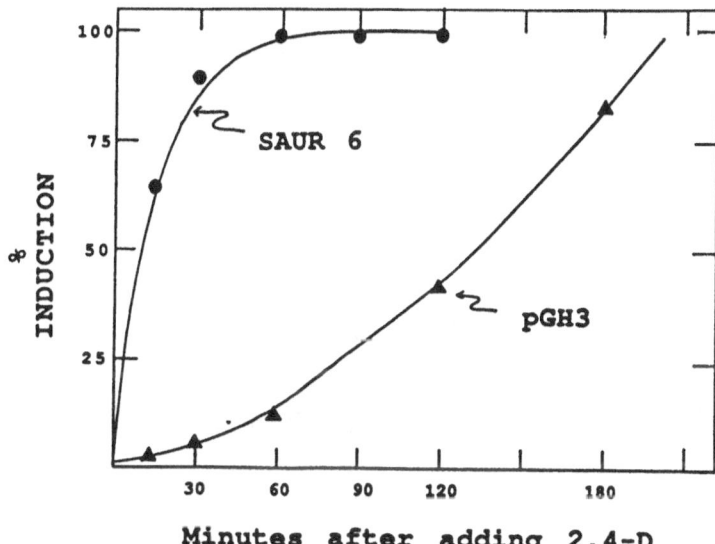

Minutes after adding 2,4-D

Fig. 3.   Kinetics for pGH3 and SAUR 6 expression after application of 50 uM 2,4-D to elongating hypocotyl sections of soybean. RNA abundance was quantitated by densitometry of slot blots for pGH3 (Hagen et al., 1984) and gel blots for SAUR 6 (McClure and Guilfoyle, 1987). pGH3 represents a rapidly induced mRNA and SAUR 6 represents a very rapidly induced mRNA.

These represent the rapidly induced class of auxin-responsive
mRNAs. The very rapidly induced class of mRNAs begin to
accumulate within about 2.5 minutes after auxin administration,
reach half maximal induced steady state levels within about 10
minutes, and achieve steady state levels at about 30 minutes
(McClure et al., 1987). These latter type of mRNA accumulation
kinetics might be expected for RNAs that could potentially play
some role in the very rapid cell extension responses (Fig. 4)
which are proposed to be mediated by auxin action (Evans, 1985;
Vanderhoef, 1980).

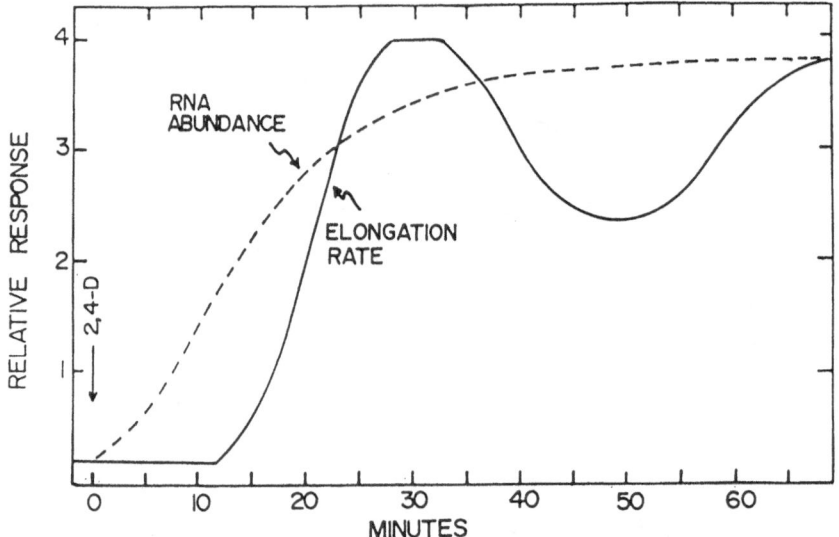

Fig. 4.   Kinetics for SAUR 6, 10A, and 15 mRNA accumulation and
           cell extension rate increase following application of
           50 uM 2,4-D to elongating soybean hypocotyl sections.
           The mRNA accumulation kinetics are from McClure and
           Guilfoyle (1987), and the growth kinetics are from
           Vanderhoef (1980).

It is possible that the rapidly induced class of mRNAs could still
play roles in cell extension, but it is more likely that these

would be roles in sustained cell elongation (which is achieved an hour or more after auxin addition) as opposed to initial cell elongation (which begins about 10-15 minutes after auxin application). At present, however, there is no direct evidence suggesting that any auxin-responsive mRNA or its encoded protein plays a role in cell extension or any other auxin-induced developmental or growth processes.

AUXIN-INDUCED TRANSCRIPTION

The mechanism(s) involved in auxin-induced mRNA accumulation could be transcriptional activation of auxin-regulated genes, posttranscriptional processing of the mRNAs, transport of the mRNAs from the nucleus to the cytoplasm, or alteration in the life times of the mRNAs. *In vitro* run-on transcription experiments with isolated nuclei, indicate that the SAURs and pGH mRNAs accumulate, at least in part, due to increased transcription rates on their genes (Hagen and Guilfoyle, 1985; McClure et al., 1989). Increased transcription rates in isolated nuclei can be detected within 5-10 minutes after auxin administration to excised organs. While nuclei isolated from both plumules and hypocotyls show a 10-fold increase in transcription rates for pGH sequences following auxin application, only hypocotyl nuclei show a several-fold increase in transcription rates for SAUR sequences after hormone addition. For the pGH probes, the kinetics for transcriptional activation is much more rapid than the kinetics for mRNA accumulation (Hagen et al., 1984; Hagen and Guilfoyle, 1985). For the SAUR probes, however, the kinetics for transcriptional induction appears to lag behind the kinetics for mRNA accumulation (McClure and Guilfoyle, 1987; McClure et al., 1989; Fig. 5). A significant change in transcription of SAUR genes occurs at about 10 minutes after auxin application, but mRNA abundance changes can be detected as early as 2.5 minutes following auxin stimulation, and half maximal steady state mRNA levels are reached in about 10 minutes after hormone addition. The lag observed for SAUR gene transcription compared to SAUR accumulation might be explained by technical difficulties involved

with transcriptional run-on experiments at very early time points
after auxin addition (e.g., insufficient radiospecific activity
of nuclear transcripts to detect transcriptional activation before

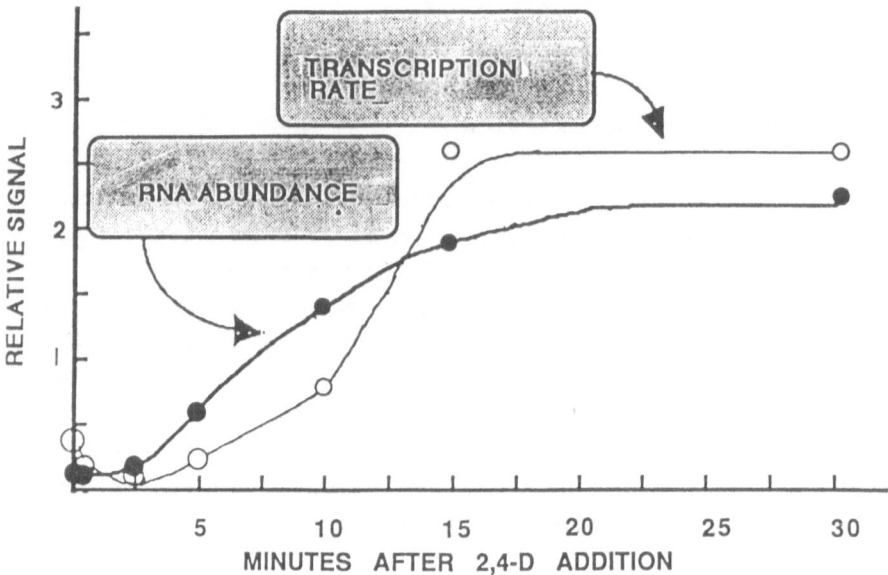

Fig. 5.   Kinetics for SAUR 6, 10A, and 15 mRNA accumulation and
            transcription rate increase following application of
            50 uM 2,4-D to elongating soybean hypocotyl sections.
            RNA accumulation kinetics and quantitation are as in
            Figure 4. Transcription rates were determined using
            run-on *in vitro* transcription with isolated nuclei.

10 minutes of auxin treatment). On the other hand if the lag is
real, this might suggest that the earliest effect of auxin on SAUR
gene expression might result from mRNA stabilization.

TISSUE- AND ORGAN-SPECIFIC GENE EXPRESSION

The relative abundance of SAURs, pJCW1, and pJCW2 in
different organs of soybean has been determined by RNA blotting
experiments. In three day old etiolated whole seedlings, the
SAURs are most abundant in the elongating zone of the soybean
hypocotyl, less abundant in basal or mature regions of the
hypocotyl, and nearly undectable in the hook region (e.g., zone of
cell division) of the hypocotyl and the root (McClure and
Guilfoyle, 1987). The mRNAs which hybridize to pJCW1 and 2 are
also most abundant in the elongating zone of the hypocotyl, but
can also be detected in the hook region as well as the basal
region of the hypocotyl (Walker and Key, 1982). In older
etiolated seedlings (e.g., 7 day old seedlings), the SAURs can
also be detected in young leaves or plumules and in epicotyls.
When the elongating zone of the three day old soybean hypocotyl is
excised from the plant and incubated in a buffered medium without
auxin, the SAURs and pJCW mRNAs decrease in abundance over a
period of a few hours (Walker and Key, 1982; McClure and
Guilfoyle, 1987). When auxin is added to the medium, the mRNAs
rapidly increase in abundance. With the SAURs, induction with
2,4-D to excised basal hypocotyl sections is at least 10-fold less
than with excised elongating hypocotyl sections (McClure and
Guilfoyle, 1987).

More recently, tissue print (McClure and Guilfoyle, 1989a and
1989b) and *in situ* hybridization experiments (M. Gee, unpublished
results) have confirmed the RNA blotting results on the organ
specificity of SAUR gene expression, and, furthermore, these
latter experiments have provided information on the tissue
specificity and and turnover of SAURs. Both tissue print and *in
situ* hybridization experiments indicate that SAURs are most
abundant in the epidermis, outer cortical cells, and starch sheath
of the elongating region in soybean hypocotyls (Fig. 6).
Addition of auxin or cycloheximide results in increased abundance
of these mRNAs in these same tissue regions and does not result in
SAUR expression in additional tissues within the hypocotyl.

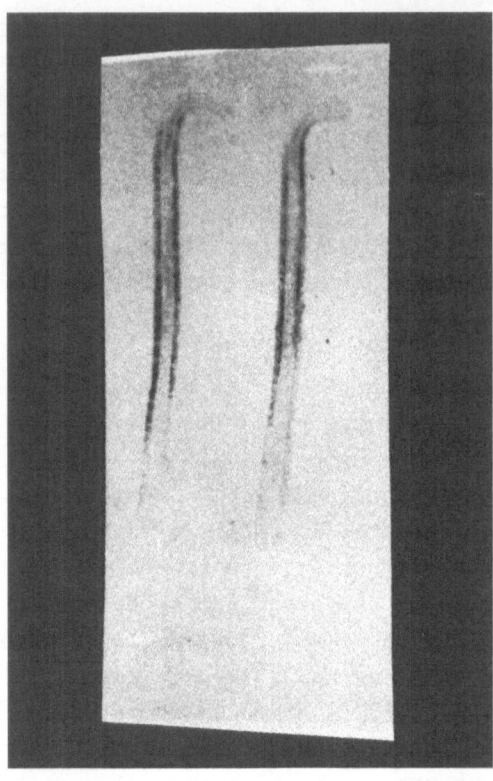

Fig. 6.   Tissue print hybridization autoradiograms (McClure and
             Guilfoyle, 1989a,b) of two 3 day old soybean seedlings
             hybridized with a ^{35}S-labeled antisense SAUR probe. A
             schematic of the seedling used for tissue printing is
             to the right of the autoradiograms.

     Tissue print hybridization has also been used to show that
prior to gravitropic bending of soybean hypocotyls, SAURs
disappear from the epidermis and outer cortex on the upper, more
slowly expanding portion of the horizontally displaced organ
(McClure and Guilfoyle, 1989a; Fig. 7). On the lower or faster
growing side of the horizontal hypocotyl, the SAURs are at least

as abundant, if not more abundant, than those observed in the
elongating zone of a vertical hypocotyl. Other tissue print
hybridization experiments indicate that the SAURs turnover very
rapidly, with a lifetime of 30 minutes or less when exposed to a
new gravity vector (McClure and Guilfoyle, 1989b). The
gravitropic experiments indicate that the SAURs are under dynamic
regulation within seedlings that have not been treated with
exogenous auxin, suggesting that either internal auxin or some
other signal regulates the expression of the SAURs in the normal
seedling. The RNA blotting experiments, along with the tissue
print and *in situ* hybridization experiments, suggest that the
SAURs may play some role in the process of auxin-induced cell
elongation.

AUXIN-REGULATED GENES

The cDNA and gene sequences have been reported for pJCW1 and
2 (Ainley et al., 1988), pGH2/4 (Czarnecka et al., 1988; Hagen et
al., 1988), and a number of independent SAURs (McClure et al.,
1989). No functions of the gene products have become apparent
from the deduced amino acid sequences, although pJCW1 and 2 are
related to each other and the SAURs are highly homologous to one
another. The 5' upstream regions of pJCW1 and 2 have three
sequence elements that are highly conserved between the two genes
(Ainley et al., 1988), and the five SAUR genes share regions of
high homology within 250 base pairs upstream from the start site
of transcription (McClure et al., 1989). Only very limited
sequence homology exists, however, within the upstream regions of
the pJCW and the SAUR genes (McClure et al., 1989). *In vitro*
mutagenesis studies will be required to determine possible auxin
inducible DNA elements and to determine whether these elements are
common to only few or many auxin-regulated genes.

At least some of the SAUR genes are clustered in the soybean
nuclear genome (McClure et al., 1989). The genes corresponding to
the 6, 10A, and 15 cDNA clones are clustered within a 5 kilobase
genomic fragment, and two additional SAUR-like genes are found
flanking this 5 kb cluster. Adjacent genes are transcribed in
alternate directions, and 5' flanking regions of adjacent genes

10

20

45

90

180 min

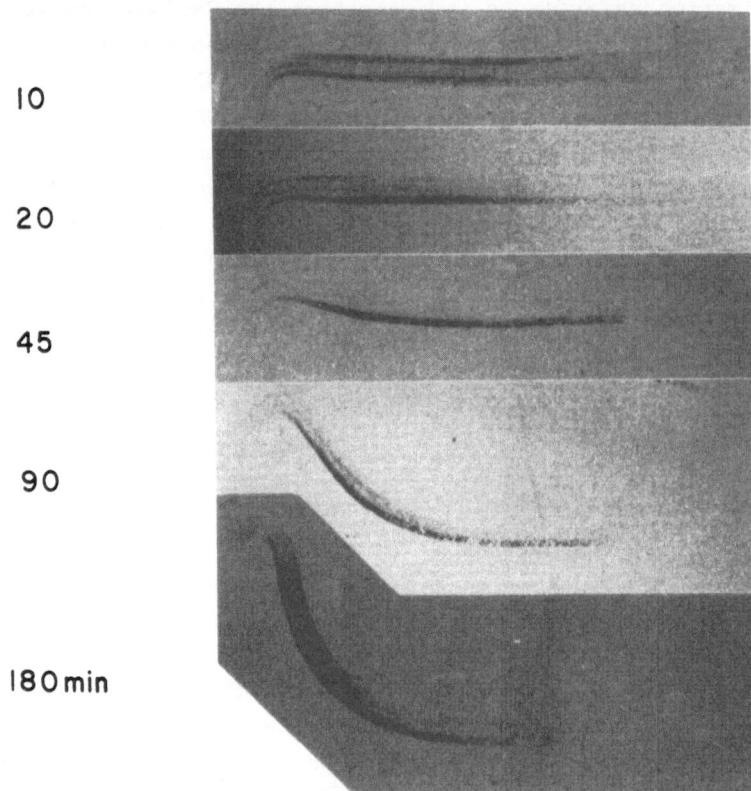

Fig. 7.   Tissue print hybridization autoradiograms of
           gravistimulated hypocotyls of 3-day old etiolated
           soybean seedlings. Hypocotyls were sectioned
           longitudinally at the different times indicated after
           being reoriented from a vertical to a horizontal
           position. The cut surface was blotted onto a nylon
           membrane and hybridized with a mixed SAUR 6, 10A, and
           15 antisense ^{35}S-labeled RNA probe. From top to
           bottom, the time of horizontal displacement is 10, 20,
           45, 90, and 180 minutes.

include only about 1.2 kilobases of DNA. This SAUR gene
organization suggests that putative auxin regulatory elements are
likely to be localized within the 1.2 kilobases of DNA that
separate any two divergent transcripts. In addition, it might be
that strong termination signals exist at the end of transcripts to
prevent read-through of transcripts into adjacent genes. A
determination of the regulatory and termination signals involved
in transcription of the SAUR genes will require detailed analysis
using both transient expression systems and transgenic studies.

To date, transgenic studies on the SAUR genes indicate that
when the entire 8 kilobase cluster of five soybean SAUR genes is
transferred into petunia, the soybean genes are still expressed
under auxin control in young leaves and petioles of regenerated
petunia (*Petunia hybrida*) plants (Wright et al., 1989). This
suggests that auxin regulatory elements are conserved between
soybean and petunia, and that the soybean *cis*-acting auxin
regulatory elements can likely be identified in a heterologous
system.

THE FUNCTION OF AUXIN-REGULATED GENE PRODUCTS

Although a number of auxin-regulated genes have been
identified, the function of the auxin-responsive gene products
remains a mystery. It is equally mysterious which cellular
proteins function in such processes as auxin-induced cell
elongation, division, and differentiation. Even less is
understood about the signal-transduction pathway(s) involved in
auxin action. It is possible that at least some of the auxin-
regulated genes which have now been identified will prove to play
some role in auxin-regulated growth processes. Identification of
the *cis*- and *trans*-acting elements that confer auxin-
inducibility to auxin-regulated genes, such as the pGH, pJCW, and
SAUR genes, should provide some insight into one or more of the
signal-transduction pathways involved in auxin regulated gene
expression.

Although it is not a trivial exercise to determine the
function of the auxin-regulated genes which have been identified,

insight into these functions might be provided by the
identification of mutants that lack or overproduce these gene
products. Such mutants may have already been identified by
classical genetic studies or might be created by molecular
approaches such as *in vitro* mutagenesis.

AUXIN-REGULATED GENES AND CROP IMPROVEMENT

The promoters of auxin-regulated genes may provide hormone-
specific and tissue-specific regulation of genes that are normally
not under these specific controls. Thus the auxin-regulated
promoters might be used in chimeric gene constructs to provide
novel types of chemically controlled gene expression.

The elucidation of functions for auxin-regulated gene
products is likely to provide insight into auxin action and the
control of growth processes in higher plants. Insight into signal
transduction pathways in plants may also result from studies on
auxin-regulated gene products. This information might then be
applied to engineering plants with improved growth habits,
herbicide resistance, or pest resistance.

ACKNOWLEDGMENTS

The research from the author's laboratory was supported by
research grants from the National Science Foundation. Melissa Gee
and Christopher Brown were supported by fellowships from Molecular
Biology and Food for the 21st Century programs at the University
of Missouri, respectively. Antonio Franco was supported by a
fellowship from the NATO scientific committee. This is journal
report number 10,825 of the Missouri Agricultural Experiment
Station.

REFERENCES

Ainley, W. M., Walker, J. C., Nagao, R. T., and Key, J. L., 1988,
     Sequence and characterization of two auxin-regulated genes
     from soybean, J. Biol. Chem., 263: 10658-10666.

Baulcombe, D. C., and Key, J. L., 1980, Polyadenylated RNA sequences which are reduced in concentration following auxin treatment of soybean hypocotyl, J. Biol. Chem., 255: 8907-8913.

Czarnecka, E., Edelman, L., Schoeffl, F., and Key, J. L., 1984, Comparative analysis of physical stress responses in soybean seedlings using cloned heat shock cDNAs, Plant Molec. Biol., 3: 45-58.

Czarnecka, E., Nagao, R. T., Key, J. L., and Gurley, W. B., 1988, Characterization of Gmhsp26-A, a stress gene encoding a divergent heat shock protein of soybean: Heavy-metal-induced inhibition of intron processing, Molec. Cell. Biol., 8: 1113-1122.

Edelman, L., Czarnecka, E., and Key, J. L., 1988, Induction and accumulation of heat shock-specific poly(A$^+$) RNAs and proteins in soybean seedlings during arsenite and cadmium treatments, Plant Physiol., 86: 1048-1056.

Evans, M. L., 1985, The action of auxin on plant cell elongation, CRC Crit. Rev. Plant Sci., 2: 317-365.

Guilfoyle, T. J., 1986, Auxin-regulated gene expression in higher plants, CRC Crit. Rev. Plant Sci., 4: 247-276.

Hagen, G., and Guilfoyle, T. J., 1985, Rapid induction of selective transcription by auxins, Molec. Cell. Biol., 5: 1197-1203.

Hagen, G., Kleinschmidt, A., and Guilfoyle, T., 1984, Auxin-regulated gene expression in intact soybean hypocotyl and excised hypocotyl sections, Planta (Berlin), 162: 147-153.

Hagen, G., Uhrhammer, N., and Guilfoyle, T. J., 1988, Regulation of expression of an auxin-induced soybean sequence by cadmium, J. Biol. Chem., 263: 6442-6446.

Jefferson, R. A., 1987, Assaying chimeric genes in plants: The GUS gene fusion system, Plant Molec. Biol. Reporter, 5: 387-405.

Kende, H., and Gardner, G., 1976, Hormone binding in plants, Annu. Rev. Plant Physiol., 27: 267-290.

Klee, H., Horsch, R., Hinchee, M., Hein, M., and Hoffmann, N., 1987, The effects of overproduction of two *Agrobacterium tumefaciens* T-DNA auxin biosynthetic gene products in transgenic petunia plants, Genes and Develop., 1: 86-96.

McClure, B. A., and Guilfoyle, T. J., 1987, Characterization

of a class of small auxin-inducible polyadenylated RNAs, Plant
    Molec. Biol., 9: 611-623.

McClure, B. A., and Guilfoyle, T. J., 1989a, Rapid redistribution
    of auxin-regulated RNAs during gravitropism, Science, 243:
    91-93.

McClure, B. A., and Guilfoyle, T. J., 1989b, Tissue print
    hybridization. A simple technique for detecting organ- and
    tissue-specific gene expression, Plant Molec. Biol., 12: 517-
    524.

McClure, B. A., Hagen, G., Brown, C. S., Gee, M. A., and
    Guilfoyle, T. J., 1989, Transcription, organization, and
    sequence of an auxin-regulated gene cluster in soybean, The
    Plant Cell, 1: 229-239.

Theologis, A., Huynh, T. V., and Davis, R. W., 1985, Rapid
    induction of specific mRNAs by auxin in pea epicotyl tissue,
    J. Molec. Biol., 183: 53-68.

Vanderhoef, L. N., 1980, Auxin-regulated elongation: A summary
    hypothesis,   in: "Plant Growth Substances 1979," F. Skoog
    (ed.), Springer Verlag, Berlin, pp. 90-96.

van der Zaal, E. J., Memelink, J., Mennes, A. M., Quint, A., and
    Libbenga, K. R., 1987, Auxin-induced mRNA species in tobacco
    cell cultures, Plant Molec. Biol., 10: 145-157.

Walker, J. C., and Key, J. L., 1982, Isolation of cloned cDNAs to
    auxin-responsive poly(A)$^+$ RNAs of elongating soybean
    hypocotyl, Proc. Natl. Acad. Sci. U.S.A., 79: 7185-7189.

Walker, J. C., Legocka,J., Edelman, L., and Key, J. L., 1985, An
    analysis of growth regulator interactions and gene expression
    during auxin-induced cell elongation using cloned
    complementary DNAs to auxin-responsive messenger RNAs, Plant
    Physiol., 77: 847-850.

Wright, R., Hagen, G., and Guilfoyle, T., 1987, An auxin-induced
    polypeptide in dicot plants, Plant Molec. Biol., 9: 625-635.

Wright, R., McClure, B., and Guilfoyle, T., 1989, Auxin-induced
    expression of the soybean SAUR locus in transgenic
    petunia, J. Plant Growth Reg., In Press.

THE MOLECULAR BASIS OF VARIATION AFFECTING GENE EXPRESSION:

EVIDENCE FROM STUDIES ON THE RIBOSMAL RNA GENE LOCI OF WHEAT

R.B. Flavell[*], R. Sardana, S. Jackson and M. O'Dell

AFRC Institute of Plant Science Research,
Cambridge Laboratory, Trumpington, Cambridge, CB2 2LQ.

[*]John Innes Institute and
AFRC Institute of Plant Science Research,
Colney Lane, Norwich, Norfolk, NR4 7UH.

INTRODUCTION

A question relevant to an understanding of plant breeding is "What is the molecular basis of allelic variation affecting gene expression?". The importance of this question is elevated if one accepts the perhaps biased belief that much of the important variation that plant breeders exploit is concerned with variation in the amounts of gene products in time and space during plant development. Many other relevant questions emerge when one wonders about variation in gene expression. For example, does a gene adopt a different structure when it is active compared with when it is silent? How does a plant turn a gene off and on? What kinds of mutations cause a gene to be more or less active.

It is with these sorts of questions in mind that we have sustained a research programme to study the control of expression of the ribosomal RNA gene (rDNA) loci in wheat. This contribution summarizes some of the findings in relation to understanding how gene structure varies in different states of activity.

CHARACTERISATION OF RIBOSOMAL RNA GENE LOCI SHOWING DIFFERENTIAL EXPRESSION

The relative activity of an array of ribosomal RNA genes can be assessed by measuring the relative volume of the nucleolus formed at the locus and the relative length of the constriction found at the nucleolus organiser (NOR) at metaphase (Martini and

Flavell, 1985). These parameters have been scored in cells of
several cultivars of hexaploid wheat, _Triticum aestivum_, and are
summarized in Table 1.

From these results it can be concluded that the rDNA locus on
chromosome 1B is more active than that at chromosome 6B in the
varieties Chinese Spring, Cappelle-Deprez and Hobbit while the
rDNA locus on chromosome 6B is more active than that on chromosome
1B in the variety Bezostaya. Studies using aneuploid stocks of
Chinese Spring have shown that when the 1B NORs are deleted the 6B
NORs become more active to compensate and _vice versa._

Table 1. Secondary constrictions visible on the 1B and 6B
         mitotic metaphase chromosomes of wheat

Cultivar examined	(a) Number of cells Scored	(b) Site of larger constriction		Summary
		1B	6B	
Chinese Spring	30	10	0	1B > 6B
Bezostaya	112	7	72	6B > 1B
Cappelle-Desprez	48	32	1	1B > 6B
Hobbit	26	18	1	1B > 6B

(a) Only cells with a complete metaphase chromosome
     complement were scored.

(b) Chromosomes 1B an 6B were distinguished by size and arm
     ratio. The numbers of cells scored is in excess of the
     number of cells in which 1B or 6B was identified as
     carrying the larger constriction. This is because in
     some cells the chromosomes were not clearly
     distinguishable, the degree of contraction obliterated
     the constrictions or there was no clearly measurable
     difference in constriction length between 1B and 6B. The
     situation in Chinese Spring has been described much more
     extensively in Martini and Flavell (1985).

When chromosome 1U from <u>Aegilops</u> <u>umbellulata</u> which carries an rDNA locus was added to the Chinese Spring chromosome complement, the wheat rDNA loci were suppressed and the 1U locus was the predominantly active locus (Martini et al, 1982; Martini and Flavell, 1985). These and other observations have led to the conclusions that the number of rRNA genes at a locus is usually in excess of the minimum required and that loci (genes) compete with one another for some limiting factor(s) which is required for their participation in an active nucleolus (Flavell, 1986).

The structure of the presumed regulatory regions of genes from each of the loci described above have been characterized. This has been achieved by:

(1) isolating specific types of genes from Chinese Spring and Bezostaya and determining the nucleotide sequences of the region between the 25S and 18S coding sequences (Barker et al., 1988; Sardana and Flavell, unpublished).

(2) using restriction enzymes, Southern blotting and hybridisation techniques on DNA isolated from the various cultivars to compare their rRNA genes with those characterised by sequencing, and

(3) comparing the rRNA genes in aneuploid and substitution lines to enable observed structural variants to be mapped to the 1B or 6B loci (Flavell et al., 1988). The different structures are illustrated in Figure 1.

A wheat rRNA gene contains a complex structure upstream from where transcription is initiated. It is assumed that as with all animal rRNA genes studied this region contains the important regulatory sequences (Clos et al., 1986; Sollner Webb et al., 1983; Jones et al., 1988). The region includes a series of 135 or 136 bp repeats that are almost identical to one another. The principal variation between the genes at the loci we have studied is in the number of these 'A' repeats. This varies between 18 in the 6B locus of Bezostaya to 7 in the 6B locus of Cappelle-Deprez (Fig. 1). Comparison of the results in Table 1, with the gene structures in Fig. 1 shows that in each case studied the locus possessing genes with more 'A' repeats is the more active locus. This suggests the 'A' repeats play a part in determining which genes participate in an active nucleolus.

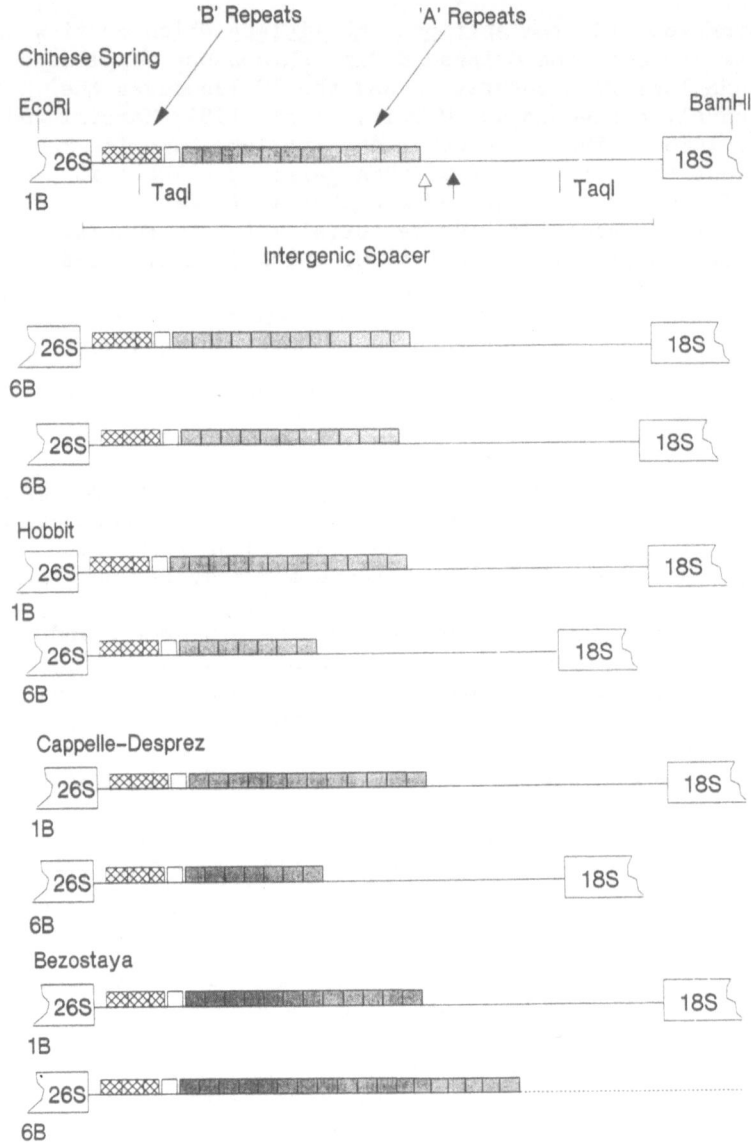

Fig. 1.   Structure of rRNA genes at specific loci
           The intergenic region between the 26S and 18S RNA
           sequences is illustrated.   The genes typical of the
           chromosomal 1B and 6B loci of cultivars, Chinese
           Spring, Hobbit, Cappelle-Deprez and Bezostaya are
           shown.   The filled arrow indicates where
           transcription is initiated.   The open arrow
           indicates the preferentially non-methylated cytosine
           in the CCGG motif in active loci.   More details and
           description of the A and B repeats can be found in
           Barker et al. (1988).

This conclusion for wheat rDNA is similar to that gained from studying transcription after injection into oocytes of Xenopus rDNA templates, differing in the number of upstream repetitive elements. As the number of upstream elements was increased so did transcription from the downstream gene (Reeder, 1984; DeWinter and Moss, 1987).

CYTOSINE METHYLATION AND DIFFERENTIAL LOCUS EXPRESSION

To discover if active and inactive loci could be distinguished by other properties of the intergenic region, the cytosine methylation status of the DNA was studied, because inactive genes are often methylated at specific sites which are not methylated in active genes (Razin and Riggs, 1980). One CCGG motif out of a large number in the intergenic DNA was found to be preferentially unmethylated in a subset of wheat rRNA genes of Chinese Spring (Flavell et al., 1988). The site lies about 165 pairs upstream from where transcription is initiated. The number of genes with this site unmethylated was considerably greater at the more active locus in all the varieties tested (Flavell et al., 1988; Sardana and Flavell, unpublished). When the 1B and 6B loci in Chinese Spring were suppressed by introduction of the 1U chromosome from Aegilops umbellulata, the number of genes at the 1B and 6B loci with unmethylated CCGG sites decreased considerably. The number in the active 1U locus was very much higher (Flavell et al., 1988). These results and others gained from a study of aneuploids in which the number of major rDNA loci varied between 2 and 6 showed that the number of genes unmethylated at CCGG sites in the intergenic region is highly regulated and correlates with the relative activity of the locus. The relative number of genes with unmethylated CCGG sites in the intergenic regions of the cells of the 1B and 6B loci of several wheat varieties were calculated. In each case the number of genes with one or more unmethylated CCGG sites is higher in the more active locus and is correlated with the number of 'A' repeats in the intergenic region of the genes in the locus.

Very few of the 'A' repeats possess CCGG motifs but they do possess GCGC motifs which are cleaved by the restriction endonuclease HhaI if the internal cytosine is not methylated. When wheat DNA is digested by HhaI the intergenic region of only some of the genes is cleaved. Those genes which are cleaved are cut at the GCGC sites in only one or a few of the 'A' repeats. These are localised preferentially towards the 3' end of the array. The number of genes cleaved again correlates with the relative activity of the locus. This is most easily observed by the reduction in the number of Chinese Spring genes cleaved in the

'A' repeats when chromosome 1U of <u>Aegilops</u> <u>umbellulata</u> is present.
The genes with an unmethylated GCGC 'A' repeat sequence include
virtually all those with an unmethylated CCGG site at -165.

How do some genes have specific cytosines unmethylated at
certain sites? The common explanation is that a protein binds to
the DNA and inhibits access to the methylases which are
responsible for addition of the methyl group, copying the
situation on the template DNA strand, at or immediately after DNA
replication. Thus one would predict that a protein binds to one
or more 'A' repeats in some genes and some of these genes also
bind a protein at a site around -165. Alternatively, one could
argue that the methylation pattern is due to unspecific errors in
the methylation process. This would not account for the
preferentially unmethylated site at -165, a major subset of genes
containing both unmethylated 'A' repeats and an unmethylated
-165 site and most importantly, the finding that the most active
locus is enriched for genes with the unmethylated cytosines. This
latter finding implies that the inhibition of cytosine methylation
is connected with gene expression.

CHROMATIN CHANGES AND DIFFERENTIAL LOCUS EXPRESSION

The binding of a protein to DNA often disturbs the chromatin
structure. This disturbance can be detected by adding DNase I to
nuclei and detecting where the nuclease has easier access to the
DNA and can consequently cut the unprotected DNA. Such studies
have revealed that a subset of the rRNA genes have DNAse I
hypersensitive sites. These sites are localised in the 'A'
repeats, the 'B' repeats (See Fig. 1) and at the two sites in the
DNA between the 'A' repeats and the site where transcription is
initiated (Thompson and Flavell, 1989). Furthermore, the genes
carrying these disturbed chromatin structures are predominantly
those with unmethylated cytosines in the same region and are
highly enriched in the more active loci. This information
suggests therefore that gene expression is correlated with
chromatin perturbations that are probably caused by binding of a
specific protein(s). Such proteins would be expected to be
involved in gene regulation.

NUCLEAR PROTEINS WHICH BIND TO REGULATORY REGIONS

Proteins have therefore been sought that bind to specific
regions of the rDNA promoter. Nuclei have been isolated from
embryos and lysed in 0.42M NaCl which is sufficient to extract
most nuclear proteins but not to take the histones from the DNA.
The proteins released were precipitated with ammonium sulphate and

redissolved. This extract was mixed with radioactively-labelled ribosomal promoter DNA in the presence of a large excess of polydI/dC which binds to general DNA binding proteins. The specific rDNA protein complexes were analyzed by two sorts of assays. The first detects slower mobility of the labelled DNA in gel electrophoresis compared with when it is free of protein. Here the 160 bp fragment from -33 to -193 has been found to be retarded under conditions where a 130 bp fragment from +12 to +141 is not. Thus a specific protein(s) binds to the -33 to -193 fragment. The fragment from -33 to +12 has also been found to form a specific high molecular weight complex when mixed with the protein extract.

The second assay defines which specific DNA sequences bind proteins. This is revealed by giving the DNA-protein complex a mild DNase I treatment before purification of the DNA and analysis of the DNA on a sequencing gel, which fractionates the cleaved DNA molecules according to their length. If DNase I has equal access to all sections of the DNA then DNA molecules of every length, corresponding to cleavage between each of the base pairs are produced. Where DNase I does not have access to the DNA, however, then DNA fragments corresponding to cleavage at these sites are unrepresented or absent leaving a "footprint". This "footprint" reveals where proteins bind to specific sequences of the DNA. Specific regions which bind have been detected at -12 to +3, -28 to -15, -86 to -67, -122 to -100 and -156 to -132 with respect to the initiation of transcription. Several protein binding sites have also been found in the 'A' repeat. It can be assumed from comparisons with analyses of animal rRNA genes that these regions are essential for rRNA transcription to occur with high efficiency (Jones et al., 1988; Clos et al., 1986). For example, for the human rRNA gene a large upstream region between -234 and -107 relative to the transcription initiation site aids transcription up to 100 fold while a core essential element between -45 and +20 affects transcription up to 1000 fold (Jones et al., 1988). For the mouse, Mus musculus, rRNA gene the region -39 to +9 contains essential information for transcript initiation but sequences extending out to -140 greatly enhance transcription (Muller et al., 1985). Proteins from human and mouse nuclei have been purified that bind to these sequences and these proteins are vital components for transcription to occur (e.g. Learned et al., 1986; Clos et al., 1986). It therefore follows that the nuclear proteins we have detected in wheat extracts are likely to play an important role in gene expression.

It is interesting to note that one protein binding region is close to the -165 site which is preferentially unmethylated in

genes at the most active loci. This protein may be that which is
postulated to bind and interfere with cytosine methylation.

REGULATORY DNA-PROTEIN INTERACTIONS AND DIFFERENTIAL GENE EXPRESSION

We stated earlier the evidence that the 135 bp 'A' repeats
are associated with differential expression of a locus. How might
this come about? One hypothesis is that important regulatory
proteins which stimulate transcription bind to these 'A' repeats
and the presence of more 'A' repeats in a gene leads to an
increased probability of attracting such proteins which would be
in limiting supply over all. We are in the process of testing
this hypothesis. The fact that not all 'A' repeats and promoters
have unmethylated cytosines in CCGG and GCGC sites or DNase I
hypersensitive sites and some genes have no unmethylated cytosines
in CCGG and GCGC sites in the 'A' repeats and promoters or any
DNase I hypersensitive sites implies that any proteins responsible
for these properties are in limiting supply or unable to bind for
some reason. The finding that the genes associated with the most
active loci are enriched with unmethylated cytosines and DNase I
sites at several sites suggests that determination of these
properties occurs co-operatively, that is, determination of a
change in gene or chromatin structure at one site greatly
increases the probability of a change occurring at a neighbouring
site on the same gene. If such changes result from the binding of
specific proteins, then it would follow that the binding of one
protein increases the probability of the binding of a second
protein nearby. The co-operative binding of transcription factors
has been described for human rRNA genes (Learned et al., 1986;
Bell et al., 1988).

The co-operative binding of protein molecules in limiting
supply to genes can theoretically provide an explanation of how
genes are selected, from the total pool available, to be the
active ones within a locus and how one locus can be differentially
expressed relative to another. This model has been described in
detail elsewhere (Flavell, 1986; 1989). It is postulated that
genes with more 'A' repeats or with a DNA sequence which provides
a greater affinity for the protein will have an increased ability
to compete for the regulatory proteins essential for efficient
transcription. The time when such competition is most critical
will be soon after fertilisation when the particular complement of
genes has come together for the first time in the life of the
plant and perhaps at DNA replication in each cell generation when
DNA helices are relatively free of protein and are transiently
available for new protein binding. Those genes which attract the

regulatory proteins will assume a chromatin structure that has
DNase I hypersensitive sites and unmethylated cytosines where the
proteins bind, and will consequently be potentially active,
produce a visible NOR and give rise to a nucleolus. Those genes
which fail to attract regulatory proteins will assume a chromatin
structure that becomes condensed into heterochromatin, be inactive
and will consequently lie outside the nucleolus.

This competition model can explain nicely how the relative
activity of an rDNA locus is decreased by an increase in the
number of rDNA loci in the cell (Martini and Flavell, 1985). It
can also explain why a locus can be relatively inactive when moved
into a new genetic background (e.g. it has to compete with a
different set of genes) or alternatively be more active.
Variation in activity of an rDNA locus when moved into a new
genetic background is commonly observed, especially in
interspecies hybrids where the regulatory DNA has probably
diverged considerably (Jessop and Subrahmanyan, 1984; Yeh and
Peloquin, 1965; Wallace and Langridge, 1971).

CONCLUDING COMMENTS

These studies have shown that genes at active rDNA loci are
different in structure from equivalent genes which are at inactive
loci in that they possess cytosine residues that are not
methylated at specific sites in the regulatory regions.
Furthermore, they are organised in different chromatin structures.
But, of special relevance to this paper, these differences are
associated with different numbers of reiterated DNA sequences in
the regions of the gene that are believed to be responsible for
determining the extent of gene expression. The results of wheat
parallel those obtained for Xenopus rRNA genes where the extent of
transcription is related to the number of upstream repeats which
act as enhancers of transcription (Reeder, 1984; DeWinter and
Moss, 1987).

Variation in gene expression is determined by variation in
protein binding domains. What kinds of mutations are responsible
for such variations? In the case of the ribosomal RNA genes of
wheat it appears to be processes such as unequal crossing over
which can create variation in the number of the tandemly arrayed
135 bp repeats (Flavell, 1986) and, of course, single base
mutations which can affect the binding affinity of a protein for
the DNA sequences. Numerous single base differences have been
detected in the 135 bp repeats of different rRNA genes (Barker et
al., 1988). The ribosomal RNA genes are perhaps a special case in
that their presumed regulatory sequences are highly reiterated but

it is now established for many genes that expression is regulated by protein binding sequences and mutations in these can greatly influence the extent of gene expression. In some cases it has also been shown that increasing or decreasing the number of such sequences can greatly influence the extent of expression of the adjacent coding sequence. Therefore we would expect that much of the variation that breeders recognise and exploit affects levels of gene expression and this results from genetic variation in regulatory DNA sequences as is described here for the ribosomal RNA genes.

Perhaps the most beautiful illustration of this in plants comes from the study of a range of promoter mutations in genes of Antirrhinum, the activity of which is essential for anthocyanin pigment formation in the flowers (e.g. Coen et al., 1986). These allelic mutations are products of excision of a transposable element that became inserted in the promoter. During the separate excision events, different sections of the promoter were deleted or rearranged. As a consequence of these mutations, the extent and position of pigment accumulation in the flower varies, illustrating that gene expression in time, space and amount in plant development can be the consequence of promoter mutations.

Understanding the quantitative basis of gene expression enables new regulatory regions to be designed for existing important plant genes. Coupled with an ability to insert the new genes into plants, this should enable much exciting new variation to be created for phenotypic characters that the breeder has been interested in for a long time.

REFERENCES

Barker, R.F., Harberd, N.P., Jarvis, M.G., and Flavell, R.B., 1988, Structure and Evolution of the intergenic region in a ribosomal DNA repeat unit of wheat, J. Mol. Biol., 201: 1-17.

Bell, S.P., Learned, R.M., Jantzen, H-M., and Tjian, R., 1988, Functional co-operativity between transcription factors UBA-1 and SL1 mediates human ribosomal RNA synthesis, Science, 241: 1192-1197.

Clos, J., Buttgereit, D., and Grummt, I., 1986, A purified transcription factor (TIF 1B) binds to essential sequences of the mouse rDNA promoter, Proc. Nat. Acad. Sci. (USA), 83: 604-608.

Coen, E.S., Carpenter, R., and Martin, C., 1986, Transposable elements generate novel spatial patterns of gene expression in Antirrhinum majus, Cell, 47: 285-296.

Dewinter, R.F.J., and Moss, T., 1987, A complex array of sequences enhances ribosomal transcription in Xenopus laevis, J. Mol. Biol, 196: 813-828.

Flavell, R.B., 1986, The structure and control of expression of ribosomal RNA genes, Oxford Surveys of Plant Molecular and Cell Biology, 3: 251-274.

Flavell, R.B., 1989, Variation in structure and expression of ribosomal DNA loci in wheat, Genome, in press.

Flavell, R.B., O'Dell, M., and Thompson, W.F., 1988, Regulation of cytosine methylation in ribosomal DNA and nucleolus organiser expression in wheat, J. Mol. Biol., 204: 523-534.

Jessop, C.M., and Subrahmanyan, N.C., 1984, Nucleolar number variation in Hordeum species; their haploids and interspecific hybrids, Genetica, 64: 93-100.

Jones, M.H., Learned, R.M., and Tjian, R.T., 1988, Analysis of clustered point mutations in the human ribosomal RNA gene promoter by transient expression in vivo, Proc. Nat. Acad. Sci (USA), 85: 669-673.

Learned, R.M., Learned, T.K., Haltiner, M.M., and Tijian, R.T., 1986, Human rRNA transcription is modulated by the co-ordinate binding of two factors to an upstream control element, Cell, 45: 847-857.

Martini, G., and Flavell, R.B., 1985, The control of nucleolus volume in wheat; a genetic study at three development stages, Heredity, 54: 111-120.

Martini, G., O'Dell, M., and Flavell R.B., 1982, Partial inactivation of wheat nucleolar organisers by the nucleolus organiser chromosomes from Aegilops umbellulata, Chromosoma, 84: 687-700.

Miller, K.G., Tower, J., and Sollner Webb, B., 1985, A complex control region of the mouse rRNA gene directs accurate initiation by RNA polymerase I, Mol. and Cell. Biol, 5: 554-562.

Razin, A., and Riggs, A.D., 1980, DNA methylation and gene
    function, Science, 210: 604-610.

Reeder, R.H., 1984, Enhancers and ribosomal gene spacers, Cell,
    38: 349-351.

Sollner Webb, B., Wilkinson, J., Roan, J., and Reeder, R., 1983,
    Nested control regions promote Xenopus rRNA synthesis by RNA
    polymerase I, Cell, 35: 199-206.

Thompson, W.F., an Flavell, R.B., 1988, DNase I sensitivity of
    ribosomal RNA genes in chromatin and nucleolar dominance in
    wheat, J. Mol. Bio., 204: 535-548.

Wallace, H., and Langridge, W.H.R., 1971, Differential amphiplasty
    and the control of ribosomal RNA synthesis, Heredity, 27: 1-
    13.

Yeh, B.P., and Peloquin, S.J., 1965, The nucleolus-associated
    chromosome of Solanum species and hybrids, Am. J. Bot, 52:
    626.

INDEX

A elements, 423
A repeats, 421, 425
abscisic acid, *227*
Ac, 117
Adh expression, 273
Adh1/GUS gene, 281
Aegilops
  sharonensis, 111
  umbellulata, 421, 423
aglycone, 395
Agrobacterium Ti plasmid
    system, 113, 117, 175,
    203, 237, 244, 251,
    320, 370, 393, 401
Agropyron, 46
albinism, 228
alien transfer, 32
alleles
  novel, 55
Allium cepa, 266
aneuploid
  methods, 145
  stocks, 420
antisense sequences, 291
Arabidopsis thaliana, 117,
    147, 238, 366, 394
Atropa belladonna, 153
autocatalytic RNAs, 321
auxin
  induced transcription,
    409
  metabolism, 406
  plant hormone, 401
   regulatory elements, 415
  responsive mRNAs, 402
  transport systems, 406

Avena sativa, 2, 22, 110,
    146, 195
5-azacytidine (AZC), 278

background noise
  genetic and
    environmental, 58
Bacillus thuringiensis,
    242, 253
bar gene, 245
barley starch, 218
barriers
  post-fertilization, 79,
    86
  pre-fertilization, 79
benzylamino-purine 218
betaglucuronidase gene
    fusion system, 148,
    175, 204, 239, 267,
    270, 280, 366, 370,
    379, 388, 392
  methylation, 275, 282
  operon, 391
  qualitative assays, 379
  quantitative assays, 379
  tissue-specificity, 380
bialaphos, 245
biochemical and molecular
    systems, 62
blast resistance, 257
Brassica, 110
  campestris, 5
  napus, 3, 65, 144, 185,
    251
  oleracea, 146
Bronze, 116

Bs1, 112
Bt2, 243
Bt4412, 243

C-banding, 65
Caenorhabditis, 366
callus induction frequency,
    229
Capsicum annuum, 146
capsid protein, 306
catalysed cleavage
    reactions, 321
cauliflower mosaic virus
    (CaMV), 306
cell accessibility, 204
cell regeneration, 152
cereal grains, 1
  quality, 28, 33
chemical response
    polymorphisms, 64
chromosome assays, 66
chromosome
  B-A, 345
  elimination, 213
  deletions, 331
  physical length, 338
  rearrangements, 331
Cin4, 112
Citrullus lanatus, 146
cloning, 56
Coix
  aquatica, 111
  gigantea, 111
coleopteran larvae, 242
conjugation-like DNA
    transfer, 237
containment considerations,
    382
control sequences, 314
conventional loci, 331
Corchorus olitorius, 146
cosmid vector, 258
cotransformation, 175
cowpea trypsin inhibitor
    gene, 253
crop losses, 289
crossability loci, 100
cucumber mosaic virus
    (CMV), 306
culture
  anther, 152, 185, 214

  wheat, 219
  cell suspension, 168
  cold pretreatment, 225
  embryo, 55
  gene control, 188
  in vitro, 164, 188
  microspore, 224
  nurse, 192
  post-injection, 192
  protoplast, 168
  totipotent suspension,
    189
cyclic improvement, 28
cycloheximide, 406
cytogenetics, 32
cytokinin, 173
cytoplasmic male sterility
    (CMS), 88

Daucus carota, 117
defective interfering
    particles (DI), 320
DGWG gene, 134
diethyl sulfate (DES), 144
DNA down excised pollen
    tubes, 177
DNA-protein interactions
    regulatory, 425
doubled haploids, 66, 152
Drosophila, 366
dsRNA complex, 323

electroporation, 175, 190,
    210, 252, 279
Eleusine coracana, 146
ELISA, 307
embryo rescue, 108
embryoids, 187
  microspore-derived, 187
En/Spm, 117
engineering
  chromosome, 55
  genetic, 43
enhancers, 314
Escherichia coli, 155, 366
ethephon, 227
ethyl methanesulfonate
    (EMS),144
European patent
    application, 215
exotic germplasm, 46

explant tissue, 203
  soybean (see *Glycine
    max*),
    cotyledons, 206
    hypocotyl, 207

Festuca arundinacea, 86
footprints, 424
Fragaria, 110
Fusarium, 153
fusicoccin, 403

gene expression
  differential, 425
  in vivo, 253
  tissue specific, 406, 411
gene fusion, 367, 394
  position effects, 376,
    378
  transcriptional, 369
  translational, 369
gene regulation, 265
gene rich, 47
gene tagging, 69
gene vectors, 298
genes
  auxin-regulated, 413
  chimeric, 306
  foreign, 55
  major, 72
  new, 55
  reporter, 368, 370
  responder, 368
  ribosomal RNA, 419
  target, 56
genetic analysis, 66
genetic diversity, 137
genetic enhancement, 49
genetic gains, 22, 24
genetic maps, 331
genetic stocks, 65
genetic vulnerability, 45
genome assay, 67
genomic library, 333
genotype assay, 67
genotype effects, 229
glucuronidation, 390
glucuronide permease, 391
Glycine max, 62, 144, 206,
    266, 402
glycosylation

N-linked, 393
glyphosate (see Roundup)
Gossypium hirsutum, 110,
    139, 208, 209
green revolution, 164
GUS (see betaglucuronidase)

hap gene, 213
haploid production, 114,
    213
Helianthus annuus, 9
Heliothis virescens, 244
Helminthosporium maydis,
    137
herbicides, 27
  resistance, 245
heterochromatic regions,
    331
highly repeated sequences,
    112
Hordeum bulbosum, 101, 213
Hordeum vulgare, 2, 22, 65,
    129, 186, 214
  x maize hybrids, 105
hormones
  abscisic acid, 403
  cytokinins, 403
  ethylene, 403
  exogenous
    dose-response profiles,
      406
  gibberellic acid, 403
hph gene, 177
hybridization
  in situ, 65, 344, 411
  intergeneric, 55
  interspecific, 55
  sexual, 55
  somatic, 185
  tissue print, 412, 413
  wide, 213

in vitro induction, 375
inbred backcross, 67
inbreeding
  recurrent, 66
incompatability, 89
insect resistance, 242
insecticidal crystal
    proteins, 242
interaction 2,4-D, 165

isoelectric focusing (IEF),
    62
isozyme polymorphisms, 331

knobs, 331
knotted, 116
Kr genes, 113

Lactuca sativa, 320
leafhopper vector, 290
lepidopteran larvae, 242
linkage map
    F2, 333
Linum usitassimum, 2, 144,
    208, 209, 210
Ln1, 145
lodging resistance, 24
Lolium
    multiflorum, 190
    perenne, 86
Lotus
    corniculatus, 86
    tenuis, 86
Lycopersicum esculentum,
    65, 113, 251, 290, 325,
    394

maize, (see *Zea mays*), 42
male sterility, 29
Manduca sexta, 243
maps
    genetic, 61
    homozygous, 341
    interval, 341, 345
    physical, 65
    saturated, 62
marker mediated location,
    72
marker systems, 61
    biochemical, 62, 68
    cytological, 64
    electrophoretic, 69
    functional, 68
    genetic, 61
    molecular, 68
    morphological, 63
    nonfunctional, 68
    selectable, 239, 279
maximum likelihood methods,
    70

multi-point (see
    MAPMAKER), 336
mechanization, 24
Medicago sativa, 192
mediums
    EDAM, 218
    FHG, 218
    liquid vs solid, 221
    N6, 218
    potato extract, 222
    requirements, 220
megachromosomes, 113
melibiose, 218
methylation
    cytosine, 423
    cyotsine residues, 280
    differential locus
        expression, 423
    DNA, 267
    hyper-, 267
    hypo-, 267
5-methylcytosine-sensitive
    restriction enzymes,
    277, 280
4-methylumbelliferyl-beta-
    D-glucuronide (MUG),
    270
microinjection, 186, 192,
    252
    systems, 190
Microseris
    bigelovii, 113
    douglasii, 113
MnSOD, 240
molecular blockers, 295
molecular drive, 112
Mu, 112, 254
multi-gene families, 369
mutagenesis, 55
    site-directed, 307
mutants
    cis-acting, 394
    conditional-lethal plant,
        148
    induced, 148
    meiotic synapsis-
        deficient, 146
    trans-acting, 394
mutant selection, 128
    mass screening, 131
    population size, 131

mutation induction, 127
  gamma fusion, 153
  gamma rays, 132
  pollen treatment, 147.
  super-mutagens, 150

N-banding, 65
naked-eye polymorphism, 340
Nicotiana
  plumbaginifolia, 240
  repanda, 153
  tabacum, 110, 153, 185,
     203, 239, 251, 266,
     320, 325, 402
  x *N. otophora*, 113
  x *N. plumbaginifolia*,
     113, 153
Nilaparvata lugens, 138
non-capsid proteins, 293
nucleolus organiser (NOR),
    419

o2-m5, 116
oilseeds, 1
opaque-2, 116
organ specificity
    variation, 384
Oryza
  australiensis, 79
  brachyantha, 79
  glaberrima, 77
  granulata, 79
  longistaminata, 77
  meridionalis, 77
  nivara, 77
  officinalis, 77
  ridleyi, 79
  sativa, 77, 163, 186,
     222, 251, 265, 266,
     289
    indica, 163
    japonica, 163
  spontanea, 77

Panicum maximum, 190
particle bombardment
    technique, 177, 192,
    193, 210, 238, 252,
    253, 266, 279
patatin, 374
pearl millet/*Sorghum*, 82

PEG-treatment, 190, 279
Pennisetum americanum, 56,
    104, 144, 188
pest management, 21
Petunia, 245
Petunia hybrida, 203, 415
Ph1 , 115
ph2b, 146
Phthorimaea oppercullela,
    244
Pieris brassicae, 243
Pisum sativum, 146, 402
plant architecture, 136
plant breeding, 21, 40,
    365
plasmid DNA, 164
plastid DNA, 228
plastid metamorphosis, 228
pNGI, 267
pollen embryogenesis, 226
pollination
  in vitro, 187
polygeneic systems, 57
potato virus X (PVX), 306
potyviruses, 309
pre-breeding, 50
primary gene pool, 113
progeny evaluation, 29
promotors
  subgenomic, 317
    RNA, 317
subgenomic, 317
protein polymorphisms, 331
protoplast
  callus formation, 173
  fusion, 55, 185
  regeneration, 163, 252
Prunus, 110
Pseudomonas
  solanacearum, 153
  syringae, 242
PstI, 333
pulse field gel
    electrophoresis, 254
Pyricularia oryzae, 257

quantitative trait loci
    (QTLs), 57, 62, 69, 72,
    115, 340

r-x1 system, 347

random intermating, 29
random reassortment, 55
recombinant DNA methods,
    366
regeneration
  cytoplasmic gene control,
      168
  frequency, 229
  in vitro plant, 188
  nuclear gene control, 165
  proporation of green
      plants, 229
  response, 208
  via somatic
      embryogenesis, 164
resistance
  CP-mediated, 307
  disease, 289
  nonconventional, 291, 295
  non-host, 296
  symptom development, 289
  targets, 289
  transformation, 289
restriction endonucleases,
    63
restriction fragment length
    polymorphisms (RFLPs),
    63, 115, 150, 254, 257,
    331
  map, 333, 345
    definitively ordered,
        349
  physical location, 344
Rht, 116, 138, 139
Rhynchoryza subulata, 79
ribosomal RNA gene (see
    gene)
ribosomes, 321
RNA blotting, 413
RNA viruses, 313
Roundup, 208, 245, 253

Salmonella typhimurium, 113
  aroA, 113
satellite RNAs, 319
satellite sequences, 291
satellites, 298
sd1, 138
SDS-PAGE electrophoresis,
    33, 62
Secale cereale, 65, 96, 130

  x maize, 105
*Secale cereale/Hordeum
    bulbosum*, 82
Secale cereale/Zea mays, 82
seed abortion, 108
selection
  co-dominant marker, 67
  efficiency, 205
  in vitro, 185, 188
  kanamycin, 175, 207
  mass, 28
  neo gene, 175
  recurrent, 28, 29, 30
  recurrent reciprocial, 29
self-pollinated, 21, 24
Sesamum indicum, 136
single seed decent, 30, 31,
    66
sodium azide, 146
*Solanum
  pinnatisectum*, 153
  tuberosum, 65, 117, 136,
      185, 208, 283, 373
Sorghum bicolor, 56, 144,
    188
sp, 143
Spartina townsendii, 110
Streptomyces hygroscopicus,
    245
subliminal infection, 293
sulfonylurea resistance,
    253
super-induced, 406
superoxide radicals, 240
supply function, 6
synergism, 297

target characters, 60
tissue print, 411
tobamoviruses, 309
tolerance
  chlorsulfuron, 155
trans-acting factors, 394
transcapsidation, 297
transcription factors
  co-opersative binding,
      425
transcriptional
  promoters, 314
  regulation, 374

transformation, 43, 56,
    164, 205, 265, 365, 376
  (see *Agrobacterium*)
  biolistic, (see particle
    bombardment)
  (see cotransformation)
  direct DNA delivery, 195
  (see DNA down excised
    pollen tubes)
  efficiency, 205
  (see electroporation)
  (see microinjection)
  (see PEG-treatment)
  monocot protoplasts, 175
  (see particle
    bombardment)
  particle gun, (see
    particle bombardment)
  pollen-mediated, 193,
    252, 253
  polyethylene glycol, 175,
    252
  potential systems, 292
  protoplasts, 206
transgenic organism, 237,
    251
transgenic rice plants, 176
translational control
    sequences, 314
translational enhancer
    sequences, 316
translocations
  B-A, 343
  reciprocal, 332
transposable elements, 112
  quiescent, 114
transposon tagging, 117,
    253
Tripsacum, 46
Triticosecale, 188
Triticum, 2, 22
  aestivum, 55, 96, 129,
    164, 186, 251, 265,
    266, 373, 419
  DNA content, 97
  durum, 101
  embryology, 98
  monococcum, 190, 318
  mRNA genes, 421
  squarrosa, 101
  thaoudar, 101

  urartu, 101
  x barley, 96
  x pearl millet, 104
  x rye, 96
  x sorghum, 103
*Triticum aestivum/Hordeum
    bulbosum*, 82
Triticum aestivum/Zea mays, 82
triton X-100, 270
tuber-disc method, 374
tungro disease, 289

unwindase activities, 323

variation
  discontinuous or
    continuous, 57
  discreate, 57
  discrete, 151, 188
  gametoclonal, 227
  genetic, 57
  heritable, 60
  phenotypic, 57
Vicia faba, 136
viral coat protein
  genes, 291
  expression, 315
viral enhancers, 316
viral promoters, 314
viral replicases, 294
virus expression, 294
virus spread, 293

Western blotting, 62
whole genome approach, 70
wide-crossing, 95
wild-type alleles, 340
Wis-2, 112

yeast, 366
Ym, 142

Zea diploperennis, 48
Zea mays, 29, 65, 130, 186, 251,
    265, 266, 331, 373, 394
  A chromosomes, 332
  B chromosomes, 332
  mexicana, 110
Zea mays/Sorghum, 82, 96
Zenopus, 423
  rRNA genes, 426